2nd International Conference on Advanced Materials (ICAM-2017)

IOP Conference Series: Materials Science and Engineering
Volume 305

Irbid, Jordan
10 – 13 July 2017

ISBN: 978-1-5108-5951-7
ISSN: 1757-8981

Printed from e-media with permission by:

Curran Associates, Inc.
57 Morehouse Lane
Red Hook, NY 12571

Some format issues inherent in the e-media version may also appear in this print version.

Printed by Curran Associates, Inc. (2018)

Institute of Physics
Dirac House, Temple Back
Bristol BS1 6BE UK

Phone: 44 1 17 929 7481
Fax: 44 1 17 920 0979

techtracking@iop.org

Additional copies of this publication are available from:

Curran Associates, Inc.
57 Morehouse Lane
Red Hook, NY 12571 USA
Phone: 845-758-0400
Fax: 845-758-2633
Email: curran@proceedings.com
Web: www.proceedings.com

Table of contents

Volume 305

The 2nd International Conference on Advanced Materials (ICAM-2017)

10–13 July 2017, Irbid, Jordan

Accepted papers received: 19 January 2018 Published online: 1 March 2018

Preface

Papers

IOP Conf. Series: Materials Science and Engineering **305** (2018) 011001 doi:10.1088/1757-899X/305/1/011001

Preface

Second International Conference of Advanced Materials (ICAM 2017)
Jordan University of Science and Technology, Irbid, Jordan
10-13 July 2017

On behalf of the Organizing Committee, it is my great pleasure to welcome you to the "Second International Conference of Advanced Materials ICAM 2017. After the success of ICAM 2015 held at Jordan University of Science and Technology (JUST), Irbid, we decided to continue the ICAM conferences and organize the second edition at JUST on 10-13 July 2017. This conference is a clear sign of the strong commitment of JUST to be a front liner in Jordan and the region.

The conference attracted over sixty contributions on new advances in the fields of chemical, physical, biological and engineering aspects of material sciences. ICAM includes representatives from academia, industry, governmental and private institutes. The plenary and invited speakers presented new and sometimes unpublished data, as well as ongoing research activities. Topics ranged from synthesis, applications, solid state to nanomaterials are discussed. Talented junior investigators presented their best ongoing research at poster sessions.

The proceeding of the International Conference on Advanced Materials (ICAM 2015) was published in IOP Conference Series: Materials Science and Engineering, Volume 92. An open access copy of all articles are found at http://www.proceeding.com/28343/html. Here, once again we are happy to publish selected manuscripts presented at ICAM 2017 at IOP. Papers submitted for publication are peer reviewed.

To organize this conference, the Faculty of Science and Arts at JUST setup several functioning committees, who have had a fruitful collaboration and in a short period of time, we have spared no effort to make this conference a success. We would like here to commend all staff and local committees for their outstanding effort, their sincere contribution, enthusiasm and all the sponsors for their support. We would also like to give very special thanks to all participants for their contributions to the Conference program. Special thank goes to the International Advisory Committee whose members gave precious inputs and were always side by side with the Organizer.

Finally, we are looking forward to continue such conferences on a regular basis. We sincerely welcoming all of you to participate in our future conferences. We hope that you had an enjoyable, unforgettable cultural and technical experience in Jordan.

Mohammad El-khateeb, Conference Chair

Mohammad-Ali Al-Omari,
Organizing Committee Coordinator

Nathir Al-Rawashdeh
Scientific Committee Coordinator

IOP Conf. Series: Materials Science and Engineering **305** (2018) 011002 doi:10.1088/1757-899X/305/1/011002

SPONSORS

Jordan University of Science and Technology	Cairo Amman Bank	Biqai Center for Comprehensive Care & Rehabilitation
Association of Arb Universities	Sedrah Palace Hotel	Helmy Abu-Sham Co
Omaya Restaurant	Poinciana for Scientfic Systems	Jordanian Chemical Society
Hijaz Electronic and Scientific Ins.	**Bab ElReda for Chemicals**	**Sydoor for Trades**
JUST Library	**Al-Fajer Library**	**Al-Haik For Housing**

IOP Conf. Series: Materials Science and Engineering **305** (2018) 011004 doi:10.1088/1757-899X/305/1/011004

Peer review statement

All papers published in this volume of *IOP Conference Series: Materials Science and Engineering* have been peer reviewed through processes administered by the proceedings Editors. Reviews were conducted by expert referees to the professional and scientific standards expected of a proceedings journal published by IOP Publishing.

Defect Functionalization of MoS_2 nanostructures as toxic gas sensors: A review

A.A. Ramanathan
Department of Physics, University of Jordan, Amman-11942, Jordan

amallahmad@gmail.com

Abstract. Toxic gas sensing plays an important role in many parts of our life from environmental protection, human health, agriculture to biomedicine. The importance of detecting toxic gases in the environment cannot be minimised in today's highly polluted world and the reality of global warming. Carbon monoxide and NO gas are highly toxic air pollutants and can cause serious health problems. Therefore, materials able to detect these toxic gases are urgently needed. Doping and defect substitution is a versatile and new tool for changing the chemical and electronic properties of 2D layered materials and boosting the applications of these materials. Molybdenum disulphide (MoS_2) as a 2D layered material has unique properties and applications due its semiconducting nature, bandgap and layered structure. In the past decade, although, extensive research of Graphene as a gas sensor was conducted, the zero bandgap limited its potential and applicability. This is overcome in MoS_2 nanostructures (MSNs) and the current focus is defect engineering of MSNs. The large surface to volume ratio, bandgap and cheapness makes MSNs very attractive for gas sensor applications. The idea is fuelled by the recent finding of Ding et al [16] of successful doping strategies on monolayer MoS_2 for enhanced NO detection. Moreover, the work of Luo et al [17] shows that substitutional doping is the new way of boosting and engineering the properties of ML MoS_2. A short and focused report in this exciting field is presented in this review.

Key words: MSNs, Defect functionalization, bandgap, sensors

1. INTRODUCTION

Toxic gases like CO, NO, NO_2 and NH_3 in the environment is a serious problem and needs to be detected and redirected or made harmless by conversion to useful products. Oxygen delivery to the body organs, mainly heart and brain can be reduced by Carbon monoxide and it is a serious health threat. NO gas can cause serious respiratory and skin problems among people and it is important to monitor its levels especially in asthma patients. A robust, portable, low cost, highly sensitive and selective sensor is in urgent need. Inspired by the well known 2D material Graphene for gas sensing applications [1-5], extensive research has been conducted on other layered materials, especially MSNs. The reduced dimensionality of 2D layered materials leads to unique properties and bandgap dependent applications. Unlike Graphene which has zero bandgap, MoS_2 has a thickness dependent bandgap and has more potential in device applications. Recently, there have been several theoretical and experimental researches on MoS_2 as gas sensors [6-10]. The desirable direct band gap in monolayer (ML) MoS_2 and large surface area to volume ratio makes it highly suitable for such an application. This has been proved by various research results for example the first principles work of Yue et al [6] which shows the effect of electric field on the adsorption of various gas molecules H_2, O_2, H_2O, NH_3, NO, NO_2, and CO on ML MoS_2. The electrical response to NO_2 and other gases has been studied experimentally by Donarelli et al [7] for a few layers of MoS_2. Sensitivity of multilayer MoS_2 transistors to NO has been shown by Li et al [8]. The detection of NH_3 down to 1ppb and the single-molecule sensitivity for NO2 detection was proposed by He et al [9]. Several experimental works have also been conducted to examine the effectiveness of ML and few layers of 2D MoS_2 for successful gas sensing applications. Multilayer MoS_2 films based on transistor sensors have been experimentally demonstrated to show stable sensitivity towards NO gas molecules by Late et al [10].

IOP Conf. Series: Materials Science and Engineering **305** (2018) 012001 doi:10.1088/1757-899X/305/1/012001

They have explored the performance of single and multilayer MoS_2 on NO_2 and NH_3 gas detection and also the effect of gate bias and light manipulations.

Chemical doping is an effective way to alter the binding configurations, modify electronic and transport properties and enhance the application of materials as a gas sensor. Chemisorption is found to be thermally more stable and enhanced by dopants. The recent years have seen some interesting theoretical investigations of doping of ML MoS_2 and absorption of various toxic gases NO, NH_3, CO. In their first principles calculation, Ma et al [11] show that the performance of NO and CO gas absorption is vastly improved by doping pristine ML MoS_2 with transition metal atoms like Au, Pt, Pd and Ni. The effect of non-metal doping of pure ML MoS_2 with Boron, Phosphorus and chlorine for NO shows strong chemisorptions, especially with B and P as reported by Ding et al [12]. Further, the adsorption of NO2 and NH_3 by ML MoS_2 doped with Al, Si and P was studied by Luo et al [13] to determine the best gas sensor materials. This article presents a short review of the role of substitutional doping on MSNs from first principles calculations over the past five years. The paper summarizes the recent advances in this field with a focus on defect production, non-metal and metallic dopants on ML of MoS_2 and the gas sensing performance related to the dopant atoms.

2. SENSING MECHANISMS

The realization of graphene field-effect transistors (FET) sensing devices and its limitations regarding selectivity has given way to the new generation MoS_2 sensors. The noticeable and specific changes in electronic properties like charge density and transport properties of the sensing material by gas molecules provide the necessary signals for detection of the gas. The gas sensing mechanisms could be resistance based, application of strain or electric field, or transistor based etc. As shown by Yue et al [6] the application of electric field on ML MoS_2 affects the adsorption of the gases NH_3, NO, NO_2, and CO by modulating the charge transfer process. The absorption energies and charge transfer for the stable configurations obtained from their first principles calculations show that all the gas species act as charge acceptors except NH_3, which acts as a donor. This is shown in Fig.1 by the charge density difference plot for the adsorption of the various gas molecules. Shown in the figure is the direction and value of the charge transfer. They also showed that the application of a perpendicular electric field can significantly modulate the charge transfer between the adsorbed molecules and the substrate.

Recently, single layer and multi-layers of MoS_2 sensors have been fabricated and investigated for performance. The relative change in resistance has been used as a sensing signal in several MoS_2 FET devices [8-14] for sensing NO, NO2 and other gases. Li et al [8] have detected NO gas with a sensitivity of 0.8 ppm using mechanically exfoliated MoS_2 FET devices. Using mechanical exfoliation they deposited one-four layers of MoS_2 on Si/SiO_2 substrate to fabricate FET devices. Their work showed that although the ML MoS_2 has a fast response it was unstable and that the multilayers were more stable with better performance for NO detection. Transistors of varying thickness of large area MoS_2 sheets synthesized by micro-exfoliation were assembled on 300nm Si/SiO_2 substrates by Late et al [10] in the search for the ideal layer structure for enhanced gas absorption. Different conditions of gate bias, humidity and light irradiation were tested for the best performance of NO2 and NH_3 absorption. Their results showed high sensitivity and recovery for few layers of MoS_2 as compared to ML MoS_2, which could be manipulated by gate bias and green light. Their DFT studies also confirmed this and the fact that resistance decreases due to charge transfer upon application of electric field.

IOP Conf. Series: Materials Science and Engineering **305** (2018) 012001 doi:10.1088/1757-899X/305/1/012001

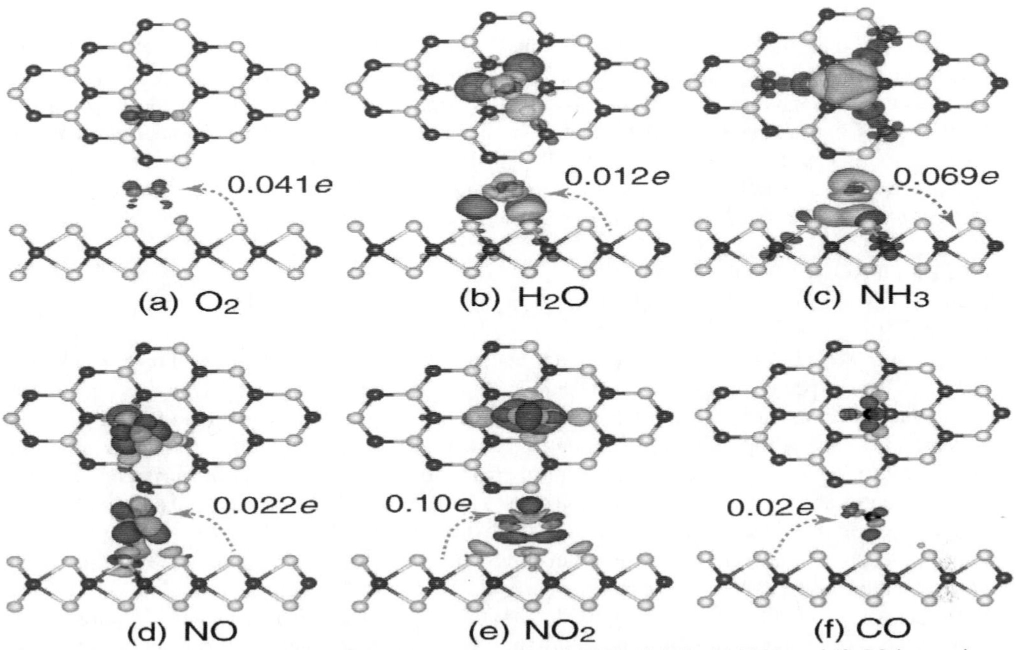

Fig. 1 Charge density difference plots for (a) O2, (b) H2O, (c) NH3, (d) NO, (e) NO2, and (f) CO interacting with ML MoS2. The red (green) distribution corresponds to charge accumulation (depletion). The isosurface is taken as 5 × 10-4e/Å3, ref [6].

Resistance type, liquid chemical exfoliated MoS_2 flake gas sensors were fabricated by Donarelli et al [7]. They investigated the response of MoS_2 flakes annealed in air at temperatures of 150°C and 250°C to NO_2 and other gases. Annealing at temperatures 250°C shows an n-type of gas sensor behavior with a detection limit of 20 ppb for NO_2. The presence of significant number of S vacancies from thermal annealing and the partial oxidation of the surface of MoS_2 flakes is the reason for this n-type as opposed to the p-type sensor annealed at 150°C, which has nitrogen substituted S vacancies. Fig. 2 shows the resistive response to 1 ppm of NO_2 in dry air for both devices.

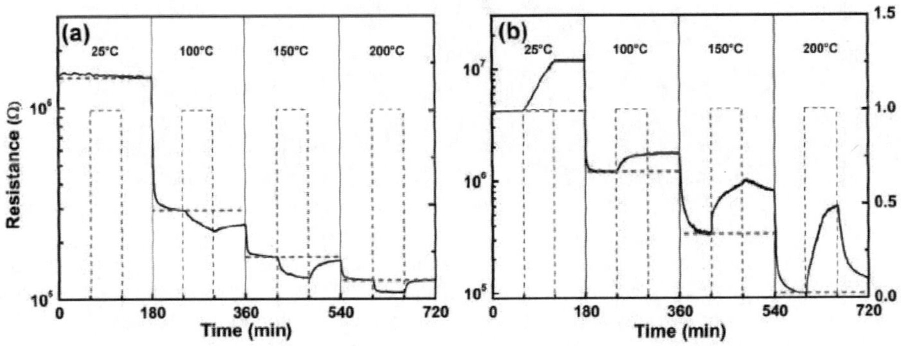

Fig. 2 Resistive responses as a function the operating temperature (a) of the "150°C annealed" device (p-type) and (b) of the "250°C annealed" device (n-type) [7]

From Fig. 2 we see that the response time for p-type is faster but the n- type has a resistance response that is much higher, almost 2.5(5) times the p-type at operating temperatures of 150°C (200°C). The ideal operating temperature is found to be 150°C considering the response intensities, time, and power consumption of the devices.

IOP Conf. Series: Materials Science and Engineering **305** (2018) 012001 doi:10.1088/1757-899X/305/1/012001

Fundamental insights of the influence of the adsorbed gas molecules on the electronic and transport properties of MSNs can be obtained by first principles calculations. These can then be used to design highly sensitive and selective gas sensors. Simulation studies help in understanding the interactions and optimizing sensor performance by indentifying the key response mechanism in the analyte-sensor interactions. Since, in reality the different methods of synthesis of MSNs produce defects, a systematic study of defects from first principles and experiment are essential and are discussed in the next section.

3. DEFECT PRODUCTION

Previous studies [4, 6] show that both pristine graphene and MoS_2 have only physisorption for gas adsorption. This can be altered by the introduction of defects that can change the electronic structure and chemical reactivity of the MoS_2 surface and increase the selectivity and adsorption performance by doping. It is expected that MoS_2 will show higher sensitivity, similar to defective and doped graphene that shows greater sensitivity for NO, NO_2, CO and other gases as reported by Ratinac et al [3], especially with the presence of edge states. Hence, information on defect formation plays an important role and both experimental and first principles works [19-24] have been carried out towards this end. Zhou et al [21] have made a systematic study of intrinsic point defects, dislocations, grain boundaries, edges etc in chemical vapor deposited (CVD) ML MoS_2. They have performed first principle calculations in addition to atomic imaging with consistent results. Their findings show the highest structural stability for S vacancies (V_S) which are very frequent compared to the antisite defects. Also, it is found that V_S is randomly distributed and V_{2S} is rarely observed as it requires twice the formation energy of V_S. This behavior is different from graphene where di-vacancies are preferred over mono-vacancy of C atoms. Fig.3 shows the image analysis of the various point defects.

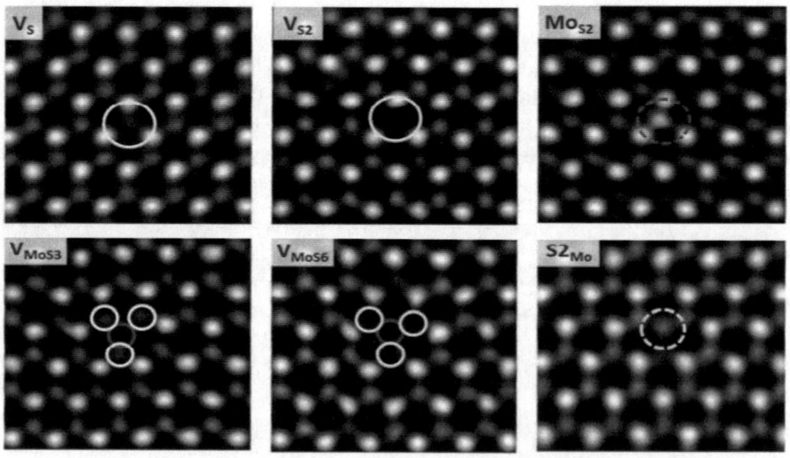

Fig.3 Image analysis for intrinsic point defects in ML MoS_2 [21]

In the filtered images the Mo atom has 2-2.5 times the intensity of S2 column and a single S atom (V_S) has half the intensity of S2 column. Najmaei et al [19, 20] have explored the sources of both intrinsic and extrinsic defects in CVD MoS_2 and ways to control production of the defects to suit the needs of a device application. Fig. 4 shows schematics of such a process.

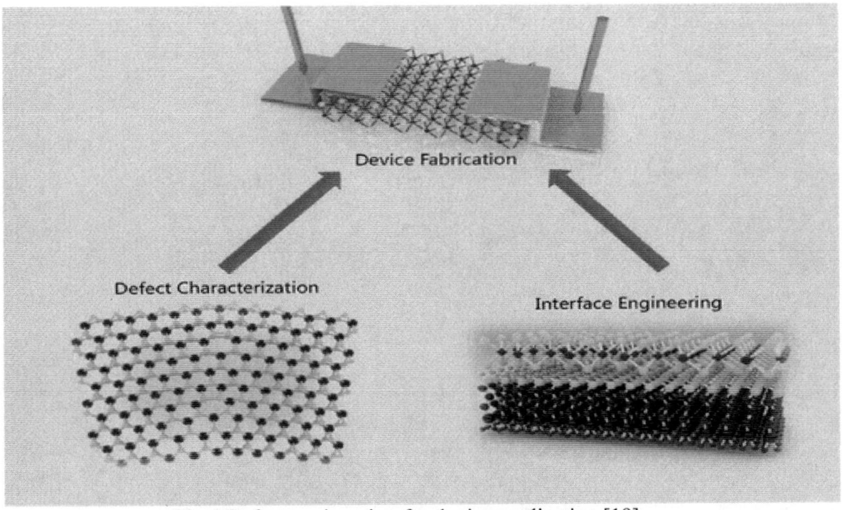

Fig.4 Defect engineering for device application [19]

The vast range of possibilities with the creation of defects and simultaneous doping with atom of another species has been realized by Komsa et al [23, 24], who studied the effect of electron irradiation on vacancy production. Using transmission electron microscopy (TEM) and DFT calculations they have shown that various structural defects can be induced in a controlled manner and the vacancies created by electron irradiation can be doped with other atomic species for tuning the electronic properties.

4. DEFECT SUBSTITUTION

The process of defect engineering and property modulation for suitable gas sensing applications with metal and non-metal atoms doping has been studied in graphene [25-31] over the last decade. In all cases it was found that the introduction of defects and dopants altered the sensitivity and selectivity of the gas sensors. Taking a cue from the encouraging results for graphene, doping MSNs is the state of art tool for the next generation of gas sensors. Several experimental studies [7, 10, 32-34] have proved the high potential of MSNs as gas sensors for NO_x and NH_3 adsorption. The recent years have also seen some interesting theoretical investigations of doping of ML MoS_2 for enhanced absorption of various toxic gases NO, NO_2, NH_3, and CO. The next section discusses this issue for non-metal and metallic dopants from first principles case studies in the last 2-3 years.

The effect of non-metal doping of pure ML MoS_2 with Boron, Phosphorus and chlorine for NO adsorption was demonstrated by Ding et al [16]. All dopants show good adsorption and the interactions of NO with MoS_2 is described by charge density changes, which is responsible for changes in the conductivity. Strong chemisorptions, especially with B and P were reported and the fact that pure MoS_2 shows only weak physisorption was also ascertained by them. Weak interactions between gas molecules and the sensing material cannot provide accurate readings and sensitivity. Doping provides the necessary sensitivity through chemical interactions and strong adsorptions. The adsorption is so strong with B with large energies that desorption time is very large and practically impossible. The best sensors are the ones that have adsorption energies in the intermediate range and the best candidate in the present scenario is the P doped MoS_2, which makes both a sensitive and reversible sensor.

In their first principles calculation, Ma et al [15] show that doping pristine ML MoS_2 with transition metal (TM) atoms like Au, Pt, Pd and Ni leads to vast improvement in the performance of NO and CO gas absorption. Using the Vienna Ab-initio Simulation Package (VASP) they have simulated the TM doped at the S vacancies of ML MoS_2 and calculated the most stable adsorption configurations. The adsorption energies, distance of gas molecules to sensor surface, the DOS, charge densities etc have all been evaluated. Fig.5 shows the side view of the atomic structure and spin density iso-surfaces for the adsorption of CO and NO gas on TM doped MoS_2. The bond length of the gas molecule and also distance between Gas molecule and TM are shown in the figure.

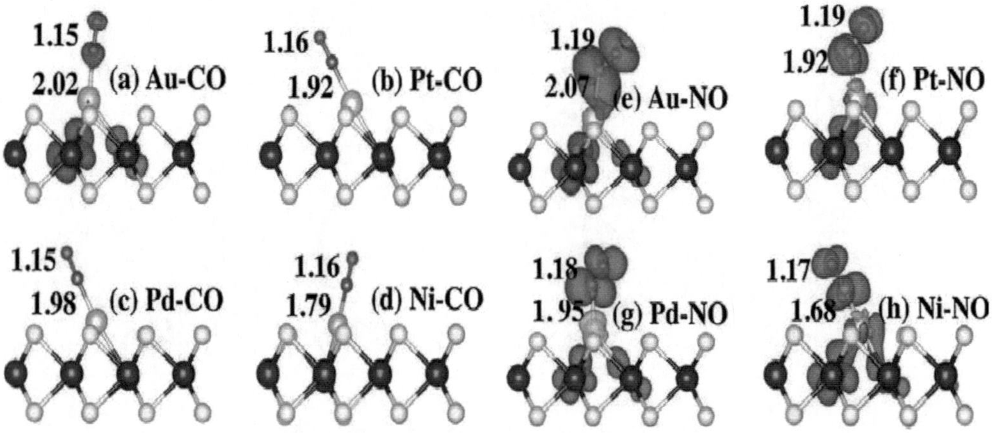

Fig. 5 Side view of the atomic structures and spin-density iso-surfaces of the CO and NO systems. The red and green regions represent positive and negative values (in 0.002 e/bohr3), respectively [15]

Again, in an attempt to determine the best gas sensor materials for the adsorption of NO_2 and NH3, Luo et al [17] studied ML MoS_2 doped with Al, Si and P. The density of states (DOS) and the charge transfer between the gas molecule NO_2 and substrate taking place is shown for Si substitution as an example in Fig. 6 from their work.

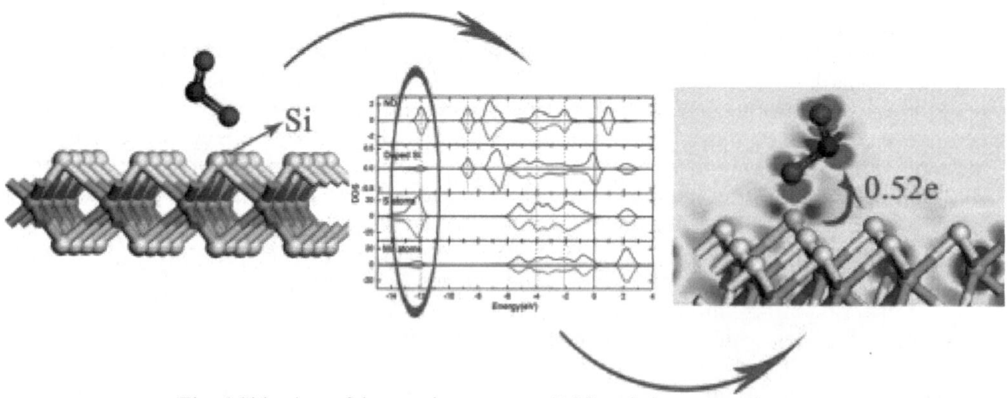

Fig. 6 Side view of the atomic structures, DOS and charge transfer

The dopants increase the stability of the adsorbing molecules by improving the orbital hybridization with the MoS_2 substrate and causing charge transfer.

The effect of Cu impurities embedded in ML MoS_2 on the adsorption of gases NO, NO2, NH3 and O2 were investigated by Zhao et al [18]. The Cu atom is strongly attached to the S vacancy and shows high adsorption energy for NH3 that acts as an electron donor, whereas, all the others act as electron acceptors. This can be seen from Fig.7 which shows side view of the stable atomic structures and charge difference densities for the adsorption of the different gas molecules. The yellow and cyan regions in the charge density plots represent the positive (electron accumulation) and negative (electron depletion) values respectively, and we see for NH3 that the cyan color is predominant indicating it to be an electron donor.

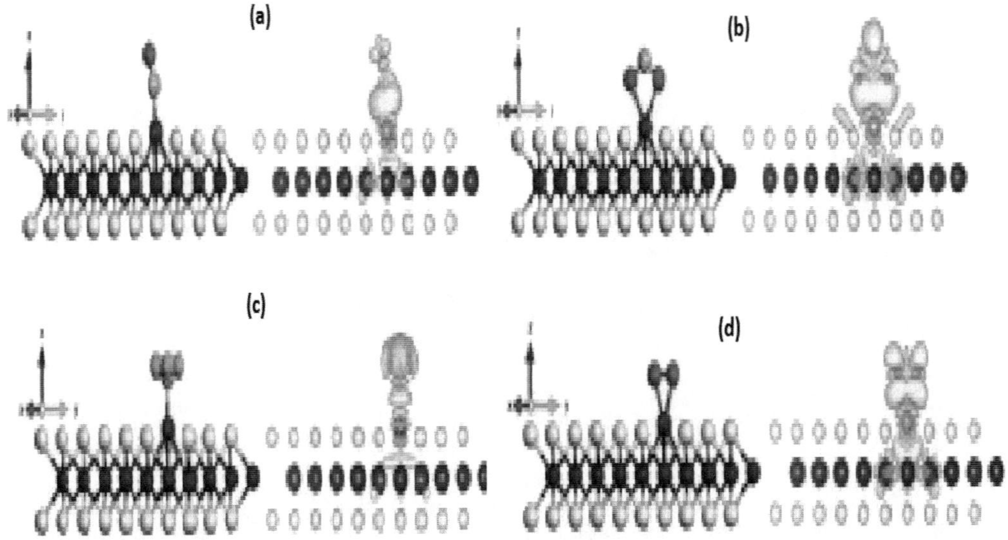

Fig. 7 Side views of the geometric structures and the charge density difference of a) NO b) NO_2 c) NH3 and d) O_2 adsorption on MoS2-Cu system, the pink, gray, violet, red, blue, and yellow spheres represent the H, N, Cu, O_2, Mo and S atom, respectively [18]

The hybridization of orbital between the TM Cu atom and the gas molecules gives rise to a change in the charge densities as observed. The environment around the dopants and nearby Mo atoms is changed by relaxations and chemical interactions to form an effective way of changing the transport properties of MoS_2 and increasing their performance as sensors. This exciting field of investigation of MSNs sensors through defect doping requires more first principles modeling for technological and feasibility assessments.

5. CONCLUSIONS AND OUTLOOK

Toxic gas sensing plays an important role in many walks of our life from environmental protection, human health to agriculture. Doping and defect substitution is a versatile and new tool for changing the chemical and electronic properties of 2D layered materials and boosting the applications of these materials. In the past decade, although, extensive research on Graphene as a gas sensor was conducted, the zero bandgap limited its potential and applicability. This is overcome in MoS_2 which is a direct bandgap semiconductor, with a large surface to volume ratio similar to graphene and can exhibit excellent sensing performance. MoS_2 as a 2D layered material has unique properties and the current focus is on defect engineering of MSNs for high sensitivity and selectivity. The large surface to volume ratio, bandgap and cheapness makes MSNs very attractive for gas sensor applications. Chemical doping is an effective way to alter the binding configurations and improve and enhance the application as a gas sensor. This promising method requires further investigations and research for industrial scale mass productions and for fabricating robust, low cost, highly sensitive and selective sensors that can be tested for best performance, reliability and repeatability to become a reality. It is hoped that this focused account of important first principles work on this new and promising research field will guide and encourage scientists and experimentalists to explore further.

References

[1] Basu S and Bhattacharyya P, 2012, *Sens. Actuators*, B, **173**, 1
[2] He Q, Wu S, Yin Z and Zhang H, 2012, *Chem. Sci.*, **3**, 1764.

[3] Ratinac K R, Yang W, Ringer S P and Braet F, 2010, *Environ. Sci.Technol.*, **44**, 1167.
[4] Leenaerts O, Partoens B, Peeters F M, 2008, *Phys. Rev.* B **77,** 125416.
[5] Dai J et al, 2009, *Appl. Phys. Lett.* **95,** 232105
[6] Yue Q, Shao Z, Chang S, Li J, 2013, *Nanoscale Res. Lett.* **8**, 425
[7] Donarelli M et al, 2013, *Sens. Actuators* B **207** Part A, 602–613.
[8] Li Hai et al, 2012, *Small* **8**, 63-67
[9] He Q et al, 2012, *Small* **8**, 2994.
[10] Late D J et al, 2013, *ACS Nano* **7**, 4879–4891.
[11] Perkins F K et al, 2013, *Nano Lett.,* **13**, 668–673
[12] Shur M et al, 2014, *IEEE Sensors 2014*, Valencia, Spain, November 2-5, 2014.
[13] Zhang S L et al, 2014, *Curr. Appl. Phys.,* **14**, 264.
[14] Huang Y et al, 2015, *Nanoscale* **7**, 19358–19376.
[15] Ma Dongwei et al, 2016, *Applied Surface Science* **383**, 98–105
[16] Ding K et al, 2016, *Vacuum* **130**, 146-153
[17] Luo Hao et al, 2016, *Chemical Physics Letters* **643**, 27–33
[18] Zhao B et al, 2016, *Applied Surface Science* **382**, 280–287
[19] Najmaei S et al, 2015, *Acc. Chem. Res.*, **48** (1), pp 31–40
[20] Najmaei S et al, 2013, *arXiv:***1301**.2812
[21] Zhou Wu et al, 2013, *Nano Lett.,* **13**, 2615−2622
[22] Sun L et al. 2015, *MRS Bulletin*, **40**, Issue 1 (Frontiers of in situ electron microscopy) 29-37
[23] Komsa H P et al, 2012, *Phys. Rev. Lett.* **109**, 035503
[24] Komsa H P et al, 2015, *Phys. Rev.* B **91**, 125304.
[25] Dai J et al, 2009, *Appl. Phys. Lett.* **95**, 232105
[26] Dai J et al, 2010, *Phys. Rev.* B **81**, 165414
[27] Gutes Albert et al, 2012, *Nanoscale,* **4**, 438
[28] Ma C et al, 2014, *Sci. China Chem.* **57**, 911–917.
[29] Liu X Y et al, 2014, *Appl. Surface Sci.*, **313**, 405–410.
[30] Niu F et al, 2013, *J. Mater. Chem.* A, **1**, 6130–6133.
[31] Zhang Y H et al, 2009, *Nanotechnology*, **20,** 18
[32] Cho B et al, 2015, *ACS Appl. Mater. Interfaces* **7**, 16775–16780
[33] Liu B et al, 2014, *ACS Nano* **8**, 5304–5314.
[34] Lee K et al, 2013, *Adv. Mater.* **25,** 6699–6702.

IOP Conf. Series: Materials Science and Engineering **305** (2018) 012002 doi:10.1088/1757-899X/305/1/012002

Olive mill wastewater treatment in Jordan: A Review

Abeer Al Bawab[1,2*], Noor Ghannam[1], Saida Abu-Mallouh[2], Ayat Bozeya[2], Rund A. Abu-Zurayk[2], Yazan A. Al-Ajlouni[3], Fida'a Alshawawreh[1], Fadwa Odeh[1], Muna A. Abu-Dalo[4*]
[1] *Chemistry Department, The University of Jordan, Amman, 11942, Jordan,*
[2] *Hamdi Mango Center for Scientific Research, The University of Jordan, Amman, 11942, Jordan*
[3] *New York University Abu Dhabi, Saadiyat Island Abu Dhabi 129188, United Arab Emirates*
[4] *Department of Applied Chemical Sciences, Jordan University of Science & Technology, Irbid-22110, Jordan*
drabeer@ju.edu.jo
maabudalo@just.edu.jo

Abstract The environmental impact of olive mill wastewater (OMW) pollution is a public concern. OMW contains high levels of phenols, organic compounds, chemical oxygen demand (COD), biological oxygen demand (BOD), microorganisms, nutrients, and toxic compounds. The treatment of OMW has been investigated by many researchers in the Mediterranean region, using several treatment techniques to remove contaminants from OMW. These techniques include chemical, biological, physiochemical, and biophysical techniques. Surfactants and some adsorbents were used in chemical techniques, anaerobic and aerobic in biological techniques, while the combined treatment methods used Electroosmosis, ozonation and electrocoagulation processes as physiochemical methods, and ultrasonic irradiation combined with aerobic biodegradation as biophysical method. The effects of OMW, whether treated or untreated, have been evaluated on both plants' growth and soil properties. The treatment methods as well as the environmental impact of OMW in Jordan were summarized in this review.

1. Introduction

Olive oil production is considered as a major industry in Jordan with approximately 21.5 thousand tons per year. The process of olive oil extraction generates around 200,000 m^3 of olive mill wastewater (OMW), which also known as Zebar [1,2].

OMW generally characterized as acidic dark colored water which contain ample amounts of organic compounds (COD, BOD), total dissolved solids (TDS), and high concentration of phenols and potassium [3-5]. In Jordan, according to a recent study carried out by Azzam et al,.2015, OMW contains high phenolic contents (360 mgL^{-1}), high COD (40,000 mgL^{-1}), high total suspended solids TSS (20,700 mgL^{-1}), high total dissolved solids TDS (27,700 mgL^{-1}), high concentrations of cations and anions (Ca 2.11 mgL^{-1}, Cu 0.94 mgL^{-1}, K 25.18 mgL^{-1}, Mg 1.90 mgL^{-1}, Na 0.68 mgL^{-1}, Zn 33.43 μgL^{-1}) and a low pH value, estimated to be approximately 4.6.

Majority of mills in Jordan discharge their OMW without any treatment due to lack of knowledge, complexity, and costs of treatment or transport to a landfill site. The treatment of OMW is not only desirable, but also necessary. This is because if untreated OMW is discharged into rivers or streams, it will deplete the dissolved oxygen, thereby causing fish death and other undesirable effects, including malodorous gases produced due to the decomposition of organic materials. Other compounds also contribute to the toxicity of OMW, these include for instance, lipids and humic-like substances. Untreated OMW also contains numerous pathogenic, microorganisms and toxic compounds (especially phenolic compounds), which affects human health negatively.

On the other hand, the disposal of OMW to the environment has beneficial effects, as they are rich in nutrient, especially potassium. Potassium can stimulate plant growth and could be used as soil conditioner, biomass fuel, compost, or as starting material to obtain valuable products such as antioxidants, enzymes, and biogas fuel [6-10]. Additionally, olive mill solid residue offers a potential for the removal of heavy metals by biosorption [11].

Several methods have been proposed to reduce the environmental impact of OMW. These methods are classified into physical, chemical, and biological methods, including floatation and sedimentation [12], sand filtration [13,14], ozonation [15], membrane filtration [16-18], neutralization with addition of acid advanced chemical oxidation (Fenton reaction) [19-20], adsorption by activated carbon, aerobic and anaerobic digestions [9,10,21], and magnetic nanoparticles [22], surfactant enhanced aquifer remediation technology [23], and modified surfactant [24,25].

In recent years, there has been growing interest to develop environmentally friendly and cost-effective treatment methods for OMW. Upon these bases, the aim of this review is to explore various methods and technologies for treatment, recovery, and reuse of OMW in Jordan.

2. Treatment methods of OMW

In Jordan, many researches in the literature have focused on attempting to find efficient and cost-effective treatment method for OMW. Previously, various methods have been tested to achieve this aim. These methods have drawn on aspects from chemical, biological, and physical sciences, and have often used an inter-disciplinary approach combining more than one method.

2.1. Biological treatment methods

Biological treatment methods used for OMW treatment are ultimately aerobic activated sludge and anaerobic digestion, focusing heavily on the ability to remove organic matter and inorganic nutrients. These processes are reliable, environment friendly, and cost-effective [26-30].

2.1.1 Anaerobic

During anaerobic processes, biomass waste is converted to biogas (by bacteria in the absence of oxygen) and compost [29]. The produced biogas (mainly methane) has an economic importance, as it can be used to generate heat and electricity [31].

Up-flow anaerobic sludge blanket reactor (UASB) is considered as the most popular bioreactors to treat agro- industrial wastewaters characterized with high organic load [32].

The impact of the removal efficiency and the observed biomass yield on nutrient requirements for anaerobic treatments of OMW was studied by Ammary, 2004 using laboratory scale anaerobic sequencing batch reactors. The researcher found more than 80% of COD removal could be achieved by varying the COD: N: P ratio. In addition, the observed biomass yield was about 0.06 kg volatile suspended solids (VSS) per kg of COD degraded. Thus, it was concluded that OMW have sufficient nitrogen and phosphorous concentrations, and the addition of such nutrients was not necessary [33].

These results were again supported when the characteristics of Jordanian OMW were investigated [34]. In this study, the use of anaerobic sequencing batch reactor (ASBR) was also tested for its suitability for onsite treatment, and a COD removal of 83% was achieved.

Furthermore, Sobhi et al. conducted a study in 2007 aiming to identify the most efficient anaerobic treatment for OMW. It was found that the (UASB) reactor achieved COD removal efficiency in the range 75–85 % at a Hydraulic Retention Time (HRT) of 5 days with an influent COD concentration of about 40 gL^{-1} and Organic Loading Rate (OLR) of 7–8 g $CODL^{-1}$. d^{-1} [35].

As the COD is a major OMW pollutant, an UASB was constructed in 2009 by Khatib et al. to reduce COD level under the specified parameters. After eight months of operation program, it was shown that 46%-84% of COD removal was achieved and that the organic load was reduced from 27,000 mgL^{-1} to less than 5,000 mgL^{-1}, which permits the direct discharge of OMW municipal wastewater treatment plants. Biogas (mainly methane) was also collected and used [36].

2.1.2 Aerobic

Naturally occurring microorganisms work as an effective factor in wastewater treatment. These microorganisms include bacteria, fungi, protozoa and other microbes. They ultimately live on wide variety of complex compounds contained in wastewater. In aerobic treatment processes, bioreactors are used to provide microorganisms with optimum growth conditions by the addition of soluble oxygen as well as organic and nitrogenous compounds. During this process, microorganisms act as decomposers that are responsible for the oxidation of complex organic compounds to return them to simple forms of carbons that could safely be returned to the environment [37]. Additionally, other microorganisms are capable of producing biofuel by oxidizing phenolic compounds [38].

For example, the fungus *Pleurotus sajor-caju* was used to treat OMW and reduce its toxicity, making it suitable as a fermentation medium by yeast to produce ethanol. The maximum ethanol production (14.2 gL^{-1}) was obtained after 48 hour of yeast fermentation using 50% diluted OMW that was thermally processed and pretreated with *P. sajor-caju* [38].

Another study using *P. sajor-caju* was performed to evaluate its effect on OMW treatment. In this study, it was reported that the OMW could be a suitable media for *P. sajor-caju* ligninolytic enzymes, which is involved in the reduction of toxic compounds in OMW. It was possible using this fungus to significantly reduce phenolic compounds content (from 5.7 to 2.1 gL^{-1}) along with enzymes secretion and action. By the end of this treatment process, a complete degradation of some phenolic and non- phenolic compounds was achieved and the color of OMW was reduced by more than 60% [39].

2.2. Chemical treatment methods

2.2.1 Adsorption

Continuous packed bed of zeolite was used to investigate the possibility of OMW treatment by Abu Al-Haija, 2008. Pilot-scale column was designed and operated under various feed flow rates using different zeolite particles sizes. Jordanian mined zeolite packing the bed was treated and activated as a part of this investigation. The results showed a high reduction in phenolics compound and COD using the smaller zeolite particles size, the removal was 98 % for both terms and decreased with increasing particle size as well as operating condition [40].

Sedimentation, filtration, and flotation experiments were conducted by Gharaibeh et al., 2008 to treat OMW, using Azraq bentonite (AB), reddish volcanic tuff (RVT), lime (CaO), aluminum sulfate (alum), ferric chloride, and sodium carbonate. Two parameters were studied; OMW turbidity removal and COD removal. The results revealed that alum, lime and AB have a high efficiency in turbidity and COD removal, while RVT, ferric chloride, and sodium carbonate did not have significant effects. The turbidity removal using alum, lime and AB was 95%, 99%, and 96%, respectively, while the COD removal was 65%, 69%, and 37.5%, respectively. However, given its high cost it was not recommended to use alum, whereas lime and AB are inexpensive and locally available and could be used easily [41].

The adsorption of phenols and organic compounds was also investigated. Different pretreatment processes of OMW such as sedimentation and filtration were used. This was followed by batch adsorption process using activated clay as the adsorbent. The maximum reduction of organic load was about 71%, while the removal of phenols was about 81% [42]. In another study, Azzam et al., using activated carbon as the adsorbent, the maximum adsorption capacity for the tested concentrations of activated carbon is reached in less than 4 hour. Phenols were about 94% removed, while for organic matter in general it reached about 83% [43]. This was followed by batch adsorption process using activated carbon for one, two and three stages process. The pretreatments showed an effective reduction of COD, phenols, and total solids content. It was more effective to use a two stage countercurrent adsorption process compared to a one-stage adsorption process, however, the use of a three-stage adsorption showed only a slight improvement compared to the two-stage process. For example, a treatment protocol composed of a three-stage countercurrent adsorption process using activated carbon of concentration 24 $gL^{-1}$of OMW was able to reduce the COD from 60000 mgL^{-1} down to 22300 mgL^{-1}, while phenols were reduced from 450 to 15 mgL^{-1} [44].

Naturally occurring Jordanian clay was also tested as an adsorbent material for OMW treatment, while COD and phenols levels were monitored. Natural clay was calcined at different temperatures (350–550°C) and some were further treated with different HCL concentrations (1, 3, and 5 M-HCl solutions) at 85°C. The treatment was done in batch experiment to identify the optimum conditions, which were then used for a continuous packed bed treatment system. In this investigation, both COD and phenols of raw OMW were reduced by 10–20% in batch experiments. Additionally, the reduction in COD in the continuous packed bed experiment reached 50%. Such reductions are considered relatively significant, considering the large amounts of organic matter involved (COD~40,000 mgL^{-1}) [45].

2.2.2 Coagulation advanced oxidation

Lafi et el., 2010 proposed the use of a combination of two treatment processes for the removal of organic pollutants from the OMW. These processes included a single coagulation step in which they used Al^{+3} and Fe^{+3} ions as coagulant, followed by a single advanced oxidation process, AOP, step. The study used UV, O_3, O_3/UV and H_2O_2/UV depending on the operating time. For the AOPs, the concentration of the chemical oxygen demand (COD) was measured in the effluent of the treated wastewater. It was found that the COD removal achieved using Al^{+3} was 54%, while for Fe^{+3} ions it reached 58% at pH 9. However, when advanced oxidation process was used alone, the percent COD removal ranged between 10 and 39%. The percent removal of the COD concentration was achieved using the combined processes, coagulation and AOPs (O_3 90%, O_3/UV 95% and H_2O_2/UV 94%) [46].

2.2.3 Surfactants

Surfactant enhanced aquifer remediation (SEAR) technology was applied to reduce phenolic compounds content in OMW using modified surfactant, which is a novel extended surfactants sodium polypropylene oxides sulphate of the type (branched hydrocarbon chain)-(propoxyle group)-(sulphate). Phase diagram was modeled using phenol as a standard compound. This study indicated that the needed amounts of surfactants to extract the phenol can be determined using the phase behavior of phenol, water, and proposed surfactant [23].

Another study in the literature proposed new modified surfactants. These included surfactants combined with cationic hydrotropes tetra butyl ammonium bromide (TBAB) in different molar ratio, namely: sodium linear alkyl- polypropylene-oxide-polyoxyethylene- sulfate (X-AES), Sodium polypropylene oxides sulfate (L167–4S), and Linear alkyl benzene sulfonate (LAS). Similarly, the amount of the needed surfactant to extract the phenol was determined using the three-dimensional phase behavior of the phenol, water, and the proposed surfactants. The results of this study showed that only polypropylene oxides sulfate (L167–4s) modified surfactant can be used for real sample which showed a 95 % COD removal [24].

Moreover, Al Bawab et al., 2018 used cost effective media of two types of granular carbons. Granular Activated Carbons non-oxidized (GAC), and Granular Activated Carbons oxidized (GAC-OX), particles impregnated with surface active materials (nonionic surfactants (Tween 80, Span 20, Span 80, Span 85, and Brij 93) at three different concentrations (10, 30, 55 mM) for OMW treatment. Experimental results for 24 hours showed that 10 mM nonionic surfactant concentration was the optimum concentration used in media preparation, which is below Critical Micelles Concentration (CMC). The results of percent phenol removal using the 10 mM impregnated media GAC-OX-Span 20, GAC-OX-Span 80, GAC-OX-Brij93, GAC-OX- Tween 80, and GAC-OX-Span 85, were 83.64, 79.55, 73.97, 66.54, and 31.96%, respectively. While, on GAC for GAC-Span 20, GAC-Brij 93, GAC-Span 80, GAC-Tween 80, and GAC-Span 85, were 82.56, 67.65, 64.68, 33.45, and 29.36%, respectively. The results of media, which was soaked with OMW for 15 days, showed an increase in percent removal of phenol using of GAC-OX-Span 20, GAC-OX-Span 80, and GAC-OX- Brij 93, were 92.94, 91.45, and 78.07% respectively. While for GAC-Span 20, GAC-Span 80, and GAC-Brij 93, percent removal was 97.03, 86.25, and 79.25% respectively. After 15 days, the percent removal was relatively reduced.

Finally, the media were tested at different temperatures (7, 25, and 35°C). The results showed that there was no change in percent removal with temperature. Additionally, the shaking condition was not significant. Therefore, these media provide an economical treatment strategy [25].

2.3. Physicochemical treatment methods

2.3.1. Electroosmosis process

Electro-osmosis dewatering (EOD), is a technique that extracts water by placing a colloidal material between two electrodes. This technique is based on the electrostatic effect of the electrochemical double layer that is formed at the particle water interface of the colloidal material. Near the solid-liquid interface the polarization of water molecules causes an electrical double layer to form [47].

Additionally, this process that OMW goes through affects its pH, voltage, current and bed height, and these effects have been investigated independently. This process resulted in converting the pH of the water from acidic to basic and reduction in COD and total dissolved solid (TDS) levels. EOD process was enhanced by increasing the voltage or the current and by the addition of alum or an electrolyte. Furthermore, it was reported that sodium chloride is the most effective electrolyte in EOD process [48].

2.3.2 Ozonation and electrocoagulation processes

OMW could be successfully treated by applying electrocoagulation accompanied with ozonation. This was manifested by Bani Salameh 2015, where the electrocoagulation process resulted in 82.5% and 47.5% removal of TSS and COD respectively after 70 minutes at 45mA/cm^2, by using coupled iron–aluminum electrodes. The subsequent ozonation reveals high levels of organic compound elimination [49].

2.4. Biophysical treatment methods

A combined ultrasonic irradiation and aerobic biodegradation treatment was used for the reduction of toxic phenolic compounds from greenish black (GB) and dark brown (DB) OMW by Al-Qodah et al., 2014. Several parameters were tested for their effect on phenols, COD and BOD degradation. Parameters tested include the duration of ultrasonic irradiation, ultrasonic power intensity, and ultrasonic frequency. The results revealed that total phenols were degraded by 81% when it was exposed to 90 min to ultrasonic field. It was demonstrated that phenol degradation followed a first-order kinetics model with rate constant, k, of 0.0083 and 0.0077 min^{-1} for (GB) and (DB) OMW samples, respectively. In the aerobic degradation step, the COD consumption followed the Grau kinetic model. The order of COD degradation rate, is about 1.13 and 1.27, whereas, the Grau kinetic constant, $k = 0.0218$ and 0.0149 h^{-1} for GB and DB OMW, respectively. The maximum removal of COD was approximately 80% [50].

2.5. Biochemical treatment methods

Different stage processes of advanced oxidation with ozone (O_3), an aerobic biodegradation treatment and photodegradation by UV radiation was used for COD removal from OMW. For both single-stage treatment of O_3 and two-stage treatment of O_3/UV, the COD remains quite high. However, a combination of biological and UV/O_3 process for the OMW treatment showed an alteration in the reduction of the COD. Biodegradation of UV/O_3 pretreated OMW had the highest removal levels while the percent of COD removal reached approximately 91%. The decay of chemical oxygen demand follows a first-order models for advanced oxidation and pseudo first-order models for biodegradation processes as the kinetic study showed [51].

3. Environmental impact of OMW

Given the high abundance of organic substances, phenols, and other toxic compounds and microorganisms in OMW, OMW has significant polluting concerns on both environmental and biological systems. Often, OMW is collected in open ponds and left to dry throughout the summer season, thus producing pollution to underground water and the soil surface in addition to unpleasant odor [52, 53].

Consequently, there is an urgent need to raise awareness, improve technical skills and change the public's attitude toward this issue. Change must be advocated for and policy-makers must focus on intervening to utilize and manage OMW, especially given its economic value [54]. However, other studies did not recommend the use of either treated or untreated OMW in agriculture because of the high dissolved solids concentration within it, and it was advised to use natural occurring materials such as bentonite and lime as a solution to treat OMW [41].

Generally, to achieve sustainable agricultural practices, OMW applications should be carefully adapted, while taking into consideration several issues, including climatic conditions, soil properties, and biological systems.

In an experiment carried out by Rousan, 2007, barley crops were treated with OMW for two years during 2005 and 2006. During this time period, several factors were studied, including dry matter, leaf area index, plant population, and grain yield at harvest. The results of dry matter and grain yield were comparable when treated and untreated OMW used, however the untreated OMW produced necrosis of barley leaves and led to slow secondary stems emission [55].

The effect of untreated OMW on soil properties, oil quality, plant performance, oil content, and fruit yield was studied by Ayoub et al., 2014. Fresh OMW was applied on the soil surface of orchards planted with 15 years old olive trees (cultivar Nabali Muhassan) at five different application rates; 0 Lm^{-2}, 5 Lm^{-2}, 10 Lm^{-2}, 20 Lm^{-2} and four equal doses of 20 Lm^{-2}. It was indicated that OMW could increase the concentrations of potassium, organic matter, phenols, and total microorganisms in treated soil when compared with control soil, but that was not considered a negative effect. Moreover, this soil could serve as an efficient fertilizer. Furthermore, and regarding plant performance, it was found that fruit set, shoot growth, fruit yield and photosynthesis were significantly enhanced [56].

Soil is considered an indicator for the health and integrity of ecosystem, thus its measurement provides valuable insights in research. Notably, any contamination or pollution in the soil will negatively affect plant growth and other vital living systems. Biological soil quality index (QBS-ar) was used as a valuable tool to study the impact of OMW on the physiochemical and biological properties of soil. Wahsha et al., 2014 used this tool to study soil samples collected from northwest Jordan. The results showed a decrease in the biological quality of soil samples polluted with OMW according to QBS-ar due the alteration in chemical and physical characteristics of soil microhabitat [57].

OMW contains a substantial amount of oil, which might affect soil thermal conductivity. Thus, an experimental work investigated the effect of OMW on thermal conductivity of three soil types (sandy clay, silty clay and clay) using heating and cooling methods under laboratory conditions. The results showed a decrease in thermal conductivity for all soil tested at different concentrations of OMW (20%-100%). Thermal conductivity ranged from 0.89-0.71, 0.77-0.63, 0.80-0.57 W/mK for sandy clay, silty clay, and clay, respectively. Higher thermal conductivities were obtained from heating methods [58].

4. Summary and Conclusion

The purpose of this exploratory review study was to provide an overview of the OMW treatment methods and the environmental impact of each method, with specific focus of the application of these methods in Jordan. In summary, there is no ideal solution in the area of treatment of OMW. Rather, several environmentally friendly and economically viable solution are presented and proposed. These methods include biological treatment (anaerobic and aerobic), chemical treatment (adsorption, surfactants), physicochemical treatment (Electroosmosis, ozonation and electrocoagulation) and biophysical treatment (combined ultrasonication irradiation and aerobic).

Previously, studies in the literature examined the efficiency of these methods by using phenols removal percentages and/or COD removal percentage as a measure of efficiency. The best removal percentages using different methods were reported in the literature as follows:

- Anaerobic: 80% COD removal.
- Aerobic: 63% Phenols removal.
- Adsorption: 98 % for both COD and phenols removal.
- Surfactants: 95% COD removal.
- Ozonation and electrocoagulation: 47.5% COD removal.
- Combined ultrasonication irradiation and aerobic: 80% for both COD and phenols removal.
- Chemical treatment methods (adsorption, surfactants) are found to show highest efficiency of treatment of OMW (98%, 95%).

As a conclusion from the literature reviewed above, it is plausible to state that OMW are important from an environmental perspective. Ultimately, OMW are currently considered to be a waste in most situations. However, with proper treatment, it can be become a recovered resource. The current study advances our knowledge of OMW treatment method by providing a rigorous review of the current methods, and providing a setting for potential combination of different methods. Future research should continue to explore various treatment methods and combine different approaches to utilize the most appropriate approach in Jordan.

References:

[1] Ministry of Agriculture 2012. Annual Report. Department of Agricultural Economy, The Hashemite Kingdom of Jordan.
[2] Ministry of Agriculture 2014. Annual Report. Department of Agricultural Economy, The Hashemite Kingdom of Jordan.
[3] Coskun T, Debik E and Demir NM 2010. Treatment of Olive Mill Wastewaters by Nanofiltration and Reverse Osmosis Membranes. Elsevier. pp. 65–70
[4] Tsagaraki E, Lazarides H N, and Petrotos K B 2007 Olive Mill Wastewater Treatment. In: Oreopoulou V., Russ W. (eds.) *Utilization of By-Products and Treatment of Waste in the Food Industry*. New York: Springer. p. 133-158.
[5] Niaounakis M and Halvadakis C 2006 Olive processing waste management: literature review and patent suurvey. 2nd Edn., Elsevier, Amsterdam, The Netherlands, pp: 514
[6] Cabrera F, López R, Martinez-Bordiú A, de Lome E D and Murillo J M 1996 Land treatment of olive oil mill wastewater. *Int. Biodeterior Biodegrad* **38** 215–25
[7] Mekki A, Dhouib A, Aloui F and Sayadi S 2006 Olive wastewater as an ecological fertiliser. *Agron Sustain Dev* **26** 61–67

[8] Schaumann G.E., Borisover M., Nasser A., Bukhanovsky N., Hasan J., Marei Sawalha A. 2010. Potential effects of olive oil production waste water on soil quality. Acta Hortic. 888, 337–344.

[9] Azbar N, Keskin T and Catalkaya E C 2008 Improvement in anaerobic degradation of olive mill effluent (OME) by chemical pretreatment using batch systems. *Biochem Eng J* **38** 379–83

[10] Azbar N, Keskin T and Yuruyen A 2008 Enhancement of biogas production from olive mill effluent (OME) by co-digestion. *Biomass Bioenerg* **32** 1195–201

[11] Martinez-Garcia G, Bachmann R, Williams C, Burgoyne A and Edyvean R 2006 Olive oil waste as a biosorbent for heavy metals. *Int Biodeter Biodegrad* **58(3-4)** 231-38.

[12] Achak M, Ouazzani N, Yaacoubi A and Mandi L 2008 Modern olive mill effluent characterization and their treatment by coagulation–flocculation using lime and aluminum sulphate. *Revue Des Sciences De L'Eau* **21** 53–67

[13] Achak M, Mandi L and Ouazzani N 2009 Removal of organic pollutants and nutrients from olive mill wastewater by a sand filter *J Environ Manage* **90** 2771–79

[14] Achak M, Ouazzani N and Mandi L 2009 Treatment of modern olive mill effluent by infiltration-percolation on a sand filter *Revue Des Sciences De L'Eau* **22** 421–33

[15] Chedeville O, Debacq M and Porte C 2009 Removal of phenolic compounds present in olive mill wastewaters by ozonation. *Desalination* **249** 865–69

[16] Paraskeva C A, Papadakis V G, Tsarouchi E, Kanellopoulou D G and Koutsoukos P G 2007 Membrane processing for olive mill wastewater fractionation. *Desalination* **213** 218–29

[17] Dhaouadi H and Marrot B 2008 Olive mill wastewater treatment in a membrane bioreactor: process feasibility and performances *Chem Eng J* **145** 225–31

[18] Akdemir E O and Ozer A 2009 Investigation of two ultrafiltration membranes for treatment of olive oil mill wastewater *Desalination* **249** 660–66

[19] Gomec C Y, Erdim E, Turan I, Aydin A F and Ozturk I 2007 Advanced oxidation treatment of physico-chemically pre-treated olive mill industry effluent *J Environ Sci Heal* B **42** 741–47

[20] Zorpas A A and Costa C N 2010 Combination of Fenton oxidation and composting for the treatment of the olive solid residue and the olive mile wastewater from the olive oil industry in Cyprus *Bioresour Technol* **101** 7984–87.

[21] Boubaker F and Ridha B C 2008 Modelling of the mesophilic anaerobic codigestion of olive mill wastewater with olive mill solid waste using anaerobic digestion model No. 1 (ADM1) *Bioresour Technol* **99** 6565–77

[22] Nassara N, Ararb L, Mareia N, Abu Ghanim M, Dwekatb M and Sawalha S 2014 Treatment of olive mill based wastewater by means of magnetic nanoparticles: Decolourization, dephenolization and COD removal *Environ Nanotechnol Monit Manag* **1(2)** 14–23

[23] Odeh F, Al Bawab A, Fayyad M and Bozeya A 2013 Surfactant enhanced olive oil Mill wastewater remediation. *APCBEE Procedia* **5**, 96 – 101

[24] Al Bawab A, Alshawawreh F, Abu-Dalo M, Al-Rawashdeh M and Bozeya A 2017 Separation of soluble phenolic compounds from olive mill wastewater (OMW) using modified surfactants *Fresenius Environ Bull* **26(3)** 1949-58

[25] Al Bawab, A., Ghannam, N., Abu-Zurayk, R., Odeh, F., Bozeya, A., Abu Mallouh, S., Al-Rawashdeh, N., Abu-Dalo' M." Olive Mill Wastewater Remediation with Granular Activated Carbon Impregnated with Active Materials" *Fresenius Environmental bulletin*. Accepted Jan, 2018.

[26] Sarika R, Kalogerakis N and Mantzavinos D 2005 Treatment of olive mill effluents Part II. Complete removal of solids by direct flocculation with poly-electrolytes. *Environ Int* **31** 297-304.
[27] Paraskeva P and Diamadopoulos E 2006 Review technologies for olive mill wastewater (OMW) treatment: a review. *J Chem Technol Biotechnol* **81** 1475-85
[28] Oktav Akdemir E and Ozer A 2006 Application of microfiltration process to the treatment of olive oil mill wastewater. *Electron J Environ Agri Food Chem* ISSN:1579-4377. 1338-1348
[29] Arvanitoyannis I S, Kassaveti A and Stefanatos S 2007 Olive oil waste treatment: A comparative and critical presentation of methods, advantages & disadvantages. *Crit Rev Food Sci* **47** 187-229
[30] Tziotzios G, Michailakis S and Vayenas D V 2007 Aerobic biological treatment of olive mill wastewater by olive pulp bacteria. *Int Biodeterio & Biodegr* **60** 209-14.
[31] Rajeshwari KV, Balakrishnan M, Kansal A, Lata K and Kishore V V N 2000. State of the art of anaerobic digestion technology for industrial wastewater treatment *Renew Sus Energ Rev* **4(2)**, 135-56.
[32] Erguder T H, Guven E and Demirer G. N 2000 Anaerobic treatment of olive mill wastes in batch reactors *Process Biochem* **36** 243–48
[33] Ammary B 2004 Nutrients requirements in biological industrial wastewater treatment. *Afr J Biotechnol* **3(4)**, 236-38
[34] Ammary B 2005 Treatment of olive mill wastewater using an anaerobic sequencing batch reactor *Desalination* **177** 157-65.
[35] Sobhiv B, Isam S, Ahmad Y, and Jacob H 2007 Reducing the environmental impact of olive mill wastewater in Jordan, Palestine and Israel. *Water Resources in the Middle East* eds Hillel Shuval and Hassan Dweik. Vol 2 pp 409-15
[36] Khatib A, Aqra F, Yaghi N, Subuh Y, Hayeek B, Musa M, Basheer S and Sabbah I 2009 Reducing the environmental impact of olive mill wastewater *Amer J Environ Sci* **5(1)** 1-6
[37] Buchanan J R and Seabloom R W 2004 Aerobic treatment of wastewater and aerobic treatment Units (Module Text). *University Curriculum Development for Decentralized Wastewater Management* 1-22
[38] Massadeh M and Modallal N 2008 Ethanol production from olive mill wastewater (OMW) pretreated with pleurotus sajor-caju *Energ Fuels* **22** 150-4
[39] Fraij A and Massadeh M 2015 Use of Pleurotus sajor-caju for the Biotreatment of Olive Mill Wastewater *Romanian Biotechnological Letters* **20(4)** 10611-17
[40] Abu Al-Haija M Azzam M O J and Al-Ghazawi Z 2008 Olives mills wastewater treatment using packed bed of zeolite. Jordan University of Science and Technology MSC. Thesis
[41] Gharaibeh S, Rababah A and Nishino H 2008 Low cost treatment and disposal of olive mill wastewater *Pollu Sol* **Sep/Oct 2008**
[42] Al-Malah K, Azzam M and Abu-Lail N I 2000 Olive mills effluent (OME) wastewater post-treatment using activated clay *Sep Purif Technol* **20** 225–34.
[43] Azzam M, Al-Malah K and Abu-Lail N I 2004 Dynamic Post-Treatment Response of Olive Mill Effluent Wastewater Using Activated Carbon *J Environ Sci Heal* A **39(1)** 269–80
[44] Azzam M, Al-Malah K, Al-Gazzawi Z and Al-Omari S 2010 Dynamic treatment response of olive mills wastewater using series of adsorption steps *Clean Soil Air Water* **38(9)** 822–30

[45] Azzam M, Al-Gharabli S and Al-Harahsheh M 2015 Olive mills wastewater treatment using local natural Jordanian clay *Desalination Water Treat* **53(3)** 627-36
[46] Lafi W, Al-Anber M, Al-Anber Z, Alshannag M and Khalil A 2010 Coagulation and advanced oxidation processes in the treatment of olive mill wastewater (OMW) *Desalination Water Treat* **24** 251-56
[47] Chen G and Mujumdar AS 2002 Application of Electrical Fields in Dewatering and Drying *Dev Chem Eng Mineral Process* **10** 429-442
[48] Al-Asheh S, Juma R, Banat F, Al-Zou'bi A and Al-Qutaish N 2011 Treatment of olive mills effluent using Electro-Osmosis Dewatering *Hydrol Current Res* **2(2)** 1-7
[49] Bani Salameh W 2015 Treatment of olive mill wastewater by ozonation and electrocoagulation processes *Glo Adv Res J Eng Technol Innov* **4(2)** 69-78
[50] Al-Qodah Z, Al-Bsoul A, Assirey E and Al-Shannag M 2014 Combined ultrasonic irradiation and aerobic biodegradation treatment for olive mills wastewaters *Environ Eng Manage J* **13(8)** 2109-18
[51] Lafia W, Shannak B, Al-Shannaga M, Al-Anber Z and A-Hasan M 2009 Treatment of olive mill wastewater by combined advanced oxidation and Biodegradation *Sep Purif Technol* **70** 141–146.
[52] Borja R, Martin M, Maestro R, Alba J, Fiestas J A 1992 Enhancement of the anaerobic digestion of olive mill wastewater by the removal of phenolic inhibitors *Process Biochem 27* 231-37
[53] Benitez F J, Beltran-Heredia J, Torregrosa J, Acero J L and Cercas V 1997 Chemical pretreatment by ozone of wastewater from olive oil mills *Technol Environ Chem* **60** 97-109
[54] Attom M and Al-Sharif M 1998 Soil stabilization with burned olive waste. *Appl Clay Sci* **13** 219-30
[55] Rousan L 2007 Farmers' perceptions and adoption of using olive mill wastewater on barley crop in Jordan. *World J Agr Sci* **3(3)** 280-84
[56] Ayoub S, Al-Absib K, Al-Shdiefata S, Al-Majalia D and Hijazeanba D 2014 Effect of olive mill wastewater land-spreading on soil properties, olive tree performance and oil quality *Sci Hortic-Amsterdam* **175** 160–66
[57] Wahsha M, Bini C and Nadimi-Goki M 2014 The impact of olive mill wastewater on the physicochemical and biological properties of soils in northwest Jordan. *Environ quality* **15** 25-31
[58] Khdair A, Abu-Rumman G 2015 Effect of olive mills wastewater (OMW) on soil thermal conductivity. *Int J Soil Sci* **10 (2)**, 84-92

The Dead Sea Mud and Salt: A Review of Its Characterization, Contaminants, and Beneficial Effects

Abeer Al Bawab[1,2*], Ayat Bozeya[2], Saida Abu-Mallouh[2], Basha'er Abu Irmaileh[2], Ismail Daqour[3], Rund A. Abu-Zurayk[2]

[1] *Chemistry Department, The University of Jordan, Amman, 11942, Jordan,*
[2] *Hamdi Mango Center for Scientific Research, The University of Jordan, Amman, 11942, Jordan*
[3] *Numeria Mixed Salts & Muds Co. Jordan,*
drabeer@ju.edu.jo

Abstract The Dead Sea has been known for its therapeutic and cosmetic properties. The unique climatic conditions in the Dead Sea area make it a renowned site worldwide for the field of climatotherapy, which is a natural approach for the provision of medications for many human diseases including unusual exclusive salt composition of the water, a special natural mud, thermal mineral springs, solar irradiation, oxygen-rich and bromine-rich haze. This review focuses on the physical, chemical, and biological characteristics of the Dead Sea mud and salts, in addition to their contaminants, allowing this review to serve as a guide to interested researchers to their risks and the importance of treatment. Beneficial effects of Dead Sea mud and salts are discussed in terms of therapy and cosmetics. Additional benefits of both Dead Sea mud and salts are also discussed, such as antimicrobial action of the mud in relation to its therapeutic properties, and the potency of mud and salts to be a good medium for the growth of a halophilic unicellular algae, used for the commercial production of β-carotene; Dunaliella.

Keywords: Dead Sea mud, Dead Sea salts, Therapeutic properties, Cosmetic properties, Climatotherapy

1. Introduction:

The Dead Sea (DS), the lowest geographical location on earth, is considered to be the biggest natural saline reserve in the world. The sea's salt content, approximated to be (348 g/L), makes its salinity 10 times the typical salinity of oceans [1]. The Dead Sea is located in the Syrian - East African rift valley and surrounded by the Moab Mountains to the east and the Judean Mountains to the West [2,3].

The atmosphere of the DS is rich in oxygen by 10% more than any other typical sea, which might be attributed to its exceptionally low altitude [4], approximated to be 396 meter below sea level [5]. However, the unparalleled salinity is not the only extraordinary characteristic of the DS, as it contains natural thermo-mineral waters, mineral muds, higher bromine content in the air, as well as high selenium content of local drinking water [6].

Due to the rarity of its atmospheric and climatic features, the DS is considered to be an attractive destination for patients who seek a medication for diseases such as psoriasis [3,7], rheumatic disorders [6, 8], and atopic dermatitis [6]. Treatments are mainly based on: (a) bathing in the DS water while exposing the skin to filtered UV radiation, and (b) mud packs prepared from highly saline black mineral mud that is rich in sulfide, which is found abundantly in the area [8, 9].

Furthermore, the distinctive combination of photobiologic characteristics and elemental properties of the DS gives this area the uniqueness that cannot be found elsewhere. The sunburn spectrum of ultraviolet light (UV) is very weak at the DS [10], because of a continuous mist that is established above water level. This mist results from remarkably high rates of water evaporation estimated at $2*10^9$ m^3/year. Consequently, most UVB sun-burning rays (290-320nm) are filtered out, which allows a better exposure to the longer wavelength UVB and penetrating natural UVA rays (320-400nm) [11]. The sun and the sea are two factors that have played a main role in the management of many different medical illnesses, especially dermatological diseases [12, 13].

Besides the essential health purposes that the DS can serve, cosmetics is a field that has not been overlooked. Over the past years, many units of DS related cosmetic products have been marketed. These natural beauty products are made from natural pure components and extracts in addition to the DS mud, salts and minerals which are extracted directly from the DS area.

Despite the great benefits associated with the dead sea products, it is important to note that risks or side effects may be associated with these products dues to the possibility of the presence of contaminants such as heavy metals, microorganisms, or radioactive elements. We believe that treatment is required to avoid risks of such contaminants, and it is considered a challenge to find a suitable technique to detect the presence of those contaminants and to remove them without affecting the benefit of DS products. Yet, barely any research is found in the literature concerning the treatment of the DS products. Only lately one study was published by our group, [14], in which we assessed some heavy metals in the DS mud and optimized a treatment methodology using chelating agent.

On these basis, this review aims to provide details regarding the characteristic of the DS natural mud and salts, as well as their contaminants. Additionally, it aims to give an overview of their beneficial effects. This work might be a valuable reference that guides interested researchers into the importance of treatment of DS mud and salts, as it is an area that is under-studied in current research, especially that the DS products are currently considered a largely emerging industry in Jordan. The wide variety of these products and the versatile importation worldwide necessitate considering a high standard of their purity and a minimum harmfulness to the end user.

2. Dead Sea Mud:

The DS mud is derived from older sediments or the red-brown soils that are usually swept back into the sea during the winter time [15]. It is well known for its therapeutic properties for skin diseases, as it contains a unique composition of minerals.

Khlaifat *et al.*, 2010 focused on studying the chemical and physical properties of DS mud samples collected from three different locations (north, middle, and south collection points). The results indicated that the chemical analysis of mud samples has revealed high CaO concentrations (20.61 - 27.86 wt.%), high CO_2 concentrations (15.47 - 25.01 wt.%) and high SiO_2 concentrations (23.74 - 33.66 wt.%) while the total soluble salts (T.S.S), chlorides and sulfates were 10.19, 4.48 and 0.056 wt.% respectively. Regarding physical properties, it was found that in individual grain size characteristic, most the samples are classified as fine grained, the liquid limit values range from 7 to 23, plasticity index range from 5 to 18, and the specific gravity ranges from 2.257 to 2.386. The variation in DS mud samples prosperities depends on the collection locations and this due to the different processes taking place at the different location [16]. Physical properties of DS mud were also studied by Arab and Alshikh, 2012 who found that the liquid limit value was about 44, plasticity index was about 15 and Plastic limit value was about 29 [17].

3. Dead Sea Salts:

The Dead Sea is the most hypersaline waterbody on earth. Its water has pH value of about 6 and contains about 348 g/L mineral salts [1].
When compared with other oceans and seas, the DS is more abundant in many elements, including chloride (212.4 g/l), magnesium (40.65 g/l), sodium (39.15 g/l), calcium (16.86 g/l), potassium (7.26 g/l), and bromide (5.12 g/l). Conversely, it has a lower concentration of sulfate (0.47 g/l), and bicarbonate (0.22 g/l) [2, 18-20].
In the period between 1959 and 1960, the DS had approximately 290 g in total of dissolved salts per liter. However, with time and due to high evaporation rates, water level decreased dramatically. This led to an increase in the salinity to become approximately 340 g/L in 1979 [21]. Since that time, the overall concentration of salts in DS has not changed significantly, with a value of 348 g/L in 2010 [1]. Despite the continuing decrease in water level and the supersaturation of DS with salts, especially sodium chloride (NaCl), the actual concentration of sodium (Na) has decreased due to the huge quantities of halite (rock salt) that precipitated in the bottom of the DS [22-24]. However, the concentrations of other soluble ions increased significantly, such as magnesium; the most prevalent divalent cation, leading to a belief that the DS is principally a magnesium chloride lake [25].
Furthermore, the DS minerals consist of specific elements that participate in regulatory activities of skin metabolism. Magnesium, potassium, and calcium are the most substantial elements present. It was proved that magnesium (Mg^{+2}) is a co-factor for phosphate transferring enzymes, and participates in balancing the regulation between cyclic adenosine monophosphate (c-AMP) and cyclic guanosine monophosphate (c-GMP). Potassium (K+) promotes CO_2 transport, and calcium is the inducer of lamellar secretion and the regulator of cell membrane permeability [26, 27].
It is also important to note that minerals have the capability in restoring moisture and enhancing intracellular water capacity due to their hygroscopic characteristics. Therefore, if absorbed into skin, they can contribute to the skin's natural moisturizing factor (NMF) [28].

4. Contaminants of Dead Sea mud and salts:

Contamination in the DS mud and salts represents a potential risk to the local population and for cosmetics producers who usually use the black sea mud which they claim to impart a relaxed feeling, nourishing the skin, activating the circulatory system and ease rheumatic discomfort [8, 29, 30]. This section will provide an overview of the most important contamination sources and how they can influence the current uses of the DS mud and salts.

4.1. Heavy metals:

Heavy metals comprise a well-known group of inorganic chemical hazards. Furthermore, it was proved that chromium (Cr), cadmium (Cd), lead (Pb), copper (Cu), zinc (Zn), mercury (Hg), nickel (Ni), and arsenic (As) are usually found at contaminated areas [31]. The abnormal accumulation of heavy metals in soils of both urban and rural environments comes because of a slow occurring geochemical cycle of heavy metals, therefore causing risks to human, animals, plants, and other ecosystems [32] .
Comparing the DS water and mud for the presence of heavy metals, DS water is found to be rich in heavy metals while the mud is found to contain very low levels of trace elements. Studies in literature show that it is more appropriate for heavy metals is to join to soluble salts in the acidic water rather than precipitating in mud [33]. These studies also revealed that the black mud usually has low content of heavy metals, thus low toxicity. Moreover, the results of sequential

extraction showed that Ni and Co were found in the carbonate fraction, Mn was joined up with iron oxide, and the residual phase contained Cr, Cu, Fe and Pb.

Abdel-Fattah and Pingitore, 2009, investigated the composition of the DS mud samples (from 3 spots in the Jordanian side of the DS) and found that toxic heavy metals present in concentrations below standard levels. In the same investigation, 16 commercial DS mud-based and mud-enhanced cosmetic products, which were marketed in Jordan and in the USA, were analyzed. Generally, cosmetic products have diluted minerals, except for cadmium, which was found in levels exceeding those in the plain DS mud samples in several of the commercial muds and in one facial mask. It was concluded that there is risk regarding mineral toxicity from DS mud or DS mud-based products [34].

Similarly, Khlaifat *et al.*, 2010, determined physical and chemical properties of 24 different DS mud samples collected from three different locations on the eastern seaside of the DS. Their results showed that the mud samples were rich in some elements (Barium, Vanadium, Strontium, lead, cadmium and zinc), although there were significant differences between mud samples collected from different locations, there was no strong correlation between the location and the elements content. The most abundant element was strontium followed by barium, vanadium and lead, with the concentration ranges of 410–810, 155–380 , 209–264 , 108–114 part per million (ppm) respectively [16].

Arab and Alshikh, 2012, conducted another investigation to measure the concentrations of trace elements in DS mud using Atomic Absorption Spectra and Polargraph instrument. The results of atomic absorption spectra approved that iron has the highest concentration 964.036 ppm, followed by Selenium 6.4 ppm, zinc 5.72 ppm, and lead 3.64 ppm. On other hand, Copper concentration were very low, with a value of 0.58 ppm, while Cadmium concentration was as low as 0 ppm. Similar results were obtained by using polagraph instrumental, confirming their validity [17].

Such studies that reveal the presence of heavy metals caused the development of numerous processes that were developed to remove high concentrations of dissolved heavy metals. These include, but not limited to, ion exchange, oxidation-reduction, precipitation-filtration, membrane separation, solvent extraction, as well as reverse osmosis [35, 36].

Moreover, much attention has been paid to applying biotechnological methods for controlling and removing heavy metal pollution in recent years. Despite this, alternative processes are available, including biosorption (metals sequestering), which utilizes various natural materials derived from bacteria, fungi, yeast, and algae. These biosorbents can decrease the concentration of dissolved heavy metal from part per million (ppm) to part per billion (ppb) level. This technique is considered to be a natural and an applicable technique to treat waste water with both low and high levels of metals due to the potency of biosorbents to sequester metal ions from diluted solutions [37].

4.2. Microorganisms:

The domestic microbial flora of DS mainly involves a limited number of microorganisms which can be divided into two main groups, obligate halophilic bacterial strains (such as Sarcina-like coccus *Halobacterium sp.*) and facultative halophytic algae (such as Dunaliella). The recorded number of microorganisms' species at DS is reported to be very low, whereas the total biomass is relatively high (about 10^5 bacteria/ml and 10^4 algal cells/ml). Two different antagonistic mechanisms are reported at DS, which are developed by both bacteria and algae. The bacteria adjust their internal inorganic ionic strength to that of the medium. On the other hand, the algae developing a mechanism for salts exclusion from the intracellular fluid using osmotic regulation and glycerol [38].

In addition to bacteria and algae, six different genera of fungi were isolated from DS water, suggesting that the DS water may have health hazard. This was concluded from a study conducted by Mbata, 2008. The study involved 100 water samples collected from DS, 68% of these samples showed fungi contamination. The recovered fungi genera included *Aspergillus versicolor* (44.1%), *Chactomium globosum* (20.6%), *Hortaea werneckii* (13.2%), *Aureobasidium pullulans* (11.8%), *Eurotium spp* (8.8%) and *Gymnascella spp* (1.5%). It was concluded that *Aspergillus versicolor* and *Chactomium globosum* are the most prevalent genera [39].

During the twentieth century, biological monitoring for the microorganisms in the DS revealed the presence of pathogenic non-halophilic *Sporohalobacter lortetii* [40]. *Haloferax volcanii, Haloarcula marismortui, Halorubrum sodomense, Halobaculum gomorrense* [1]. These are present alongside the different types of cyanobacteria, and several Dunaliella species which could grow in DS water such as *Dunaliella viridis* and *Dunaliella parva* [41, 42]. On the other hand, protozoa were not found at all due to the extreme salinity of the water [1].

Throughout the years, there was a negative balance between DS water level and salinity. On one hand, this created an environment for salt-tolerant algal genus. On the other hand, Dunaliella cells cannot survive in DS water due to this negative balance [43].

Recent studies in the literature directed their focus towards the metabolic potentials of halophilic archaea and bacteria, as well as the possibilities to use them in bioremediation applications [44]. Other studies focused on the potential of halophilic bacteria in biotechnological applications such as PHA production, extracellular protease production, halocin production and bioemulsifier production [45].

5. Beneficial effects of Dead Sea products:

5.1. Therapy

Since ancient times, DS mud has been used for treatment of various skin disorders, as it contains high concentration of minerals, allowing it to retain heat for hours and be highly absorbent [6, 46-48]. Thus, DS mud can stimulate blood circulation, enhance lymphatic flow, cleanse the skin from dead cells, and help in wounds healing and soothing irritation [47, 49].

DS salt solution, which is rich in magnesium, has many therapeutic uses. It was proved that bathing in this salt solution improves skin hydration, enhances functions of skin barriers, and reduces dry skin inflammation [50]. In addition to magnesium, other elements are present. This includes zinc, which plays a role in wound healing and epidermal regeneration [51].

Wound healing potential of natural and compounded facial masks prepared from DS black mud was tested on wounds created on the dorsum region of BALB/c mice by Abu-Al-Basal, 2012. Test mice were treated once a day for two consecutive days with 0.1% natural or compounded DS black mud or 0.2% nitrofurazone, in addition to untreated mice group used as a control. Wound healing was evaluated at day 3, 7, 14 and 21 from wounding day, by measuring weight and percentage of tissue granulation. The results showed that wound healing process (which includes wound contraction, granulation, epithelialization, angiogenesis, and collagen deposition) were accelerated when DS black mud was used compared with nitrofurazone, with note that compounded facial masks have healing property greater than that of natural black mud [52].

In addition, other medicinal properties of DS have been known for thousands of years. The major diseases that are frequently treated by DS balneotherapy are musculoskeletal and dermatologic diseases, such as rheumatoid arthritis, osteoarthritis, ankylosing spondylitis, low back pain, psoriasis, atopic dermatitis, and other joint diseases [2, 53-54]. This success in treatment comes due to the high mineral content in the DS [1, 10], and to the UVB radiation which is an attenuated radiation that leads to elevate the concentration of atmospheric oxygen and other relaxation factor [29].

Furthermore, Psoriasis is considered to be one of the most important diseases that are often treated by DS spa therapy. Several studies have proved that DS mud is efficient in Psoriasis treatment [2, 12, 29, 55-66], in addition to DS salt which also help in Psoriasis treatment due to the effects of magnesium bromide or magnesium chloride [19, 67, 68].

Psoriasis is highly affected by exposure to the sun, which acts as the main factor in the treatment process. However, the effect of solar radiation could be enhanced by bathing in DS water. This was concluded from a study conducted by Even-Paz *et al.*, 1996, when eighty-one patients with Psoriasis (plaque type) underwent treatment at the DS for four weeks. They were divided into three groups: DS water bathing only; sun exposure only; and sun exposure combined with DS water bathing. Psoriasis area and severity index (PASI) was used to measure the reduction percentage in the psoriasis area, which was scored 28.4% for the first group (when patients only bathed in DS water), 72.8% for the second group (when patients exposed only to sunbathing), and 83.4% for the third group (for patients who bathed in DS water and exposed to sunbathing). This study was also found that there were no significant seasonal differences in the results related to the sun-exposure groups [69].

Similarly, Elkayam *et al.*, 2000 used balenotherapy (treatment of diseases by bathing) and phototherapy (treatment of diseases by radiation) to treat forty-two psoriatic arthritis patients at DS area. From these forty-two, twenty-three patients were receiving additional treatment with mud packs and sulfur baths, while nineteen patients did not receive any type of additional treatment. The results revealed that both groups showed significant improvement in right and left grip, patient self assessment, morning stiffness and axial skeleton movements. Over the time, better results were observed in the group that received mud packs and sulfur baths [70].

Furthermore, Fibromyalgia (musculoskeletal disorder) is another disease that can be treated with balneotherapy at the DS. This has been extensively studied by Sukenik *et al.*, 2001 when twenty eight patients with fibromyalgia and psoriatic arthritis were treated with balneotherapy at the DS area, and were followed-up by assessment of several clinical indices such as number of active joints, duration of morning stiffness, a point count of eighteen fibrositic tender points, and determination of tenderness threshold in nine fibrositic and in four control points using a dolorimeter. The results revealed a reduction in both number of active joints and number of tender points which was reduced from 18.4 to 9 for active joints, and from 12.6 to 7.1 for tender points in men, while there was also a decrease from 13.1 to 7.5 for tender points in women [71].

A clinical trial optimized the use of combined mud bath treatment to treat fibromyalgia patients who have a poor response to drugs. This study was performed by Fioravanti *et al.*, 2007. Forty patients were submitted to a cycle of twelve mud packs and thermal baths, while forty other patients were employed as a control group. Then, patients were evaluated by FIQ (fibromyalgia impact questionnaire), tender points count, VAS (visual analogue scale) for minor symptoms, HAQ (health assessment questionnaire), and AIMSI (arthritis impact measurement scales). After sixteen weeks of treatment, the results proved the efficiency of mud bath treatment of all evaluation parameters [72].

DS spa has a therapeutic potential in atopic dermatitis, which was evidenced by Shani *et al.*, 1997 and Giryes *et al.,* 1997 when they reported 90% clearance of lesions in a study sample (n=1408) after 4-6 weeks of therapy at the DS area [65, 66].
Mud packs and sulfur baths (each separate or in combination) can be used as an effective treatment for rheumatoid arthritis patients, which was evidenced through randomized controlled trials for two weeks therapy at the DS area [8]. On the other hand, employment of mud compresses on the hands of patients suffered from swollen and tender joints could relives pain [73].
Osteoarthritis is a joint disease that affects the articular cartilage. This may cause bone to grow at the margins of joints, and leads to changes in the synovial membrane. Over the time, joints loss its normal motion which later causes swollen and pain [30]. Osteoarthritis is one of arthritis and occurs equally in men and women, and could be treated by pharmacological mediations, physiotherapy treatments or balneotherapy.
Flusser *et al.,* 2002 reported a comparison between compresses prepared from unaltered DS mud and mineral-poor DS mud. Fifty-eight knee osteoarthritis patients were selected randomly and treated as follow: forty patients were treated with compresses prepared from unaltered DS mud, and eighteen patients were treated with compresses prepared from mineral-poor DS mud. Using Lequesne index, the results indicated that knee pain can be reduced. Data suggested a better outcome for patients treated with unaltered DS mud packs compared to those treated with mineral-poor mud packs [74].
Ma'or *et. al.,* 2003 prepared compresses, bandages, warps, and dressings from small magnetic particles added to Nano-powders of DS salts. Magnetic particles were absorbed to epidermal layer of skin and provided a magnetic field that helps in pain relief when those compresses, bandages, warps, or dressings placed in a direct contact with the patient's skin, above painful region [75].
Additionally, it is important to finally note that one of the main therapeutic uses of the DS is the role of DS minerals in anti-aging. The effect of minerals on keratinocytes cultures and human skin were investigated by Soroka *et al.,* 2008. The obtained results revealed that DS minerals can decrease the expression of some aging markers, enhance proliferation, stimulate mitochondrial activity and limit apoptotic damage after UVB irradiation [76].

5.2. Cosmetics
Cosmetics products that contain either DS mud or salts have been used for long years and many companies offer product lines that feature DS minerals. Such products include bath salts, mineral mud soaps, mineral peeling soaps, hand and body lotions, eye cream, cleansing mud masks, body butter, body exfoliates, acne lotions, sunscreens, lightening cream with sun protection factor (SPF), collagen firming creams with SPF, firming night creams, scalp masks, antidandruff and numerous other shampoos, and products that have an 'anti-wrinkle effect'.
DS water is widely used in cosmetics, because of its moisturizing and smoothing properties. This is ultimately due to the high magnesium content that enhances water retaining in the skin [77, 78], and $CaCl_2$ which provides the DS water its oily feel [2, 18-20].
To prove the efficiency of DS water in skin smoothing, Ma'or et al., 1997 tested three liquid gels prepared from DS products. Those gels were applied on twenty women twice daily over four weeks. Computer-aided laser profilometry, in accordance with standard testDIN 4768 ff (German Institute of Standardisation (Deutsches Institut für Normung)) were used to determine skin roughness parameter for women at the beginning and at the end of the study. By the end of

this study, it was found that skin roughness parameter could be reduced by 40.7% when using liquid gel containing 1% of a DS mineral solution, 27.8% when using liquid gel without mineral additives, and 10.4% when using control gel without anti-wrinkle agents or the additives [77].

Zeng et. al., 2004 prepared a skin caring product that contained DS mud and salts, hydrolyzing collagen, Ginseng Radix, and optionally one or more materials selected from soybean isoflavone, sea snake bile and bamboo charcoal. The resulting product has proven improvement in skin cleansing, supplying skin with necessary nutritional ingredients and removing dead skin [79].

Fleischmann, 2004 prepared bathing solutions that contained DS salt, silica, and bicarbonate; those preparations showed effectiveness in dehydration, weight loss, improving bowel movement [80]. Another bathing product contains DS salts prepared by Hasunuma et. al., 2000, this product has the property to form bubbles when components were mixed and have an influence on skin moisturizing and conditioning [81].

Moreover, an invention was recorded in 2005 by Braun et. al. This invention concerned the preparation of skin formulations containing urea and DS mud or salts that helps in reducing skin irritation and skin damage [82].

A cosmetic preparation (formulated as cream, paste, milk or face mask) comprises mainly of DS salt, coenzyme Q10 and natural active ingredients was prepared by Beckermann 2001 as anti-wrinkling products that reduce skin wrinkles without causing irritation. The composition also comprised colloid former(s), with high molecular weight organic thickener, and was present as a stable colloidal formulation. The preparation contained also other additives such as organic solvent, inorganic thickener, surface active agent, antioxidant, antibacterial agent, gelling agent, fat, polysaccharide, oil, colorant and/or odorant [83].

Beckermann 2004 prepared face and body oil that composed of two phases; phase A (water, allantoin, DS salt, pentylene glycol) and phase B (soy oil, octyl decanol, perfume, jojoba oil), in addition to other cosmetic and dermatological formulations prepared from aqueous phase and oily phase (vicinal diol with 3-6 carbon atoms) without using of surfactants or emulsifiers [84].

Furthermore, and with regards to hair cosmetics, Robert et. al., 1997 focuses on alopecia patients and used the DS mud to prepare formulation that helps in retardation of hair loss and restoration of hair growth [85].

For aesthetic characteristics, Hwang et. al., 1998 prepared formulation containing DS mud which have an excellent impact on the prevention of dandruff, excluding itching, increase hair gloss and conditioning [86].

5.3. Additional benefits:

In addition to previously mentioned therapeutic properties of DS mud, antimicrobiocidal action is another property which is probably attributed to the combination of high salt and sulfide concentrations in plus low pH. This anitimicrobiocidal efficiency may explain the antiacne effect of facial DS mud masks because of the inhibitory effect on skin (Ma'or, unpublished observations, 1998). This was proved by Ma'or. et al., 2006 when suspensions of *P. acnes*, *E. coli*, and *C. albicans* were mixed with DS mud, then it found that the number of colonies that could be recovered declined rapidly [87].

There are also beneficial eukaryotic algae exist in the DS area, such as Dunaliella, which was reported for the first time at DS in 1940 by Elzari-Volcani. Dunaliella, is a unicellular biflagellate alga, and considered as the richest natural source of the β-carotene, the most widely used and commercially important carotenoid. β-carotene is generally used as an antioxidant, food coloring agent and source of pro-vitamin A. Thus, there is a global trend to develop anticancer

medicine from β-carotene based on its antioxidation feature. The halophilic species of Dunaliella could be also used as a source of glycerol, since it can accumulate high concentrations of glycerol [88]
In 2013 Emeish S. performed a study in Jordan aimed to investigate the probability of β-carotene production from Dunaliella isolated from the at the laboratory scale cultures, followed by subjecting the produced β-carotene to enzymatic oxidation to produce tretinoin (vitamin A). The amount of β-carotene produced during this study was 3-6% of the dry weight of cultured Dunaliella [89].

6. Conclusion:

The Dead Sea is the biggest natural saline reserve in the world. Characterization of DS mud and salts showed high concentrations of sulfates and mineral salts, which cause low pH value (approximately 6.0) of its water. DS mud and salts have been used for a long time for treatment of various disorders such as wound healing, rheumatoid arthritis, joint diseases, skin disorders and aging effects. In addition, it has been used as a constituent of several cosmetic products.
Despite the wide uses of DS mud and salts, it was found that they contain some contaminants which adversely affect the efficiency of their use such as heavy metals and microorganisms. Literature revealed that heavy metals are concentrated more in DS water, whereas microorganisms are found both mud and salts.
Thus, for safety purposes we recommended that care must be taken for the treatment of DS mud and salts before use in any therapeutic or cosmetic product.

Acknowledgement

The authors of this study would like to thank The Arab Potash Company in Jordan for their support in initiation the idea of this review and their interest in research. We also extend our gratitude to Hamdi Mango Center for Scientific Research (HMCSR) and the Deanship of Academic Research (DAR), both located in the University of Jordan (UJ). The authors would also like to thank Yazan A. Al-Ajlouni for contributing to this manuscript.

References:

[1] Oren A 2010 Lakes & Reservoirs: *Research and Management* **15** 215-222.
[2] Even-Paz Z and Shani J 1989 *International Journal of Dermatology* **28 (1)** 1-9.
[3] Abels D and kipnis V 1998 *A Dead Sea perspective clinics in Dermatology* **16** 695-698.
[4] Oumeish Y 1996 *Clinics in Dermatology* **14** 659-664.
[5] Falk B, Nini A, Zigel L and Blau H 2006 *Pediatric pulmonology* **41 (3)** 234.
[6] Halevy S and Sukenik S 1998 *Ar-chives of Dermatology* **134 (11)** 1416-1420.
[7] Even-Paz Z 1996 *Israel journal of medical sciences* **32** 511-515.
[8] Sukenik S, Buskila D and Neumann L 1990 *Annals of the Rheumatic Diseases* **49** 99-102.
[9] Sukenik S, Buskila D and Neumann L 1992 *Clinical Rheumatology* **11** 243-247.
[10] Kushelersky A P and Slifkin M A 1975 *Israel journal of medical sciences* **5** 588-590.
[11] Goldberg L H and Kushelevsky A P 1977 Proceedings of the Second International Symposium, 1976 (ed EM Farber, New York, Yorke Medical Books) p 461-463.
[12] Avrach W W 1976 Proceedings of the second international symposium on psoriasis, Stanford University. New York, NY: Yorke Medical Books, pp. 258-261.
[13] David J Abels D J and Kattan-Byron J 1985 *Journal of the American Academy of Dermatology* **12** 639-643.

[14] Abu-Zurayka R A , Bozeya A, Abu-Irmaileha B, Abu-Mallouha S, Al-Bawab A and Al-Dujaili A H 2017, Soil and Sediment Contamination An International Journal, DOI: 10.1080/15320383.2017.1313193
[15] Neev D and Emery K 1967 *Geol Survey (Israel) Bull* **41** 147.
[16] Khlaifat A, Al-Khashman O, and Qutob H 2010 *Materials Characterization* **61** 564-568.
[17] Arab S and Alshikh A 2012 *New York Science Journal* **5(1)** 112-115.
[18] Schamberg IL 1978 *International Journal of Dermatology* **17 (6)** 524-525.
[19] David M, Efron D, Hodak E and Even-Paz Z 2000 *The Israel Medical Association journal* **2 (3)** 232-234.
[20] Kudish AI, Abels D and Harari M 2003 *International Journal of Dermatology* **42 (5)** 359-365.
[21] Oren A, Gurevich P, Anati DA, Barkan E and Luz B 1995 *Hydrobiologia* **297** 173-185.
[22] Niemi T, Ben-Avraham Z and Gat JR 1997 The Dead Sea – The Lake and its Setting (Oxford University Press, New York).
[23] Gavrieli I, Beyth M and Yechieli Y 1999 Microbiology and Biogeochemistry of Hypersaline Environments, ed A Oren (CRC Press, Boca Raton) pp 121–7.
[24] Gavrieli I and Oren A 2004 Dying and Dead Seas. Climatic Versus Anthropic Causes, ed JCJ Nihoul, PO Zavialov, and PP Micklin PP (Kluwer Academic Publishers,Dordrecht) pp 287–305.
[25] Post FJ 1977 *Microbial Ecology* **3** 143-165.
[26] Collober I, Noel-Hudson M S and Wepierre J 1994. *International Journal of Cosmetic Science* **16** 149-160.
[27] Menon G K, Price L F, Bommannan B, Elias P M and Feingold K R 1994 *Journal of Investigative Dermatology* **102** 789-795.
[28] Jacobi O K 1967 *Journal of Cosmetic Science* **18** 149-152.
[29] Sukenik S, Giryes H, Halevy S, Neumann L, Flusser D and Buskila D 1994 *Journal of Rheumatology* **21 (7)** 1305-1309.
[30] Solomon D H, Bates D W, Panush R S and Katz JN 1997 *Annals of Internal Medicine* **127 (1)** 52-60.
[31] GWRTAC 1997 Tech. Rep. TE-97-01, GWRTAC, Pittsburgh, Pa, USA, *GWRTAC-E Series.*
[32] D'Amore J J, Al-Abed S R, Scheckel K G and Ryan J A 2005 *Journal of Environmental Quality* **34 (5)** 1707-1745.
[33] Momani K, El-Hasan T, Auaydeh S and Al-Nawayseh K 2009 *Jordan Journal of Earth and Environmental Sciences* **2 (1)** 50-59.
[34] Abdel-Fattah A and Pingitore N E 2009 *Environmental Geochemistry and Health* **31 (4)** 487-492.
[35] Hubicki Z, Jakowicz A and Łodyga A 1999 Adsorption and its applications in industry and environmental protection: Studies in surface science and catalysis, ed A Dąbrowski (Elsevier, Amsterdam, New York).
[36] Dąbrowski A, Hubicki Z, Podkościelny P and Robens E 2004 *Chemosphere* **56** 91-106.
[37] Wang J L and Chen C 2006 *Biotechnology Advances* **24** 427-451.
[38] Nissenbaum A 1975 *Microbial Ecology* **2** 139-161.
[39] Mbata T 2008 *Sudanese Journal of Public Health* **3 (4)** 170-172.
[40] Loret ML 1892 *The Palestine Exploration Fund* 48-50.

[41] Volcani B E 1944 The microorganisms of the Dead Sea, in: Papers Collected to Commemorate the 70th Anniversary of Dr. Chaim Weizmann. Collective volume: 71–85. Rehovoth: Daniel Sieff Research Institute
[42] Oren A 1988 Advances in Microbial Ecology, ed KC Marshall (Plenum Publishing Company, New York, NY) **Vol 10** pp 193–229.
[43] Oren A 2014 *Journal of Biological Research-Thessaloniki* **21** 23.
[44] Oren A 2015 *Current Opinion in Biotechnology* **33** 119-124.
[45] Satbhai A, Kasodekar A, Pachuau L and Bharambe N 2015 *International Journal of Current Microbiology and Applied Sciences*, Special Issue-**2** 1-17.
[46] Hodak E, Gottlieb A B and Segal T 2003 *Journal of the American Academy of Dermatology* **49 (3)** 451-457.
[47] Comacchi C and Hercegova J 2004 *Journal of the European Academy of Dermatology and Venereology* **18** 372-374.
[48] Moses S W, David M, Goldhammer E, Tal A and Sukenik S 2006 *The Israel Medical Association Journal* **8 (7)** 483-488.
[49] Carretero M I 2002 *Applied Clay Science* **21** 155-163.
[50] Proksch E, Nissen H P, Bremgartner M and Urqu-hart C 2005 *International Journal of Dermatology* **44 (2)** 151-157.
[51] Iwata M, Takebayashi T, Ohta H, Alcalde R E, Itano Y, and Matsumura T 1999 *Histochemistry and Cell Biology* **112 (4)** 283-290.
[52] Abu-Al-Basal M 2012 *Pakistan Journal of Biological Sciences* **15 (7)** 306-315.
[53] Shani J, Barak D and Levi D, 1985 *Pharmacological Research Communications* **17 (6)** 501-512.
[54] Nasermoaddeli A and Kagamimori S 2005 *Environmental Health and Preventive Medicine* **10** 171-179.
[55] Dostrovsky A, Sagher F and Even-Paz Z 1959 *Harefuah* **57** 143-145.
[56] Dostrovsky A and Shanon J 1963 *Harefuah* **63** 127-129.
[57] Montgomery B J 1979 *Journal of the American Medical Association* **241** 227-231.
[58] Azizi E, Kushlevsky A, Avrach W and Schewach-Millet M 1982 *Israel journal of medical sciences* **18** 267-270.
[59] Abels D J and Kattan-Byron J 1985 Journal of the American Academy of Dermatology **12** 639-643.
[60] Giryes H, Halevy S and Sukenik S 1994 Sixth International Psoriasis Symposium; July 20-24, Chicago, III.
[61] Abels D J, Rose T and Bearman J E 1995 *International Journal of Dermatology* **34** 134-137.
[62] Giryes H, Sukenik S and Halevy S 1995 *Journal of the European Academy of Dermatology and Venereology* **5** 44-46.
[63] Even-Paz Z, Efron D, Kipnis V and Abels D J 1996 *Journal of Dermatological Treatment* **7** 17-19.
[64] Halvey S, Giryes H, Friger M and Sukenik S 1997 *Journal of the European Academy of Dermatology and Venereology* **9** 237-242.
[65] Giryes H, Friger M, Sarov B and Halevy S *1997 International Symposium at the Dead Sea*; November 2-6, Dead Sea Israel.
[66] Shani J, Seidel V, Hristakieva E, Stanimirovic A, Burdo A and Harari M 1997 *International Journal of Dermatology* **36** 481-492.

[67] Shani J, Sharon R, Rand K and Even-Paz Z 1987 *Pharmacologist* **35 (6)** 339-347.
[68] Levi-Schaffer F, Shani J, Politi Y, Rubinchik E and Brenner S 1996 *Pharmacologist* **52 (5)** 321-328.
[69] Even-Paz Z, Gumon R, Kipnis V, Abels D J and Efron D 1996 *Journal of Dermatological Treatment* **7** 83-86.
[70] Elkayam O, Ophir J, Brener S, Paran D, Wigler I and Efron D 2000 *Rheumatology International* **19 (3)** 77-82.
[71] Sukenik S, Baradin R, Codish S, Neumann L, Flusser D, Abu-Shakra M and Buskila D 2001 *Israel Medical Association Journal* **3** 147-150.
[72] Fioravanti A, Perpignano G, Tirri G, Cardinale G, Gianniti C, Lanza C, Loi A, Tirri E, Sfriso P and Cozzi F 2007 *Rheumatology International* **27** 1157-1161.
[73] Codish S, Abu-Shakra M, Flusser D, Friger M and Sukenik S 2005 *Rheumatology International* **25** 49-54.
[74] Flusser D, Abu-Shakra M, Friger M, Codish S and Sukenik S *Journal of Clinical Rheumatology* **8** 197-203.
[75] Ma'or Z and Michael R *Pat.* 2003, **WO 2004000244 A1 20031231**
[76] Soroka Y, Ma'or Z, Leshem Y, Verochovsky L, Neuman R, Bregegere F M and Milner Y 2008 *Experimental Gerontology* **43 (10)** 947-957.
[77] Ma'Or Z, Yehuda S and Voss W 1997 *International Journal of Cosmetic Science* **19 (3)** 105-110.
[78] Riyaz N and Arakkal F R 2011 *The Indian Journal of Dermatology Venereology & Leprology* **77 (2)** 128-134.
[79] Zeng Q, Hu Z and Li Y 2004 *Faming Zhuanli Shenqing* **CN 1543927 A 20041110**
[80] Fleischmann E 2004 *Pat.* **DE 10259825 A1 20040708**
[81] Hasunuma, K, Hanaoka, H, Morita, K and Saito M 2000. *Pat.* **JP 2000229841 A20000822**
[82] Braun Y B and Braun E 2005 *Pat.* **WO 2005007071 A2 20050127**
[83] Beckermann 2001 *Pat.* **DE10020874 A1 20010517**
[84] Beckermann W J 2004 *Pat.* **DE 10308774 A1 20040909**
[85] Robert F R 1997 *Pat.* **WO 9722348 A1 19970626**
[86] Hwang S R and Lee I H 1998 *Pat.* **KR 154348 B1 19981116**
[87] Ma'or Z, Henis Y, Alon Y, Orlov E, SØrensen K, and Oren A 2006 *International Journal of Dermatology* **45** 504-511.
[88] Chidambara K N 2005 Thesis submitted to the University of Mysore, India.
[89] Emeish S 2013 *Journal of Environment and Earth Science* **3 (10)** 6-15.

Modeling the effect of Rose Bengal on growth and decay patterns of Pseudomonas aeruginosa, Escherichia coli and Staphylococcus aureus

M-Ali H. Al-Akhras[1,*], Mohammad Al Shorman[2], Majed M. Masadeh[3], Khaled Aljarrah[1,4], Zaid Ababneh[5]

[1]*Department of Physics, Bio-Medical Physics Laboratory, Jordan University of Science & Technology (JUST), P.O. Box 3030, Irbid 22110, Jordan.*
[2]*Tafila Technical University, Department of Physics, P.O.Box 179 Tafila 66110 Jordan Irbid 009622 Jordan*
[3]*Department of Pharmaceutical Technology, Faculty of Pharmacy, Jordan*
[4]*College of Applied Medical Science, King Saud Bin Abdulaziz University for Health Science, Al-Ahsa*
[5]*Department of Physics, Yarmouk University*

alakmoh@just.edu.jo

Abstract: Most infections caused by (Pseudomonas aeruginosa, Escherichia coli and Staphylococcus aureus) are hospital and community acquired infections in patients. Bacterial growths incorporated with photosensitizing material (Rose Bengal) with and without light were investigated. The results demonstrated that the viable counts are increasing in absence of light (in dark) for all samples incorporated with Rose Bengal. Variation in growth phases were noticed as expected, but there is no significant change in decay phases. Convenient and adequate mathematical modeling is in very good agreement with the experimental results and showed to be a very good approach of characterization the growth behaviors of the bacteria.
Bandwidths are independent of bacteria group (gram-positive or gram-negative) but it seems totally dependent on the oxygen requirements; an anaerobic bacterium takes broader bandwidths than aerobic bacteria. This concludes that the growth and lethal rates of anaerobic are much greater than aerobic.

Keywords: Pseudomonas, E coli, Staphylococcus, growth curve modeling, Rose Bengal

1. Introduction

Several studies and mathematical modeling on bacterial growth phases which have been conducted by many researcher, lead to significant results in understanding the bacterial growth behaviors. In spite of significant achievement in understanding the bacterial behaviors, there is still no mathematical strong evidence about the correlation between the strain or type of bacteria and growth and/or mortality rates. Up to our knowledge, this is the first time Gaussian function is applied to express the bacterial behaviours in terms of bandwidths and type of bacteria. Three of most important bacteria named Pseudomonas aeruginosa, Staphylococcus aureus, and Escherichia coli are used in this work. In spite of the fact that they are common colonizers of human skin and mucous membranes, they are responsible for a wide spectrum of infections targeting a broad range of organs and deep-seated infections and surgical site infections like wound infections and endophthalmitis [1, 2].
Pseudomonas aeruginosa and Escherichia coli are Gram-negative and Staphylococcus aureus is a Gram-positive bacteria. Those types of Bacteria are considered to be the most virulent human pathogens which rarely cause infection in the healthy host. They are found in a variety of habitats including soil and fresh water streams as well as on plant and animal tissues [3, 4]. Members of this genus are characterized by their ability to grow in water because of their simple nutritional requirements: they will even grow in distilled water and natural mineral water [5].

Pseudomonas aeruginosa is of particular interest as a human pathogen which is adept at infecting different tissues and organs. It is considered to be an opportunistic and human pathogen as it causes infections in individuals with neutropenia or whose normal immune status is compromised in some way. Although most infections caused by Pseudomonas aeruginosa are hospital acquired, community acquired infections in patients have also been reported [6, 7]. Epidemiological statistics indicate significant frequencies of occurrence of Pseudomonas aeruginosa infections and capable of transiently colonizing skin and mucosal surfaces, but deficient in its ability to carry out initial steps of infection [8]. The normal non-specific defenses of the human body are sufficient to prevent the bacteria from causing infections but once the opportunist, it could override the primary defense systems and may cause several fatal systemic diseases in the human body [9 - 13]. Mortality rates attributed to this bacterium in hospital acquired infections and patients suffering from cystic fibrosis are high [14]. The capacity of this organism to cause such high rates of illness and death is primarily due to a diverse range of virulence factors [15, 16]. Many outbreaks of food-borne illness are reported worldwide in which Staphylococcus spices were a common cause of food poisoning (enterocolitis). Some strains also release superantigens, which interact with the immune system causing the serious systemic disease, Toxic Shock Syndrome [17]. Staphylococcus aureus is a primary and major cause of nosocomial infections which is considered to be the second leading cause of nosocomial blood stream infections [18, 19]. Escherichia coli are commonly found in warm-blooded organisms and can cause serious food poisoning in humans and preventing the establishment of pathogenic bacteria within the intestine. Virulent strains of Escherichia coli can cause gastroenteritis, urinary tract infections, and neonatal meningitis. Virulent strains are also responsible for haemolytic-uremic syndrome, peritonitis, mastitis, septicemia [20]. Some other disturbance of cell-wall synthesis and the appearance of a multilamellar structure near the septum of dividing cells, along with loss of potassium ions from the cells were reported by [21].

In general, those bacteria remain a threat to public health with a high rate of morbidity and mortality. The goal of this study is to get better understanding and explain the behavior of the bacteria growth stages with and without additives in presence and absence of light. Those stages described by numerous mathematical modeling that varies considerably in structure and number of parameters. Therefore, accurate parameter of the growth curves such as the relationship between the bandwidth and bacteria group at half maximum value is still needed and up to our knowledge it has not been calculated. Furthermore, investigation the effect of irradiation on the bacterial growth of different strains of bacteria namely Pseudomonas aeruginosa, Staphylococcus aureus and Escherichia coli with the presence of Rose Bengal dye were studied.

2. Mathematical Modeling

Several sigmoidal functions such as (Logistic, Gompertz, and Richards) were compared to describe growth curve or lethal damage of several organisms [22 - 25].

A new calculation of the two main stages of bacterial growth and decay were fitted with Gaussian equation. In addition, a Logistic model for bacteria growth with and without additives were applied to experimental data obtained in this study. The new model successfully describes Gaussian and sigmoidal growth curve of Escherichia coli, Pseudomonas aeruginosa and Staphylococcus aureus under various initial conditions.

In general, the growth curves of bacteria are typical sigmoidal curves [26]. Original logistic model can be expressed as:

$$\frac{dN}{dt} = rN(1 - N / N_{\max}) \qquad \text{.....................} (1)$$

Where **N** is the population of the organism at time **t** and **r** is rate constant, or the maximum specific growth rate. N_{max} is the maximum population (at the stationary phase), the term $(1 - N / N_{\max})$ which suppresses the growth rate at a high population.

When N is small during the lag phase, the value of this term is almost one, which does not affect the growth rate. During the stationary phases, the values of N_{max}, approaches zero with growth rate almost zero. Sigmoidal growth curves described in this study could be explained on the bases of Gibson et al.

modified logistic model to fit bacterial growth data by the following equation:

$$N = a / (1 + \exp(-\alpha (t - \beta))) \quad \text{.................... (2)}$$

Where **N**: is the number of colony, **a** is the maximum number of colonies (always normalized to one), **α** is the growth or lethal rate, **t** is the bacterial growth time, and **β** is the time required to reduce the maximum number of colonies to one half of its maximum value.
This study is focused on a mathematical modeling for the bacterial growth and lethal curves behaviors due to the importance of such bacteria which has increased dramatically in the last two decades. On the other hand, the following Gaussian Equation was used to fit the entire stages of the bacterial growth curves:

$$N = a\, e^{\frac{-(t-b)^2}{2c^2}} \quad \text{.................... (3)}$$

where **a** is the maximum height of the curve peak, **b** is the position of the center of the peak, and **c** is the full width at the half maximum of the peak

3. Materials and Methods

3.1. Materials

The following materials were used in this study are Rose Bengal (received from NEWTECH LTD, Brixworth. Northants. UK), Muller Hinton Agar (received from HiMedia Laboratories Pvt. Ltd, India) and Muller Hinton broth (received from HiMedia Laboratories Pvt. Ltd, India).
The broad-band light source used for bacteriolysis measurements was a 200-W high-pressure Hg/Xe arc. The lamp was housed in an Oriel Research Arc Lamp Housings model 66903, with the Oriel Digital Arc Lamp Power Supply Model 68907. For irradiation of the samples, the light source was generated from a 1.5 inch Oriel Research Arc Lamp Housings, (Copyright © 1996-2008 Newport Corporation) with an output power of (50-500 W) and light output intensity of 200 watts. The sample was located at 17 cm from the light source and the output intensity reach the samples were about $60 J/cm^2$ as measured by FiledMaxII Laser power/Energy Meter/Coherent/USA. Appropriate spot size to cover the sample area was created with an objective lens.

3.2. The Growth Characteristics

In order to estimate the rate of microbial reproduction, it is necessary to determine the numbers of microorganisms present. The growth of bacteria can be estimated by several different methods, including the total count, the viable count and turbidometric procedures. The viable plate count method is one of the most common procedures for the enumeration of bacteria. It is used to determine the viable population in a bacterial culture. In this procedure, a series of dilutions of a bacterial suspension are plated onto a suitable solid growth medium in order to determine the number of colonies. It is assumed that each colony arises from an individual bacterial cell. Therefore, by counting the number of colonies that develop and by taking into account the dilution factors, the concentration of bacteria in the original sample can be determined as Colony-Forming Units (CFU). A major limitation of the viable plate count procedure is its selectivity. It measures only those cells that are capable of growth on the given plating medium under the set of incubation conditions that are used. Sometimes cells are viable, but nonculturable unless steps are taken to acclimate the microorganisms to laboratory culture conditions. Bacteria can also be enumerated by direct counting procedures (total count or microscopy), that is, counting without the need to first grows the cells in culture. However, the difficulty in establishing the metabolic status of the observed bacteria, that is, whether the cells are living or dead, is a major limitation of this procedure.

3.3. Experimental Approach

The overnight culture of Pseudomonas aeruginosa ATCC 27853, Escherichia coli ATCC 25922, and

Staphylococcus aureus ATCC 29213 were performed as follows:
25 mL of a sterile medium Muller Hinton Broth (MHB) in a 100 mL Erlenmeyer flask were inoculated with a pure culture of each bacteria (single colony, from a streak plate). The flasks were then incubated in total darkness overnight in an incubator at 37°C. For the test culture of each bacteria a fresh aliquot (75 mL) of medium (MHB) was inoculated with 250 μL of each bacteria mentioned above for overnight culture. This was performed in duplicate (flasks A & B). The flasks were incubated at 37°C in the incubator.
For Preparing Culturing Media: 38 grams of Muller Hinton agar were suspended in 1000 ml of distilled water. The suspension was boiled and dissolved completely. The medium was sterilized by autoclaving at 15-lbs pressure and 121°C for 15 minutes with well mixing before use.
Similarly, 21 grams of Muller Hinton agar were suspended in 1000 ml of distilled water and heated to dissolve the medium completely dispense as desired and sterilize by autoclaving at 15-lbs pressure and 121°C for 15 minutes.

3.4. Viable Count

An aliquot (1 mL) of the identical test culture (flask B) was aseptically removed and serially diluted in the range of 10^{-2} to 10^{-9}. Aliquots (100 μL) of the dilutions were spread plated onto appropriate agar plates Muller Hinton Agar (MHA). These plates were prepared in triplicate (three plates for each dilution). The plates were incubated overnight at 37°C. After 24 hours, colonies were counted in each plate and viable counts were calculated according to the following equation:

$$CFU\ mL^{-1} = N\ (1/DF) \qquad (4)$$

Where; **CFU** = The number of colony forming units, **N** = Mean of counts on the plates and **DF** = Dilution factor.

3.5. Sample Preparation

Fresh bacteria samples with seventeen hours old (Pseudomonas aeruginosa, Staphylococcus aureus and Escherichia coli) were used throughout this study.
Three samples from each kind of bacteria were prepared and kept inside the incubator. 100μl of bacteria was taken every one hour and cultured in a Petri dish for 24-hour long.

3.6. Bacteria Induced by Rose Bengal

Fresh bacteria samples with seventeen hours old (Pseudomonas aeruginosa, Staphylococcus aureus and Escherichia coli) were used throughout this study.
Three different mixtures of bacteria (Pseudomonas aeruginosa, Staphylococcus aureus and Escherichia coli) with drugs at different concentrations were prepared. Samples are mixed with photosensitizes drugs (Rose Bengal) and split every sample into two samples and irradiate the first one with light for continuous exposure times and the second one was incubated at total darkness. 100μl of bacteria were taken every 15 minutes and cultured in a Petri dish up to about 2 hours.
The concentrations range of the photosensitizing drug in the sample are (0.2mg/ml -0.6mg/ml). The samples were irradiated with 200-watt high pressure arc lamp.

4. Results

4.1 Samples with No Additives

The number of colonies in Petri dishes was measured and plotted for all the growth stages of three kinds of bacteria, Pseudomonas aeruginosa, Staphylococcus aureus and Escherichia coli.

Figures 1, 2, 3 and Table 1 showed the relationship between the numbers of colonies versus growth time (0-26 hours).
The typical growth curves of Escherichia coli, Staphylococcus aureus and Pseudomonas aeruginosa are

shown in Fig.1. The figure represents the living phases of the bacteria growth for Escherichia coli, Staphylococcus aureus, and Pseudomonas aeruginosa. The first phase represents the lag phase which takes about 3, 4, and 10 hours respectively. The second phase, which is called log (or exponential) phase, takes 11, 10, and 8 hours respectively. The third phase is the stationary phase and its duration is about 3 hours for the three kinds of bacteria. The fourth phase is called the death phase (decay) which takes 10, 10, 6 hours respectively.

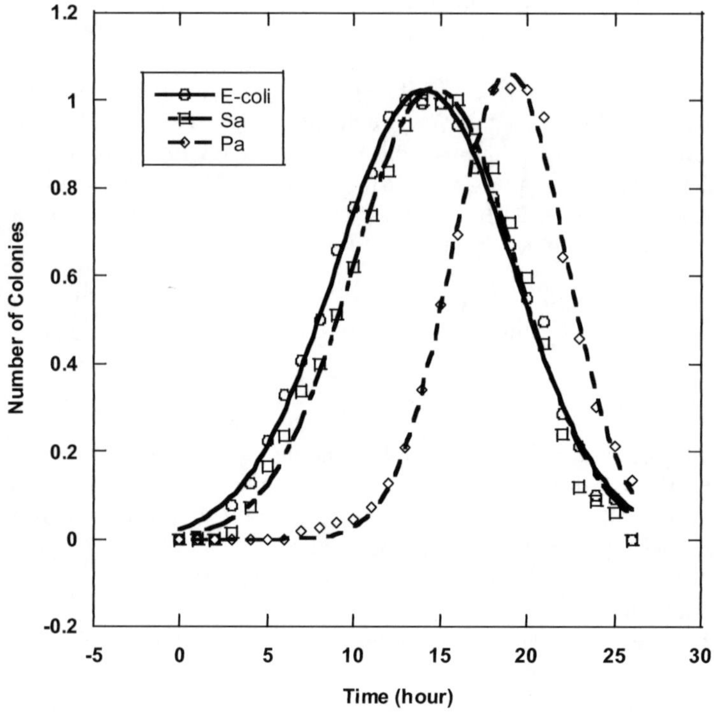

Figure 1: A typical growth stages of Escherichia coli, Staphylococcus aureus and Pseudomonas aeruginosa at temperature of (37^0 C) without any additives induced to the samples. The symbols denote the experimental values and lines represent the fitting of Gaussian Equation no. 3.

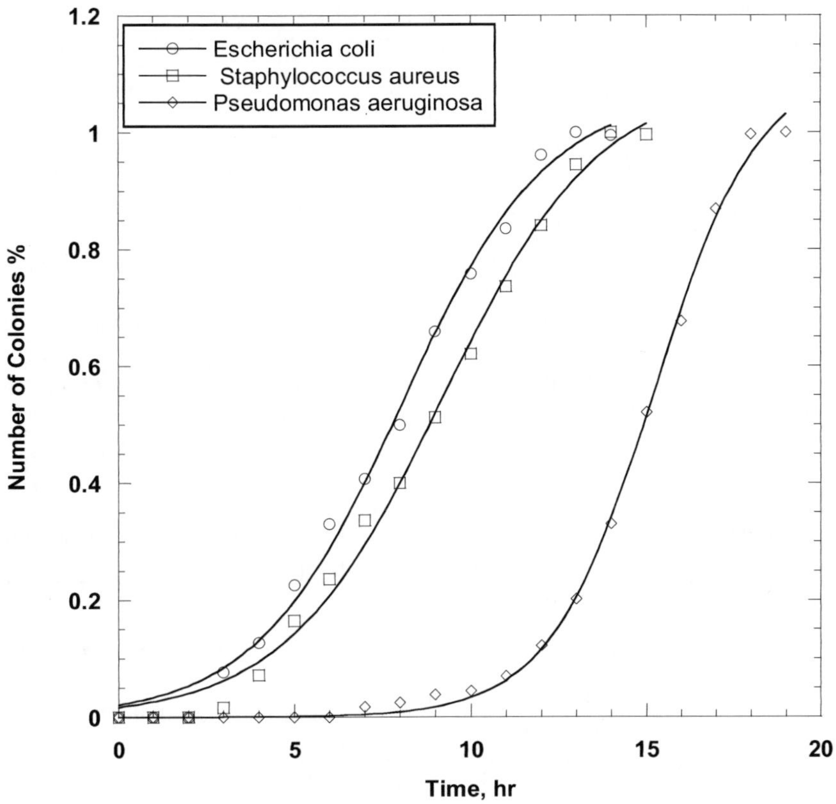

Figure 2: Growth curves for Escherichia coli, Staphylococcus aureus, and Pseudomonas aeruginosa samples without induced any additives. The symbols denote the experimental values and the lines represent the fitting function.

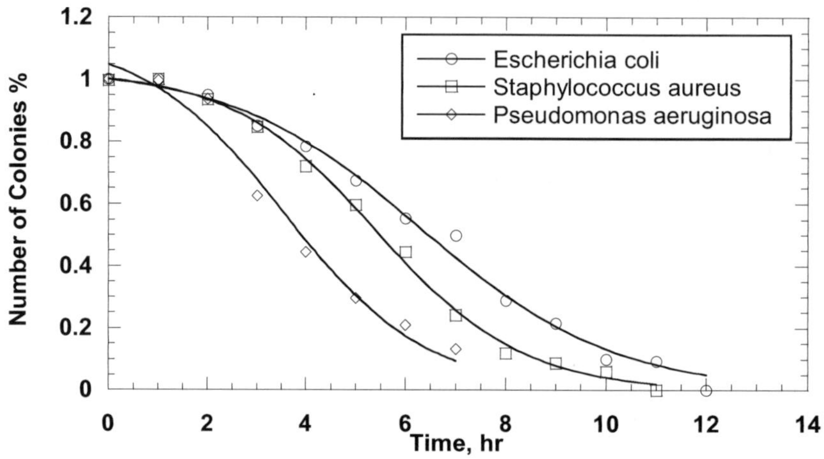

Figure 3: Lethal curves for Escherichia coli, Staphylococcus aureus, and Pseudomonas aeruginosa samples without induced any additives. The symbols denote to the experimental values and the lines represent the fitting function.

Table 1: The values of the model parameters (**a, β, α, and R^2**) represented in the modified logistic function and (**a, b, c, and R^2**) represented in Gaussian fit for the Pseudomonas aeruginosa, Escherichia coli, and Staphylococcus aureus bacterial without additives at temperature of 37^{0}C:

		Modified Logistic				Gaussian Fit			
		a	**α(hr^{-1})**	**β(hr)**	**R^{2*}**	**a**	**b**	**c**	**R^{2*}**
E. coli	Growth	1.07±0.033	0.49±0.03	8.04±0.20	0.995	1.03±0.01	14.08±0.08	5.12±0.08	0.996
	Lethal	1.04±0.032	-0.52±0.04$^{@}$	6.31±0.20	0.993				
S.a	Growth	1.09±0.036	0. 45±0.03	9.20±0.23	0.995	1.03±0.02	14.67±0.10	4.71±0.10	0.994
	Lethal	1.03±0.012	-0.69±0.04$^{@}$	5.40±0.10	0.997				
P.a	Growth	1.11±0.02	0.67±0.03	15.23±0.10	0.998	1.06±0.01	18.94±0.05	3.29±0.05	0.998
	Lethal	1.13±0.10	-0.70±0.12$^{@}$	3.55±0.375	0.981				

Results are expressed as means ± SD.
The values of the parameters were obtained by fitting the data to Eq. 2 and Eq. 3.
*Correlation Factor.
$^{@}$ -Ve signs means the bacteria are in death phase (curves decaying down)

For simplicity and comparison, Gaussian fitting to those curves can be approximately considered as a two phases, growth and decay only. Growth curves of Escherichia coli and Staphylococcus aureus are almost behaving the same and it takes about 14 to 15 hrs to reach the maximum height while Pseudomonas aeruginosa takes longer time about 19 hours to reach the maximum height. Decay curves of Escherichia coli and Staphylococcus aureus are also behaving the same and it takes about 11.5 hours to reach the minimum values while Pseudomonas aeruginosa takes shorter time (about 7 hours) to reach the lowest value. Moreover, over all behaviors there is a lag time of about 4.5 hours in growth and decay phases as compared (Escherichia coli and Staphylococcus aureus) with Pseudomonas aeruginosa. Interestingly, there is a variation in growth phases in all types of bacteria, as expected, with no significant change in decay phases. The bandwidths at the half-maximum values of the average spectrum were calculated from the Gaussian fitting and found to be 12.9 hrs, 11.8, and 8.0 for Escherichia coli, Staphylococcus aureus, and Pseudomonas aeruginosa, respectively. Interestingly, another approach of calculations was performed by fitting the exponential log phase (Figure 2) and death (decay) phase (Figure 3), separately. Growth parts, represent the number of colonies, increase with time (data points in Figure 2) and lethal parts, represent the number of colonies, decrease with time (data points in Figure 3). Data was normalized to one and fitted (solid lines) with logistic function Equ. 2 using KaleidaGraph 4.0.

Figure 2 and Figure 3 represent the survival and death phases of the bacteria lifecycle without inducing any additives. Data points in the figures was fitted using the theoretical model Equ.2, and the values of the model's parameters (**a**, **α**, and **β**) for survival phase and death phase are summarized in Table 1.

Table 1 shows the parameters of Gaussian and logistic models. The last column of the table shows the correlation factor between experimental data points and theoretical model symbolized in solid lines. Comparing exponential phase with lethal phase, we can notice that the rate parameter **α(hr^{-1})**, in case of growth phase, is lower than that of lethal phase, where the minus sign means the viable cell population declines. In the same patterns, the time **β** required to reduce the maximum numbers of colonies to one half of its maximum value were found to be higher in case of growth phase as compared with the lethal phase. Fitting the growth curves of Pseudomonas aeruginosa with induced Rose Bengal at concentration of (0.2 – 0.6) mg/ml in the presence of light showed best fit with best values of parameters. Figure 4 represents the lethal phase of the Pseudomonas aeruginosa with induced Rose Bengal at concentrations of (0.2 - 0.6) mg/ml, respectively. The experimental curves were fitted using the logistic theoretical model and the values of parameters (**a**, **α**, and **β**) are respectively shown in Table 2.

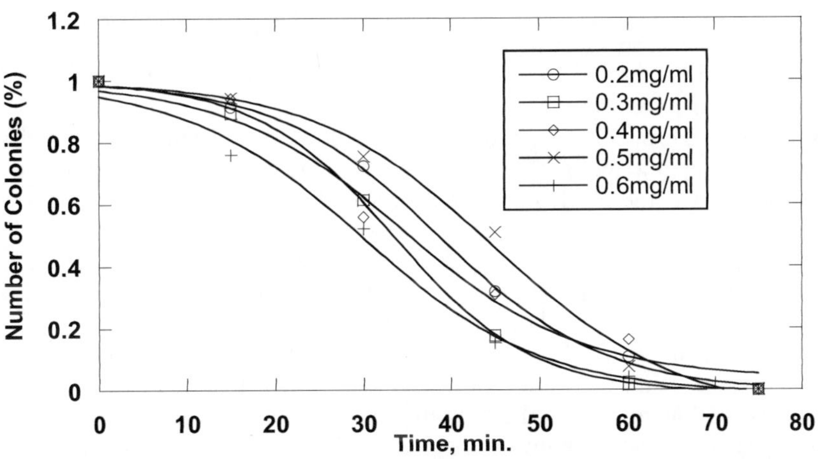

Figure 4: The lethal curves of Pseudomonas aeruginosa bacteria with induced Rose Bengal at concentrations of 0.2 mg/ml – 0.6 mg/ml at temperature 25 °C in the presence of light. The symbols denote to the experimental values and the lines represent the fitting function.

Table 2: The values of the variables (**a**, **β**, **α** and **R^2**) which represented in the logistic model for the irradiated (in light) samples of the Pseudomonas aeruginosa bacteria with induced Rose Bengal at different concentration from 0.2mg/ml to 0.6 mg/ml at temperature of 25°C:

Concentration of Rose Bengal	**Parameters**			
	a	**α (hr^{-1})**	**β (hr)**	**R^{2*}**
0.2mg/ml	1.03±0.007	0.107±0.004	38.5±0.46	0.999

0.3mg/ml	1.04±0.005	0.128±0.004	33.3±0.33	0.999
0.4mg/ml	0.99±0.028	0.092±0.013	35.0±1.86	0.987
0.5mg/ml	1.01±0.021	0.106±0.013	43.3±1.34	0.993
0.6mg/ml	1.05±0.016	0.099±0.001	29.6±1.10	0.994

Results are expressed as means ± SD.
The values of the parameters were obtained by fitting the data to Eq. 2.
*Correlation Factor.

Table 2 shows the fitting parameters of the exposed sample of the Pseudomonas aeruginosa with different concentrations of the Rose Bengal as a photosensitizing additive. Exposure light source with output intensity reaches the samples with about 60J/cm^2. The table show independent relation between the concentration and the time of the 50% control. The values of the growth rates shown in Table 2 are nearly about 0.1 in the present of Rose Bengal and light. While the values of the growth rates shown in Table 1 are nearly about 0.5 to 0.7 without any additives. The last column shows excellent correlation between the experimental data and theoretical model.

5. Discussion

5.1. Bacteria Growth Curve

Bacterial growth over time can be plotted as a cell number versus time, this is called Growth Curve. The cell number is plotted as the log of the cell number, since it is an exponential function. Regardless of the generation time in a growing culture, the plot of the log of the cell number versus time gives a characteristic curve. In this research, we plotted and fitted the growth curves for three bacteria; Escherichia coli, Pseudomonas aeruginosa and Staphylococcus aureus. These curves have typically four distinct phases: Lag phase (First part of growth, the cells prepare for synthesis of DNA and enzymes needed for cell division with no increment in cell number and this process extended to few hours); Escherichia coli takes 3 hours, Pseudomonas aeruginosa takes 10 hours, and Staphylococcus aureus takes 4 hours. However, the cells are probably active in metabolizing while preparation for cell division. Depending on the growth medium, the lag phase maybe short or very long. For example, if a culture is a rich growth medium that supplies most of their own amino acids and vitamins, the lag phase will be very long. The cells must activate the metabolic pathways for amino acid and vitamin synthesis and make enough of these nutrients to begin active growth. Exponential (log) phase is the second part of growth, where the culture reaches its maximum rate of growth for specific conditions. In this part, as shown in Figure 1, Escherichia coli takes about 14 hours, Pseudomonas aeruginosa takes 15 hours, and Staphylococcus aureus takes 19 hours to reach its maximum rate, starting from the beginning of lag phase. Stationary phase (third part of the growth curves) takes about 3 hours for this three type of bacteria. In this part, no net increase in numbers and the growth rate equals to the death rate. The death phase (fourth part of growth or the final part). Eventually there was a decline in cell number in this phase. Similarly, we found that Escherichia coli takes 10 hours, Pseudomonas aeruginosa takes 6 hours, and Staphylococcus aureus takes 10 hours in this phase. In this phase, the cells also quickly lose the ability to divide even if they are placed in fresh medium [23]. In our previous study, pre-incubation of MCF-7 cells with iron oxide nanoparticles followed by a static magnetic field exposure significantly ($P<0.05$) increased doxorubicin-induced cytotoxicity [27].

5.2. Modeling of Growth Curves

The present part reports typical results on the bacterial growth curves fitted to a Modified logistic function model. A new Modified Model for bacterial growth was developed in this study as shown in Figures 1 - 3. The model is described by equations 2 and 3 which consist of useful parameters that characterized the bacterial behaviors in most interesting and important phases; exponential growth and exponential decay, table 1 and 2 [28]. The Modified Logistic model successfully describes the sigmoidal growth curve of Escherichia coli, Pseudomonas aeruginosa and Staphylococcus aureus under various conditions.

Comparing exponential phase with lethal phase, we can notice that the rate parameter **α(hr^{-1})** in case of growth phase is lower than that of lethal phase, while the time **β** required to reduce the maximum numbers of colonies to one half of its maximum value were found to be higher in case of growth phase as compared to lethal phase.

The values of the growth rates shown in Table 2 are nearly about 0.1 in the present of Rose Bengal and light. While the growth rates shown in Table 1 are nearly about 0.5 to 0.7 without any additives. This result is consistent with similar finding by Kim, et. al., for wild-type strain MM294 of E. coli cultivated under various conditions, the growth rate without TiO2 and light (normal culture) was 0.43 and reduced 0.40 with TiO2 and light intensity of 12.5 Wm^{-2} [26]. Similarly, the fitted growth curves of Pseudomonas aeruginosa with induced Rose Bengal in presence of light are shown in Figure 4 with very good agreement as those reported by [29, 30]. From the fitted lines with Gaussian Equation, we notice that the faster bacteria to reach the maximum values the broader bandwidths at the half-maximum values could occur. For example, Escherichia coli takes about 14 hours to reach the maximum value with a bandwidth of 12.9 hours, Pseudomonas aeruginosa takes about 15 hours reach the maximum value with a bandwidth of 11.8 hours, and Staphylococcus aureus takes about 19 hours to reach the maximum value with a bandwidth of 8.0 hours.

Bandwidths are independent on bacteria group (gram-positive or gram-negative) but it seems totally dependent on the oxygen requirements. An anaerobic bacterium (Escherichia coli and Staphylococcus aureus) takes broader bandwidths than aerobic bacteria (Pseudomonas aeruginosa). This concludes that the growth and lethal rates of anaerobic are much greater than aerobic. This finding is consistent with what has been found by other researchers were Pseudomonas aeruginosa had better degradation rate when there was increase in aeration rate, the biodegradation rate increase too [31].

5.3. Bacteria (E. coli, Pseud. a. and Staph. a.) with Induced Rose Bengal

In this part, the viable counts (Number of Colonies) of Pseudomonas aeruginosa start decreasing with presence of light with induced drug (Rose Bengal) (See Figure 4). Reduction in the bacteria cells are increasing in absence of light (dark) were noticed and recorded. In normal situation, without additives, the number of colonies increases with time. This is probably because the bacteria have been frequently used for the determination of the photodynamic capabilities of sensitizers. Due to the general structural and physiological differences between the cell walls of gram-positive and gram-negative microorganisms, the behavior of photodynamic systems may vary depending on the abilities of the sensitizers to penetrate membranes and diffuse through the cell walls.

Illumination of Rose Bengal with light produces singlet oxygen ($^1O_2$) which is known to inactivate bacterial cells efficiently and led to cell death. The membrane of the cell was found to be a likely target for the $^1O_2$ reactions in a separated surface-sensitizer system. In contrast with some other sensitizer, the dye did not have to necessarily be bound to the membrane [32]. In addition, there are some clear confirmation that anionic compound penetrates through outer membrane wall and bound inside. Nitzan and co-workers used the polycationic peptide polymyxin B nonapeptide (PMBN), which increased the permeability of the Gram- negative outer membrane and allowed photosensitizer that are normally excluded from the cell to penetrate to a location where the reactive oxygen species generated upon illumination can cause fatal damage [21]**.** PMBN does not release lipopolysaccharide (LPS) from the cells and expands to the outer leaflet of the membrane allowing photosensitizer such as deuteroporphyrin to penetrate and permitting PDI of Escherichia coli and Pseudomonas aeruginosa. Nitzan, et. al., also reported that an interaction between PMBN and photosensitizer in solution could occur and speculated that this binding assisted the penetration. They found that DP seemed to work much better in concert with PMBN than many other photosensitizers, including porphyrins, phthalocyanines and merocyanine 540 [33]. In comparison with some other finding, polylysine chain of 20 lysine residues did not allow PDI with DP [34]. On our previous study, the bacteria growth curve was compared with control depending on inhibition percentage and it was found higher than control for all different bacteria strains at all concentrations [35, 36].

There is some evidence that photosensitizer can more easily intercalate into double-stranded DNA and cause damage and also Guanine residues have been shown to be the most easily oxidized [37]. The damage may be able to be repaired by various DNA repairing systems [38].

6. Conclusion

In conclusion, viable counts of bacteria incorporated with or/and without photosensitizer such as Rose Bengal are investigated. Typical curves expressed the behavior of the bacterial growth in presence and absence of light. Convenient and general mathematical modeling is obtained to characterize the growth behaviors for selected bacteria.

The conclusions of this work can be summarized as follows:

- Rose Bengal with concentration of 0.2 to 0.6 mg/ml plus enough dose of light (60 J/cm^2) are required to cause bacterial destruction.
- The viable counts decrease in presence of light with induced drug (Rose Bengal).
- These models could be considered as the suitable models for describing the growth of bacteria under certain conditions.
- Interestingly, there is a variation in growth phases in all types of bacteria as expected, while no significant change in decay phases.
- Intercomparison study on bandwidths at the half-maximum values were measured and found to be independent of bacteria group (gram-positive or gram-negative) but it seems completely dependent on the oxygen requirements; an anaerobic bacterium takes broader bandwidths than aerobic bacteria. This concludes that the growth and lethal rates of anaerobic are much greater than aerobic.
- Finally, in this study, very good agreements with acceptable correlation were found between the experimental data and theoretical modeling that express the bacterial behavior in general and bacterial growth and lethal in particular.

ACKNOWLEDGEMENTS: This work was supported by a grant from the Deanship of Scientific Research at Jordan University of Science and Technology (JUST) No. 113/2009.

References:

[1] Casey AL, Lambert PA, Elliott TS. 2007. Staphylococci. Int J Antimicrob 3: S23-32.

[2] M. Masadeh, G. Karasneh, M-Ali Al-Akhras, B. A Albiss, Khaled M. Aljarah, Sayer I Al-azzam, Karem H. Alzoubi Cerium Oxide and Iron Oxide Nanoparticles Abolish the Antibacterial Activity of Ciprofloxacin against Gram Positive and Gram Negative Biofilm Bacteria. Cytotechnology 67 (3), 427-435, 2014.

[3] Hardalo C, Edberg SC. 1997. Pseudomonas aeruginosa: assessment of risk from drinking water. Critical Reviews in Microbiology 23: 47-75.

[4] Boyd A, Chakrabarty AM. 1995. Pseudomonas aeruginosa biofilms: role of the alginate exopolysaccharide. Journal of Industrial Microbiology 15: 162-168.

[5] Legnani P, Leoni E, Rapuano S, Turin D, Valenti C. 1999. Survival and growth of Pseudomonas aeruginosa in natural mineral water: a 5-year study. International Journal of Food Microbiology 53: 153-158.

[6] Arancibia F, Bauer T, Ewig S, Mensa J, Gonzalez J, Niederman MS, Torres A. 2002. Community-acquired pneumonia due to gram-negative bacteria and pseudomonas aeruginosa - incidence, risk, and prognosis. Archives of Internal Medicine 162: 1849-1858.

[7] Weingarten JA, Paterson DL, Yu VL. 2003. Pneumonia Caused by Pseudomonas aeruginosa. Current Treatment Options in Infectious Diseases 5: 159–169.

[8] Jarvis WR. 2003. Epidemiology and control of Pseudomonas aeruginosa infections in the intensive care unit. Boston: Kluwer Academic.

[9] Appelgren P, Bjornhagen V, Bragderyd K, Jonsson CE, Ransjo U. 2002. A prospective study of infections in burn patients. Burns: 39-46.

[10] Beckett G, Williams D, Giberson G, Gensheimer K.F, Gershman K, Shillam P, Hoffman R.E, Merry R, Savalox H, Fawcett L. 2001 Pseudomonas dermatitis/folliculitis associated with pools and hot tubs - Colorado and Maine: 1999-2000 (reprinted from Morbidity and Mortality Weekly Report, vol 49, pp 1087-1090, 2000 . Archives of Dermatology 137: 522-523.
[11] Qarah S, Cunha BA, Dua P, Lessnau KD. 2004. Pseudomonas aeruginosainfections, eMedicine: http://www.emedicine.com/med/topic1943.htm.
[12] Wenzel RP. 1997 Prevention and control of nosocomial infections. Nosocomial pneumonia. 3 ed. Baltimore. Williams and Wilkins.
[13] Mandell GL, Benett JE, Dolin R. 1995. Principles and practice of infectious diseases: Pseudomonas aeruginosa. 4th. Churchill Livingstone.
[14] Van Delden C, Iglewski BH. 1998. Cell-to-cell signalling and Pseudomonas aeruginosa infections. Emerging Infectious Diseases 4: 551-560.
[15] Chatzinikolaou I, Abi-Said D, Bodey GP, Rolston KVI, Tarrand JJ, Samonis G. 2000. Recent experience with pseudomonas aeruginosa bacteraemia in patients with cancer-retrospective analysis of 245 episodes. Archives of Internal Medicine 160: 501-509.
[16] Maschmeyer G. Braveny I. 2000. Review of the incidence and prognosis of Pseudomonas aeruginosa infections in cancer patients in the 1990s. European Journal of Clinical Microbiology & Infectious Diseases 19: 915-925.
[17] Parsonnet J, Hansmann MA, Seymour JL, Delaney ML, Dubois AM, Modern PA, Jones MB, Wild JE, Onderdonk AB. 2010. Persistence survey of Toxic Shock Syndrome toxin-1 producing Staphylococcus aureus and serum antibodies to this superantigen in five groups of menstruating women, BMC Infect Dis.10, 249.
[18] Anon. 1996. Ethical issues in perinatology. Natl Med J India 9: 32-7.
[19] Lessa FC, Edwards JR, Fridkin SK, Tenover FC, Horan TC, Gorwitz RJ. 2009. Trends in incidence of late-onset methicillin-resistant Staphylococcus aureus infection in neonatal intensive care units: data from the National Nosocomial Infections Surveillance System, 1995-2004. Pediatr Infect Dis J 28: 577-8.
[20] Simon W, Daniel M, Deirdre D, Robinson C. 2006. Erythrosine is a potential Photosensitizer for the photodynamic therapy of oral plaque biofilms. Journal of Antimicrobial Chemotherapy 57: 680–684.
[21] Nitzan Y, Gutterman M, Malik Z, Ehrenberg B. 1992. Inactivation of gram-negative bacteria by photosensitized porphyrins. Photochem. Photobiol 55: 89–96.
[22] Zwietering MH, Jongenburger I, Rombouts FM, Van RK. 1990. Modeling of The Bacterial Growth Curve. Applied and Environmental Microbiology: 1875-1881.
[23] Al-Akhras M, Grossweiner LI. 1996. Sensitization of Photohemolysis by hypericin and Photofrin. Journal of Photochemistry and Photobiology B: Biology 34:169-175.
[24] Al-Akhras M. 2006. A New Application of Gompertz Function in Photohemolysis: The Effect of Temperature on Red Blood Cell Hemolysis Photosensitized by Protoporphyrin IX. Medical & Biological Eng & Computing 44: 703-710.
[25] Al-Akhras M. 2006. Effect of Additives and Photosensitive Drugs on Osmotic Fragility. Int. J. Sci. Res 16: 363-369.
[26] Kim SY, Nishioka M, Taya M. 2004. Promoted Proliferation of an SOD-deficient mutant of Escherichia coli under oxidative stress induced bt photoexcited TiO2. FEMS Microbiology Letters 236: 109-114.
[27] K. Aljarrah, N. Mhaidat, M-Ali H Al-Akhras, A. N Aldaher, B. A Albiss, Khaled Aledealat, and Fawzi M Alsheyab. Magnetic nanoparticles sensitize MCF-7 breast cancer cells to doxorubicin-induced apoptosis. *World Journal of Surgical Oncology* 2012, 10:62
[28] Al-Akhras MA, and Al-Khalili D. The Influence of Iron Oxide Nanoparticles on the Red Blood Cells Photohemolysis Sensitized with Photofrin: Temperature Effect. Jordan Journal of Biological Sciences 2015; 8(1): 55.
[29] Hiroshi F, Akemi, Satoshi M. 2003. A new logistic model for bacterial growth. J. Food Hyg. Soc. Japan 44: 155-160.

[30] Oboirien BO , Amigun B, Ojumu TV , Ogunkunle OA, Adetunji OA, Betiku E, Solomon BO. 2005. Substrate Inhibition Kinetics of Phenol Degradation by Pseudomonas aeruginosa and Pseudomonas fluorescence. Biotechnology 4: 56-61.
[31] Banks JG, R.G. Board RG, Paton J. 1985. Illuminated Rose Bengal causes adenosine triphosphate (ATP) depletion and microbial death. Letters in Applied Microhioloyy 1: 7-1.
[32] Schafer M, Schmitz C, Horneck G. 1998. High sensitivity of Deinococcus radiodurans to photodynamically-produced singlet oxygen. Int. J. Radiat. Biol 74: 249–253.
[33] Nitzan Y, Balzam-Sudakevitz A, Ashkenazi H. 1998. Eradication of Acinetobacter baumannii by photosensitized agents in vitro. J. Photochem. Photobiol B 42: 211–8.
[34] Malik Z, Ladan H, Nitzan Y. 1992. Photodynamic inactivation of Gram-negative bacteria: problems and possible solutions. J. Photochem. Photobiol, B 4: 262–266.
[35] M. Aldahoun, M-Ali Al-Akhras, M.S. Jaafar, M. Bououdina. Enhanced anti-cancer and antimicrobial activities of curcumin nanoparticles. *Artificial Cells, Nanomedicine, and Biotechnology,* 45, (2017) pp. 98-107
[36] Aldahoun, M. Jaafar, M., Al-Akhras, M-Ali, Bououdina, M. Enhanced nanocurcumin toxicity against (PC3) tumor and microbial by using Magnetic field in vitro. *Artificial Cells, Nanomedicine, and Biotechnology* **2017** *45* (4), 843-853.
[37] Hass BS, Webb RB. 1979. Photodynamic effects of dyes on bacteria. III. Mutagenesis by acridine orange and 500-nm monochromatic light in strains of Escherichia coli that differ in repair capability. Mutat. Res 60: 1–11.
[38] Imray FB, MacPhee DG. 1973. The role of DNA polymerase I and the rec system in survival of bacteria and bacteriophages damaged by the photodynamic action of acridine orange. Mol. Gen. Genet 123: 289–298.

Rapid Magnetic Nanobiosensor for the detection of *Serratia marcescen*

Alaa A. A. Aljabali[1], Emad Hussein[2,3], Omar Aljumaili[2], Mazhar Al Zoubi[2], Bahaa Altrad[2], Khaled Albatayneh[2] and Mutaz A. Abd Al-razaq[2]

[1]*Faculty of Pharmacy, Yarmouk University, Jordan*
[2]*Department of Biological Science, Faculty of Science, Yarmouk University, Jordan*
[3]*Department of Food Science and Human Nutrition, College of Applied Sciences, A'Sharqiyah University, Ibra, Oman.*

alaaj@yu.edu.jo

Abstract The development of rapid, sensitive, accurate and reliable bacterial detection methods are of keen interest to ensure food safety and hospital security. Therefore, the development of a fast, specific, low-cost and trusted methods is in high demand. Magnetic nanoparticles with their unique material properties have been utilized as a tool for pathogen detection. Here, we present a novel iron oxide nanoparticles labeled with specific targeting antibodies to improve specificity and extend the use of nanoparticles as nanosensors. The results indicated that antibody labeled iron oxide platform that binds specifically to *Serriata marcescenst* in a straightforward method is very specific and sensitive. The system is capable of rapid and specific detection of various clinically relevant bacterial species, with sensitivity down to single bacteria. The generic platform could be used to identify pathogens for a variety of applications rapidly.

Keywords (Bacterial detection, Iron Oxide Nanoparticles, Antibody Detection, Sensor)

1. Introduction

In recent years, bacterial detection from various sources is of great interest in the medical, environmental and food manufacturing. Among these bacteria that present a significant threat to human health are; *Campylobacter, Salmonella, Listeria monocytogenes, Escherichia coli (E. coli)* O157: H7, *Staphylococcus aureus*, and *Bacillus cereus* (1-5). Therefore, it is of great importance to developing reliable and sensitive methods for bacterial detection. Many approaches have been reported in the literature for the bacterial detection, including the culture and colony counting, polymerase chain reaction (PCR), and immunology-based methods (6, 7). The conventional detection mainly relies on colony counting method. This is a practical way for the detection and identification of pathogens from different resources, including bacterial culturing and isolation of the pathogen. Moreover, bacterial detection requires further identification by the confirmation using biochemical and serological experiments, which takes up to 5–7 days to get a result (5, 8). Although the methods above can obtain a reliable result, it is labor intensive and time-consuming, which cannot satisfy the request for on-the-spot bacterial detection. The PCR and enzyme-linked immunosorbent assay (ELISA) is a lot less time-consuming than the traditional culture and colony counting method (8). However, and to overcome some of the difficulties mentioned with the conventional methods the search for new alternative more reliable and sensitive methods lead to the development of the field of nanotechnology.

Nanomaterials offer an exceptional opportunity to revolutionary impact on biology and medicine because of the size-dependent physical and chemical properties of the nanoparticles.

Nanomaterials have been exploited extensively to provide targeted drug delivery, diagnostics, tissue regeneration, sensing tools, antimicrobial agents, and molecular imaging. Various platforms like, liposomes, dendrimers, quantum dots (QD), fullerenes, virus-like particles, and carbon nanotubes are being developed to improve human health (9-12).

Magnetic nanoparticles (NPs) hold great potential and currently is under investigation in magnetic hypertherthermia treatment (13), drug delivery (14) and cell sorting (15). Paramagnetic iron oxide NPs used as contrast agent in Magnetic Resonance Imaging (MRI) and photothermal therapy against cancer (16). Fe_3O_4 magnetic NPs used for liver imaging, a contrast agent for MRI, immunoassay, biosensing and drug delivery (17, 18). Silica-coated iron oxide PEG-coated NPs with contrast element Gadolinium (Gd) used to access a specific area of the brain to detect a tumor(15, 19).

Iron oxide coated NPs (IONPs) exhibit excellent biocompatibility, and they are considered a very stable colloid. IONPs labeled with antibodies have been used to detect breast cancer cells in vitro (20). In addition, IONPs conjugated with luteinizing hormone-releasing as breast cancer cells used to detect breast cancer in vivo (21). NPs are promising agents for antibacterial applications because of the selective toxicity against bacteria like silver NPs. Furthermore, some antibacterial nanoparticles can be degraded by lysosomal fusion and thus appear non-toxic to mammalian cells (22). In addition, a variety of moieties has been examined as targeting agents, including carbohydrates, peptides, receptor ligands and proteins such as transferrin and lectins (23, 24).

Bacterial infections are the fifth cause of mortality in the US with considerable economic impact on healthcare sector (25, 26). Traditionally, the presence of a bacterium in any sample is detected microscopically, usually after growth in culture. Confirmation is based on growth patterns in various media and biochemical tests. These methods, although highly specific, have some drawbacks. First, they usually take more than 24 hours and lack the desired sensitivity due to some bacteria needed, while not all microorganisms can grow in culture. In addition, other methods involving polymerase chain reaction (PCR) and fluorescent immunoassays have been developed for bacterial detection. However, despite their enhanced sensitivity and specificity, these methods are expensive, time-consuming, and need to be performed by trained and experienced personnel.

One of the methods that have been developed which holds a promising potential is that the direct immobilization of antibodies on the surface of an electrode for the detection of bacteria called impedimetric immune-sensors. This relies on the immobilization of antibodies on the electrode surface and then probing the attachment of the bacterial cells by measuring the change in electrical properties of the electrode (8). Antibodies have long been the most popular recognition elements because of their sensitivity and selectivity (27, 28).

Nanotechnology and nanoparticles (NPs) offer an attractive alternative for the identification of molecular targets *in vitro* and *in vivo*. Biosensors based on nanomaterials have been used to amplify detection signal and achieve lower detection limit due to their high surface area (29, 30). The use of NPs requires smaller samples volumes and less preparation time, without compromising the detection sensitivity (31, 32). Over the past decade, superparamagnetic iron oxide nanosensors have been designed to quantify biomolecular targets in cell lysates and tissue extracts, demonstrating high sensitivity (33, 34). The underlying detection method utilizes various molecular targets such as nucleic acids (DNA and mRNA), antibodies, peptides and other chemical moieties.

The interaction between the NPs and the microorganisms are often toxic and has been exploited for various antimicrobial applications. However, investigating the interactions and the selectivity of specific bacterial detection is of great importance in the field of bionanotechnology. The existing

IOP Conf. Series: Materials Science and Engineering **305** (2018) 012005 doi:10.1088/1757-899X/305/1/012005

detection methods for bacteria that are available relied on ELISA, Western Blots, surface plasmon resonance (SPR) biosensors, antibody microarrays Ref. The mentioned methods require high doses for the sample detection although some are sensitive the need for faster and more reliable method still in need. The method that is reported here is prompt, fast, reliable and relatively inexpensive. The development of nanotechnology offers many technological advances for the detection of bacterial. This paper acknowledges the need for the development of fast and sensitive bacterial detecting biosensor. In addition, we investigate the use of IONPs for the detection of the bacterium in solution. Modifying NPs with the target antibody as shown in Figure 1, in this case, will lead to the decoration of the bacterium surface with the modified IONPs, the high number of NPs that are expected to bind to the surface of the bacterium would lead to higher sensitive's, particularly al low target bacteria concentrations.

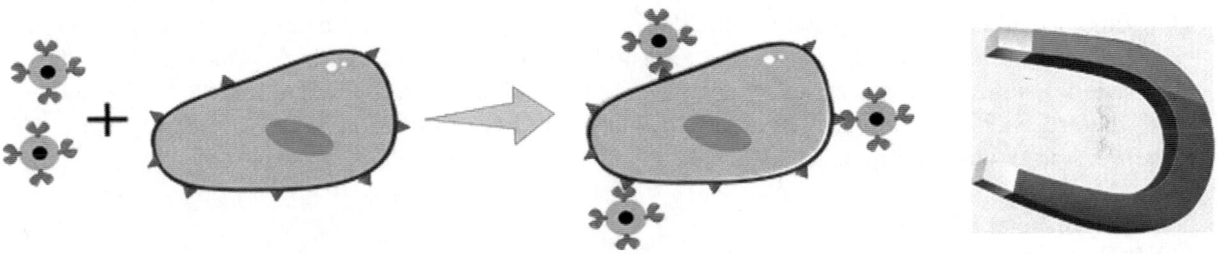

Figure 1 Schematic representation of the modification methods approach. The IONPs were decorated with antibodies that are unique to the bacterial model.

As a proof of this hypothesis, we used *Serratia marcescens* as our model organism. We selected this bacterium as a proof-of-principle model since it belongs to the family, Enterobacteriaceae, which is a large family of Gram-negative bacteria that includes, several pathogens, including *Salmonella, Escherichia coli, Yersinia pestis, Klebsiella, Shigella, Proteus, Enterobacter, Serratia,* and *Citrobacter*. This family belongs to the order Enterobacteriales of the class Gamma proteobacteria in the phylum Proteobacteria. It was originally considered to be a harmless, non-pathogenic, organism and it was frequently used as a biological marker because of its easily distinct red colonies (Merlino, 1924). *Serratia marcescens* has been recognized as an important opportunistic pathogen combining a propensity for healthcare-associated infection and antimicrobial resistance. It is a widely distributed saprophytic bacterium and has been found in food, particularly in starchy variants which provide, an excellent growth environment (35). Currently, 14 species of *Serratia marcescens* are recognised within the genus, eight of which are associated with human infection (36). Intensive care units are frequently involved in the epidemics of infection with *Serratia marcescens*. It accounts for less than 1-2% of the hospital-acquired infections which are frequently limited to the respiratory tract, the urinary tract, surgical wounds and soft tissues (37, 38). The nanosensor used in this study was prepared by the conjugation of anti- Serratia antibody (Details) to superpara magnetic IONPs. The antibody labeled nanoparticles acted as very stable months, after storage at 4°C. Apart from sensitive and fast detection, the detection method is independent of the samples optical properties.

2. Materials and Methods

All general reagents, Iron salts ($FeCl_2$. $4H_2O$ and $FeCl_3$. $6H_2O$), dialysis membrane 14 kDa, NaOH, epichlorohydrin, Purchased from Sigma-Aldrich, Dextrans (10 kDa) were received from Amersham Biosciences/Pharmacosmos and Sigma-Aldrich, succinic anhydride, N-hydroxysuccinimide, Sodium Hydroxide (NaOH), epichlorohydrin, NH_4OH, Standard concentrates 10.00 g Fe (Sigma CAS No 10421-48-4), Anti-*Serratia marcescent* antibody B/N4N was obtained from Abcam, Distilled water, DMSO, EDC: 1-ethyl-3-(3-dimethyl aminopropyl) carbodiimide hydrochloride, NHS: N- hydroxysulfosuccinimide sodium salt.

2.1. Bacterial culture

Serratia marcescens (ATCC 27117) pure culture provided by the laboratory of microbiology and biotechnology, Department of Biological Sciences, Yarmouk University-Jordan; were grown on nutrient agar media plate. The bacterial strain was inoculated into the nutrient broth, and on nutrient agar media, the resulting pure colonies were used to prepare bacterial suspensions. A loop full of the bacterial strain was used to inoculate nutrient broth and was incubated for 24hrs at 30 °C. The bacterial suspension turbidity was adjusted to 0.5 McFarland standards. One millilitre of this bacterial suspension was transferred to a new tube of broth and incubated at 30 °C for six hours before each experiment. The serial dilutions of bacterial culture were prepared using autoclaved broth water before each experiment.

2.2. Synthesis of IONPs

IONPs were synthesis following adapted protocol adapted from reference (39). Briefly, a mixture of $FeCl_2$ (3 mM) and $FeCl_3$ (6 mM) was dissolved in 100 mL Milli–Q or DD-H_2O water and prepared freshly before use. Dextran 10 kDa (2.5 g) of technical grade dextran (Sigma-Aldrich, Cat. Nr. D9260, lot Nr. BCBK8656V) dissolved in 5 mL Milli–Q water and was placed on a rotary spinner until a clear solution is formed. Iron chloride solutions with a molar ratio of 2:1 as described earlier was then mixed with a freshly prepared aqueous solution of dextran and stirred vigorously for 30 minutes under constant N_2 gas circulation. The pH of the solution was adjusted slowly up to pH 12 using 10 mM NaOH solution by drop wise addition. As an indication of the IONPs formation, the solution colour changed accordingly from yellow to dark brown to black. The solution was incubated at 80° C (Thermoblock or water-bath) for 2 hours. Samples were then transferred over a magnetic stirrer and mixed using the magnetic stirrer for one hour at ambient temperature. The solution was purified from excess dextran by dialysis overnight (17 hours) against DD-H_2O using dialysis membrane 14 kDa (Sigma-Aldrich) while exchanging water every 2 hours (3 times) and then left overnight at ambient temperature. The resultant purified NPs were concentrated using spin columns, and the iron content was determined spectrophotometrically (DAR800-Diagnostic automation, INC, USA).

2.3. Synthesis of amine terminated dextran coated NPs

The particles were prepared following adaptation from (40) 20 mL of the previously prepared dextran covered NPs were placed into a 250 mL round flask equipped with a magnetic stirrer at the ambient temperature. 36.7 mL of NaOH (5M) was added as a drop wise (roughly at a rate of 1 drop per second). After that, 13.5 mL of epichlorohydrin was added to drop by drop rate. The mixture was stirred for 7 hours at ambient temperature followed by the addition of 20 mL of NH_4OH (25%) was added at drop wise fashion, and then placed on a magnetic stirrer for 14 hours at ambient temperature. The mixture was dialyzed against water through a dialysis membrane 14 kDa Spectra-Por Float-A-Lyzer G2 bags (Sigma-Aldrich). The mixture was stirred for overnight and then dialyzed against DD-H_2O.

2.4. Synthesis of carboxyl terminated dextran coated NPs

20 mL of dextran-coated iron oxide nanoparticles were placed in 100 mL flasks equipped with stirrer bar for 5 minutes at an ambient temperature to homogenize the solution. A freshly prepared aqueous solution of 30 mg of succinic anhydrate was added in two stages to maintain pH (~ 5) of the reaction. The mixture was stirred for 1 hour at an ambient temperature, after that, another aqueous solution containing 30 mg of the succinic anhydrate was added. The mixture was stirred overnight and stored at 4°C until needed.

2.5. Preparation and Purification of bacterial suspension for TEM images

Bacteria were grown overnight in 10 mL of nutrient broth in a shaker incubator at 30° C. Followed by the addition of 1.5 mL of the bacterial broth addition an Eppendorf tube and centrifuged for 5 minutes at 10,000 g. The supernatant was discarded, and the pellet was washed twice with sterile distilled H_2O and resuspended using vortex. This bacterial suspension was viewed and examined as well as for the control under TEM.

2.6. Examination and Imaging Bacterial Suspension under TEM

5–10 µl of 0.1 mg ml^{-1} suspended particles were allowed to settle briefly (1 minute) onto pyroxylin and carbon–coated copper grids (300 mesh, Agar Scientific), and then blotted dry with filter paper. Grids were viewed at 60 kV in an EM10CR TEM (41). TEM images were performed by CCD microscopic camera (Optikam B9 Digital Camera).

2.7. Sectioning and Imaging under TEM

Bacterial cells were incubated with the antibody labeled NPs. Cells were fixed in a solution of 2.5% glutaraldehyde dissolved in 0.1 M sodium cacodylate buffer (pH= 7.3) for two h at 4 °C. Fixed cells were washed three times with 0.1 M sodium cacodylate buffer each wash for 5 minutes. Post-fixation staining was done using 1 % stock solution of osmium tetroxide for 1 hour. This was followed by washing with 0.1 M sodium cacodylate for 1.5 hours for three times each wash is for 5 minutes at 4 °C. Cells were dehydrated in acetonitrile (30, 50, 70, 90 and 100%) and infiltrated in Spurr's low viscosity (3:1, 1:1 and 1:3) resins. Then, cells were infiltrated in pure Spurr's resin (Low Viscosity Embedding Media Spurr's Kit, Electron Microscopy Sciences) for 12 hours, embedded in Beam capsules containing pure resin blocks, and finally hardened at 70 °C for 17 hours (overnight). Ultrathin sections (70-90 nm) were cut by Ultra microtome (Reichert-Jung) and the sections were loaded on copper grids (300 mesh, Agar Scientific) and stained using 5% (w/v) aqueous uranyl acetate. Samples were then analyzed under a 60 Kv ZeissEM10CRTEM microscope, TEM images were performed by CCD microscopic camera (Optikam B9 Digital Camera).

2.8. Dynamic light scattering (DLS)

DLS was measured on a DynaPro Titan, Wyatt Technology Corporation (laser wavelength 830 nm, scattering angle 20 degrees) with Dynamics software Version 6.9.2.11. Particles at a concentration of 0.5–1 mg^{-1} cm^{-1} in 10 mM sodium phosphate buffer pH 7.0 were filtered through 0.1 – 0.4-micron filters (Millipore) before analysis. Four independent measurements were performed; each single measurement presents an average of ten measurements. Data were recorded at 20 °C.

2.9. Surface charge measurement

Zeta potential was measured on a Malvern Instruments Zetasizer–Nano ZS where 1 ml of 0.5 mg ml^{-1} particles were suspended in 10 mM sodium phosphate buffer pH 7.0. Zeta cells were equilibrated at 21°C before recording three measurements each of 12 runs. The data were fitted using the Smoluchowski approximation assuming Henry's function f(Ka) of 1.5.

2.10. Preparation and purification of bacterial and IONPs.

The prepared bacterial culture as described earlier was then spun at 10,000 g for 5 minutes. The supernatant was discarded, and mixed the pellet with 300 µL of IONPs and suspended using vortex. This bacterial-IONPs suspension was viewed and examined under TEM.

2.11. Determination of Iron content

Spectroscopic measurements are most easily performed with liquid samples, and the protocol is adapted from (42). To convert the insoluble Fe_3O_4 in the unknown (NPs) to a soluble species, an acid digestion was performed. They have used a solution of Fe^{3+}. The spectroscopic methods used here are based on a molecular spectroscopy procedure that is based on forming complexes between the irons with an organic ligand and measuring the absorbance of the resulting complex.In this experiment, measurements were recorded by generating a calibration curve of prepared (known) standards of Iron atomic spectroscopy standard concentrates 10.00 g Fe (Sigma CAS No 10421-48-4) and the formation of a linear calibration model between the absorbance (X-axis) and the concentration (Y-axis) (mg/mL). The calibration curve is designed with the unknown concentration falls somewhere in the middle of its range. Ten µl of aqueous 3% freshly prepared an H_2O_2 solution and ten µl of each standard solution was added to 1 mL of 5 M HCl. The NPs particles to be tested were prepared by taking the ratio of 1/3 of IONPs solution to 2/3 Milli-Q water, in triplicate two µl of the nanoparticles from the dilution and incubate with two µl of aqueous 3% freshly prepared an H_2O_2 solution and 200 µl of 5 M HCl. The NPs particles were incubated at 50 °C for 1 hour in a Thermoblocks with the same amount of 1 mL of 5 M HCl. The standard curve and the sample absorbance were measured on 96 well plates in triplicate at 410 nm. The measured absorbance was subtracted from the background, and the standard curve was plotted on excel sheets. The iron content was calculated from the slope of the standard curve generated.

2.12. Antibody labeling NPs

Carboxylic acid groups were activated by the carbodiimide method using 1-ethyl-3-(-3-dimethylaminopropyl) carbodiimide hydrochloride (EDC-Thermofisher Scientific) and NHS (N-hydroxysulfosuccinimide-Sigma) or sulfo–NHS. Carboxyl terminated IONPs suspended in either 10 mM or 0.1 M sodium phosphate buffer pH 7.4 were incubated with a freshly prepared Milli–Q water solution of EDC (1 molar equivalent) for 5 minutes followed by the addition of a freshly prepared dimethylsulfoxide solution of NHS dissolved in DMSO (4 molar equivalents). The reaction was allowed to proceed for 2 hours at room temperature with gentle stirring. The resulted in O–succinimide esterified IONPs were purified by 10-14 kDa MWCO dialysis membrane. The esterified particles were then incubated with the anti-*Serratia marcescens* monoclonal antibody

IOP Conf. Series: Materials Science and Engineering **305** (2018) 012005 doi:10.1088/1757-899X/305/1/012005

(100 μL at 2 mg/mL stock concentration in which 20 μL of antibody were added directly to solution mixed contain EDC, NHS and carboxy NPs) and the reaction was left to proceed overnight while gently shaking at 4°C. Antibody-labeled IONPs were then purified on PD–10 desalting columns (Amersham) were used according to the manufacturer's instruction and were pre–equilibrated with ten mM sodium phosphate buffer. Eluted samples were collected and further analysed.

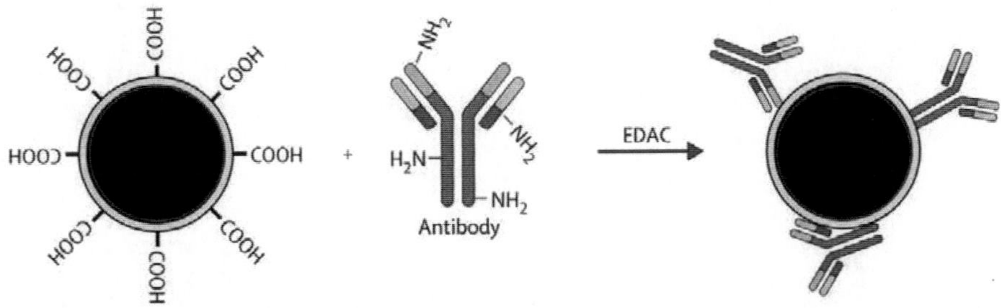

Figure 2 schematic representation of the antibody conjugation strategy on the exterior surface of the IONPs.

2.13. QuantiPro Bicinchoninic acid (QPBCA)

Bicinchoninic acid (BCA) protein assay was chosen as the second protein quantification method. The technique relies on the reduction of Cu^{2+} to Cu^{+} by proteins leading to the formation of the purple-blue copper protein complex in alkaline conditions that can be monitored spectrophotometrically at λ=562 nm. Protein concentrations in the presence or absence of MNP were measured with a BCA Protein Assay Kit as described by the manufacturer instructions (Sigma-Aldrich). Briefly, 200 μl of BCA reagent was added to each sample. Samples were incubated at ambient temperature overnight while gently shaking at 100 rpm. The absorbance of the reaction products was measured at λ=562 nm using a microplate reader (Tecan).

2.14. Bradford protein assay

Protein quantification was determined following Bradford assay method relying on the binding of Coomassie brilliant blue G-250 to the conjugated protein the IONPs resulted in the shift in the absorption maximum of the compound dye from 465 to 595 nm. The increase in the absorption at 595 was determined photometrically as described by Bradford (43). In essence, a 1 ml aliquots of freshly prepared Bradford reagent was added to 100 ml of IONPs and antibody labelled IONPs; samples were carefully mixed. A 200 μl aliquots of the solution above was transferred to a 96-well-plate. The reaction left to proceed for 45 minutes at ambient temperature, the sample absorbance was determined spectrophotometrically at λ 595 nm (Tecan).

3. Results and discussion:

3.1. The characterisation of IONPs

The particles were characterised for every step of the synthesising procedure and before modifying the NPs. The synthesized particles were characterized by TEM and DLS instrumentations.

The dextran coated DLS determined NPS with an average of ~60 nm from five independent syntheses, and each set of measurements includes ten replicates. In addition, TEM observations revealed that the morphology and the size of the iron core are ~ 20 nm surrounded by the dextran coating up-to 60 nm in diameter as shown in Figure 3. The polydispersity index of the synthesized IONPs is 0.002 % which indicates 'monodisperse' distribution of the particles. It has been reported that nanoparticles with < 0.20 % polydispersity index are considered "highly monodisperse (44, 45). Also, the correlation plot as shown in Figure 2, confirms the absence of aggregation in the sample and indicating the particles are a stable colloid.

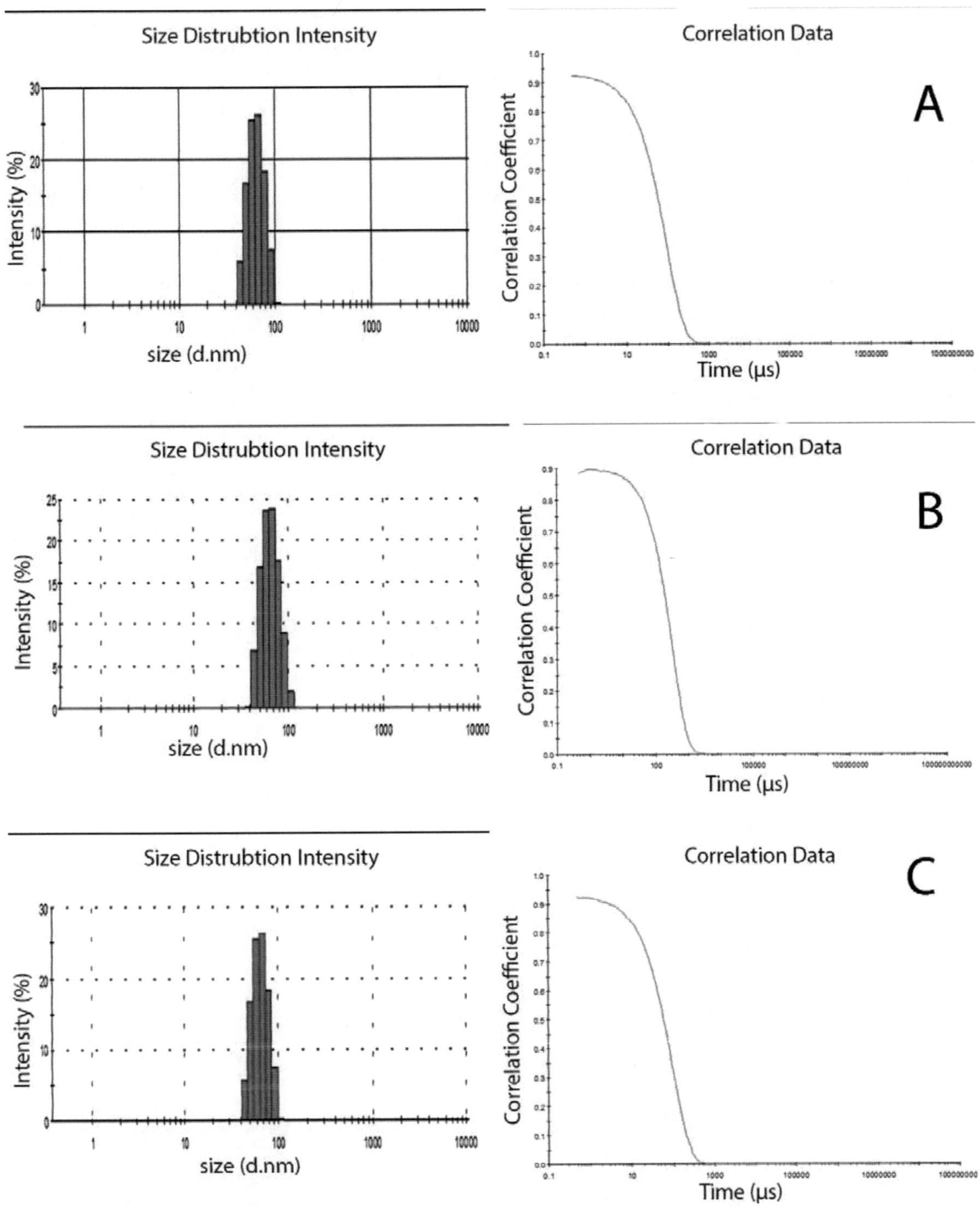

Figure 3 Size distribution intensity graph of IONPs nanoparticle as revealed by DLSwith different surface functionalities with its corresponding correlation fit (A) dextran coated IONPs (B) Carboxyl terminated IONPs (C) antibody labeled IONPs

IOP Conf. Series: Materials Science and Engineering **305** (2018) 012005 doi:10.1088/1757-899X/305/1/012005

Zeta potential values are another useful tool for the stability of colloidal particles and confirming the successful surface modification. The absolute values reflect the net electrical charge on the particle surface. The suggested theoretical limit of stability is $|25|$ mV. Particles are considered to exist as stable colloids of their zeta potential is higher than 25 mV or lower than –25 mV (46, 47).

The zeta potential analyses were conducted for all the modification steps to confirm successful modification of the NPs surface charge. The Z-potential of the dextran-coated nanoparticles were be found on average -0.165 (see Figure 4-A), while the carboxyl terminated nanoparticles (negatively charged) were found to be ranged from -10.7 to -26 as shown in Figure 4-B. In addition, Z-potential were utilized to confirm the surface modification of the IONPs with the antibody; the surface charge values indicated a value of -53.8 ± 8.0 mV see Figure 4-C. In addition, the zeta potential values suggested that the particles are very stable colloid and no propensity to aggregate, which is in agreement what the visual inspection of the samples as no visible aggregation have been seen on any of the prepared samples. Particles were stable for 2-3 months with no visible particles aggregation.

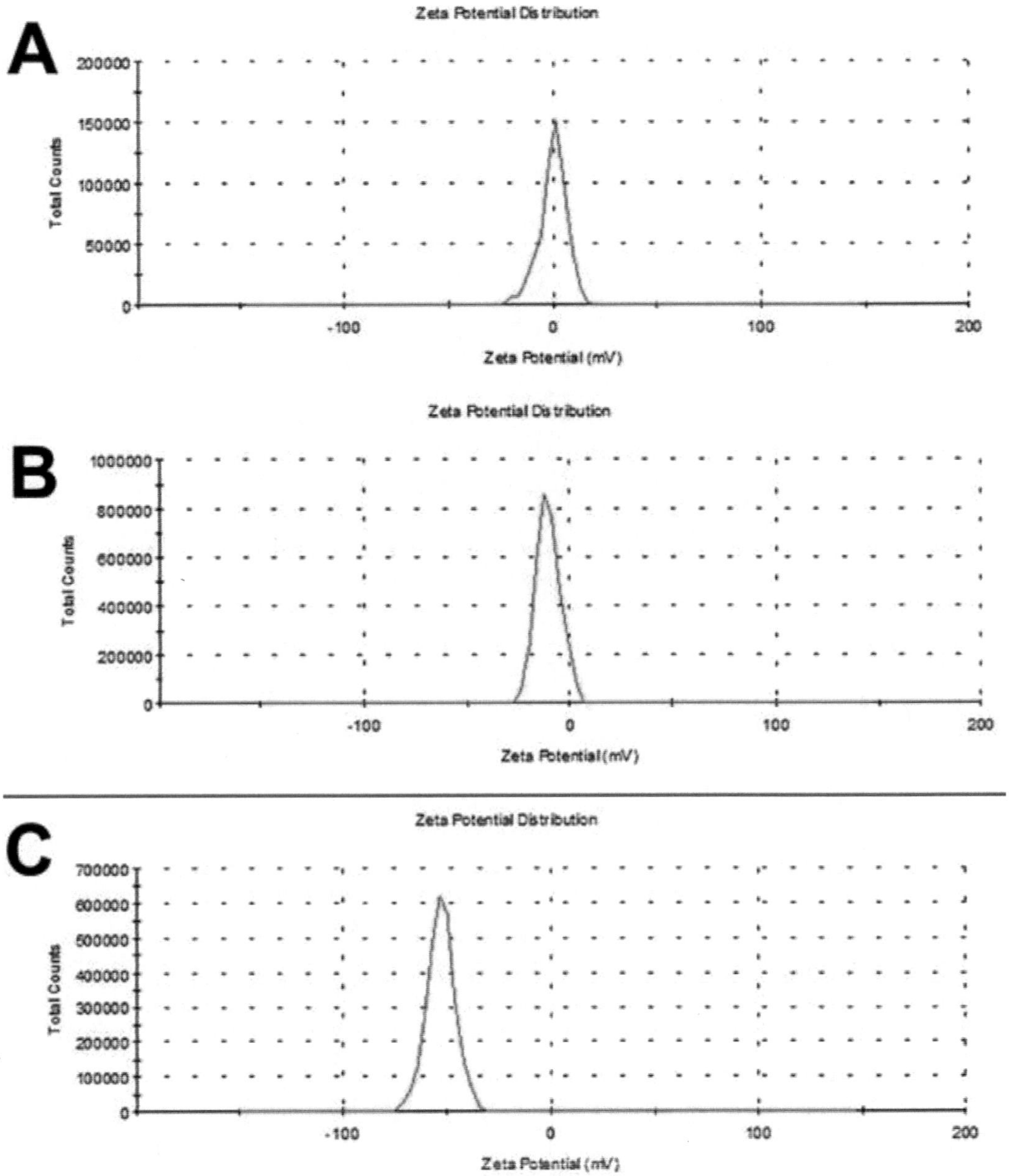

Figure 4 Zeta potential measurements of the functionalized IONPs (A) Dextran coated IONPs (B) carboxyl terminated IONPs (C) antibody labeled IONPs

TEM analyses of the synthesized IONPs revelled a solid core of the size 10-12 nm with a monodisperse size distribution and no visible aggregation in all samples synthesised as shown in as shown in Figure 5.

IOP Conf. Series: Materials Science and Engineering **305** (2018) 012005 doi:10.1088/1757-899X/305/1/012005

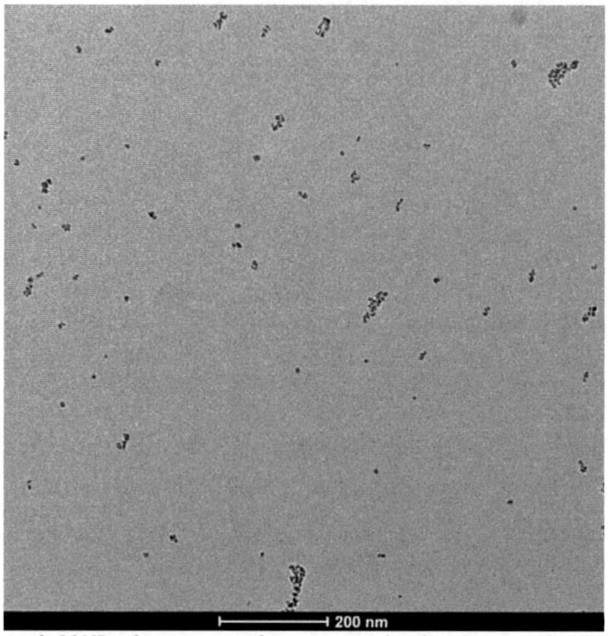

Figure 5 TEM image of unstained IONPs showing monodisperse particles distribution and solid metallic core of 10-12 nm in diameter.

TEM samples for this work were prepared by a typical process of fixation, dehydration, and embedding in a resin matrix has been reported in the literature (48, 49). TEM was utilized as a tool to examine non-sliced and sliced bacterial cells before the incubation with the IONPs and then examine the binding profile of antibody labeled IONPs elimination the electrostatic interactions as the only mean for the specific bindings as shown in Figure 6.

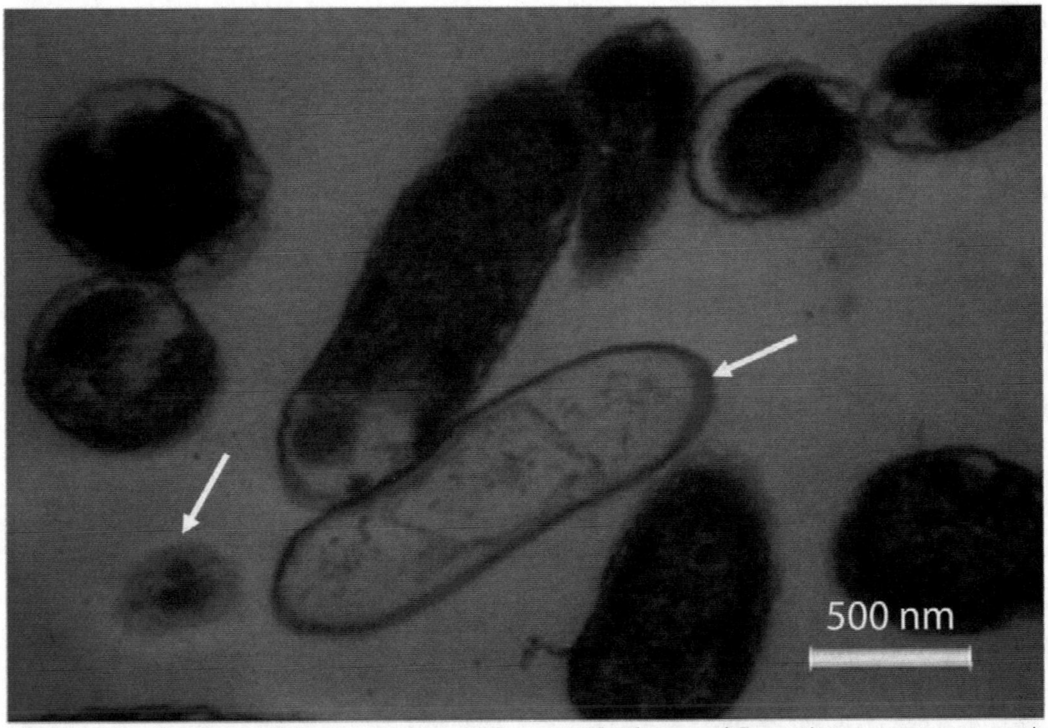

Figure 6 unstained and fixed section of 80 nm in thickness of Serratia marcescens as viewed by TEM. The yellow arrow indicates lysed or empty bacterial cell. White arrow indicates cell wall-free cytoplasmic content

The TEM images gave valuable information on how and where the NPs were about the bacterial cells. To visualize the surface interactions between the NPs and the bacterial model, sectioned TEM images were acquired to confirm the binding of the antibody labeled NPs on the bacterial surface. TEM images of Serratia in the presence of 10 µg of iron as determined by the iron content. The lower magnification images show the overall morphology of cells and NPs, while Figure 7 at higher magnification reveal more specific interactions between cells and NPs. Overall, for all the nanoparticles interaction work, no internalization of IONPs was observed into bacterial cells.
The TEM shows that upon incubating IONPs with the bacteria, Serratia cells remain intact with very minimal to the non-noticeable extent of IONPs on the bacterial cell. The lack of nonspecific association (binding) between the antibodies labeled IONPs was confirmed. The TEM reveled non-specific binding of the main carboxyl terminated IONPs in comparison to the amine terminated ones as seen in Figure 9. During TEM sample preparation IONPs-bacteria was pelleted after 10 minutes' incubation period. Followed by multiple steps of buffer rinse, the majority of antibody labeled IONPs that were not cell-bound were washed off in the supernatant, leaving the Serratia pellet free of unbound particles before embedding and sectioning. Figure 7, demonstrate that where cell surface binding occurred, a small cluster of antibody-IONPs was partially attached to the cell surface at various points, without compromising the integrity of the cell. The majority of the visible IONPs formed uniformly packed with a thin layer of the IONPs around the cells. An interesting antibody labeled IONPs interaction pattern was observed. IONPs labeled particles remain well separated, neatly lining the cell surfaces while keeping a small gap between the IONPs and the cell wall. Although the reason for this phenomena is not clear, one hypothesis that has been

published in the literature is that the NPs surface chemistry that leads to aggregation also plays a vital role here. The uniform gap between the attached IONPs and cell walls could be either the result of LPS that does not give significant TEM contrast or a double-layer (50). Overall, no internalization of IONPs was observed into bacterial cells from any of the TEM images. It is clear that IONPs were attached to the thick peptidoglycan layer of the cell wall, far from the buried lipid membrane layer. Furthermore, to investigate whether the observation from the EM images is actual IONPs. The EM grids were further analyzed by Energy-dispersive X-ray spectroscopy (EDX-INCA Energy 200Premium attached to a JEOL JSM 5900LV scanning electron microscope-SEM), the data confirmed that the layer that is surrounding the EM images are composed of iron as shown in the inset of Figure 8.

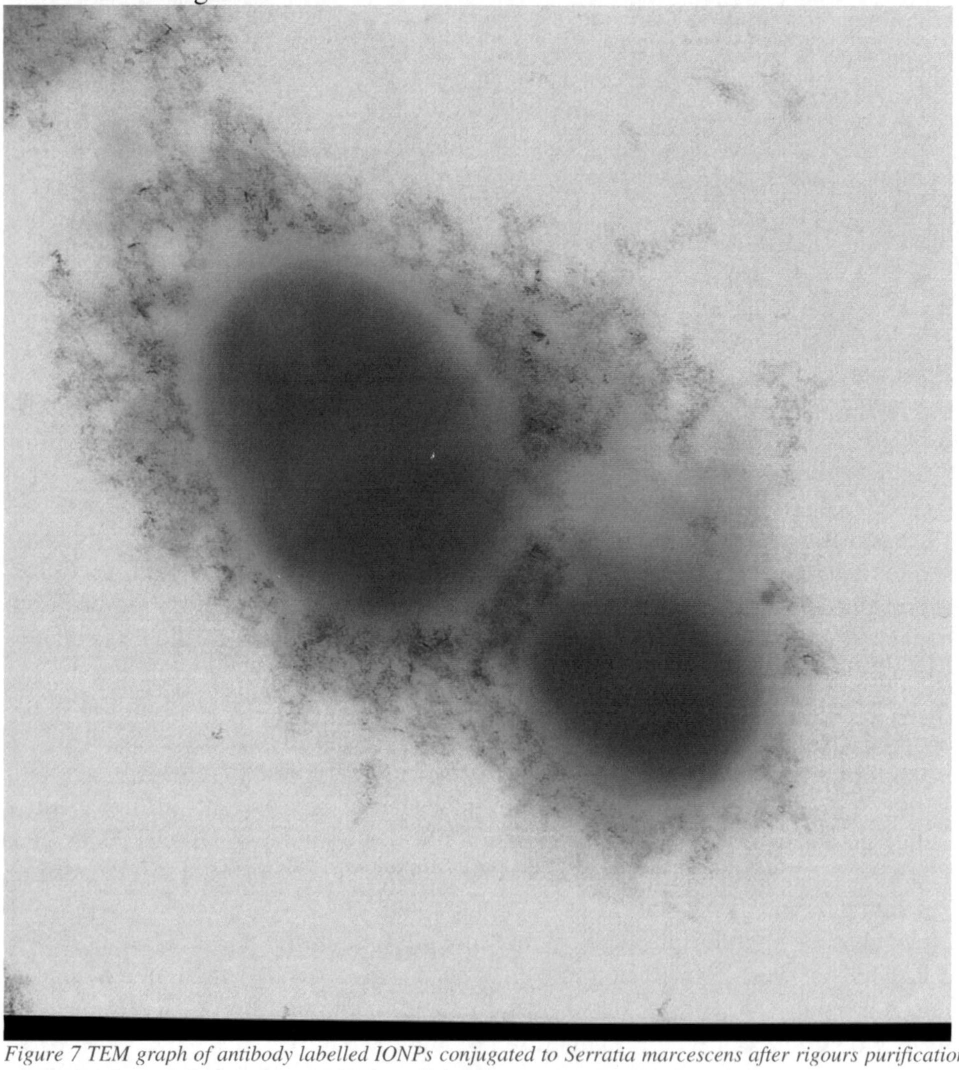

Figure 7 TEM graph of antibody labelled IONPs conjugated to Serratia marcescens after rigours purification. The particles are attached to the carbohydrate layer on the bacterial surface.

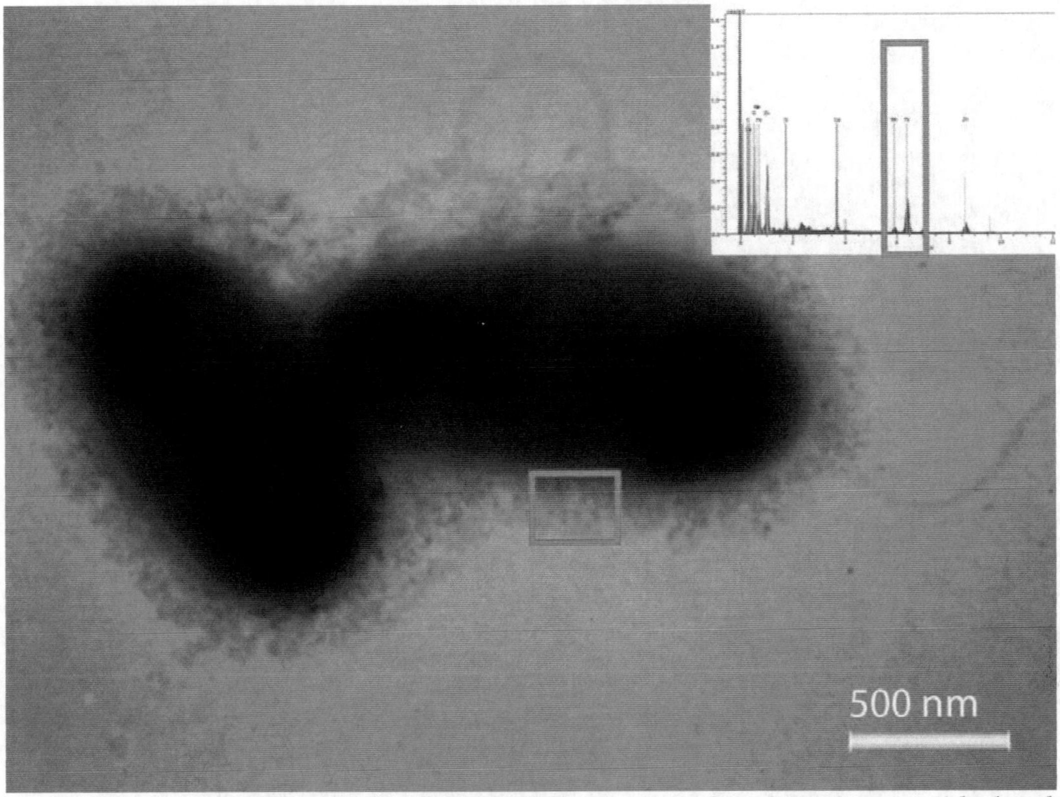

Figure 8 TEM of the antibody lableed IONPs conjugated to Serratia spp with the elemental analysis in the inset. The EDXs confirmed that the layer surrounding the bacteria consists of the iron signal in the selected area.

Whole cell imaging of Serratia with an antibody labeled IONPs showing the attachment of the IONPs on the outer layer of the bacteria. EDX analysis confirmed the presence of iron which is in agreement of the successful binding of the IONPs to the bacterial surface

TEM studies of the antibody labeled IONPs provided a snapshot of the localization interactions. The non-specific binding of the amine terminated-IONPs, carboxy terminated-IONPs, and the antibody labeled IONPs were investigated. The amine terminated IONPs some affinity towards the cells. The cationic coating on the IONPs surface induced limited non-specific binding. The non-specific binding could be attributed to the presence of negative charge on the surface of the IONPs based on the comparisons with control samples as seen in Figure 9 and 10. This could be hypothesized through the electrostatic interaction between the cationic IONPs and the negative cell surface, leading to cell wall attachment to higher degrees than the anionic IONPs. Also, the TEM sample preparation procedure involving repeated pelleting steps may have artificially enhanced the attachment observed in these images. Furthermore, the ultra-thin (70-90 nm) sections of the same antibody labelled bacterial samples that were used for the whole particles studies revelled that the IONPs were present attached to the cells or outside. None of the examined sections did show any visible IONPs within the bacterial cells confirming that the binding is to the exterior of the cells.

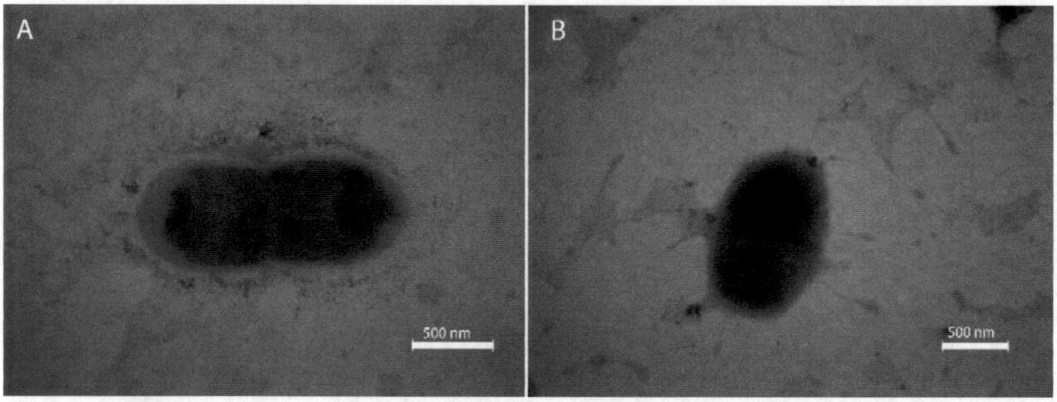

Figure 9 TEM graph of non-specific binding to Whole cell imaging of Serratia marcescens (A) carboxyl terminated (B) amine terminated IONPs

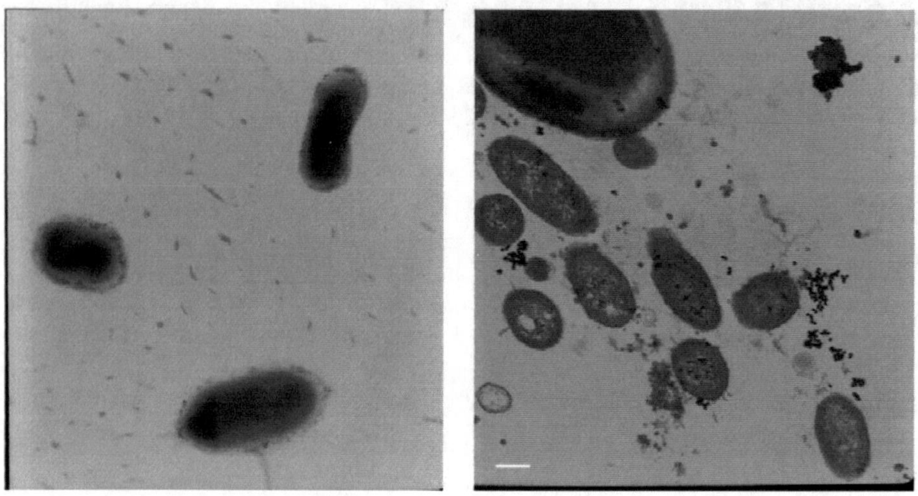

Figure 10 TEM images of the ultra-thin cross sections showing that the particles are either attached to the bacterial surface or are present on the outside of the cells. No particles were observed at any of the sections within the cells. Scale bar is 100 nm

Furthermore, the antibody quantification was determined spectrophotometrically in the presence and the absence of the antibodies at $\lambda = 562$ nm, which was quantified via BCA and determined to be 1.5 ± 0.2 mg/ml iron. Bradford method for the estimation of the protein content showed no net absorbance change at λ=595 nm. All measurements on both methods were in agreement and within the experimental errors of the protein conjugated to the IONPs as shown in Figure 11. To demonstrate the magnetic properties of the IONPs-labelled with the antibody the labelled particles were put close to Neodymium Magnet which resulted in pulling the labelled bacteria to the side of the tube as shown in figure 12.

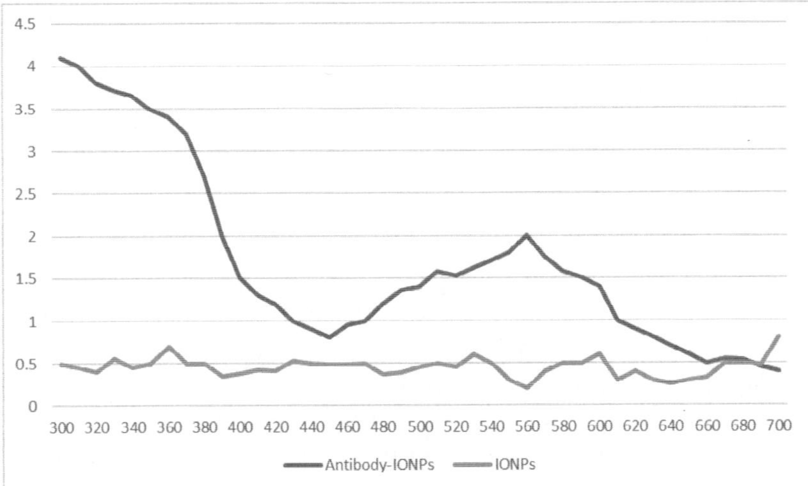

Figure 11 Absorbance spectra of antibody conjugated to IONPs after the incubation with the Bicinchoninic acid (BCA) reagent overnight. Spectra were recorded at λ= 562 nm. All assays were performed in triplicates; absorbance has been registered with the standard deviation as the error in the mesuremtns.

Figure 12 image of the IONPs labelled bacteria (A) that move with an applied external magnetic field (B).

It has been reported in the literature that, a TEM study examining non-sliced bacteria interactions with six nm-diameter cationic AuNPs reveled AuNP clusters on the Bacillus surface could be dispersed upon the removal of surface proteins by trypsin (51). These results may contribute

IOP Conf. Series: Materials Science and Engineering **305** (2018) 012005 doi:10.1088/1757-899X/305/1/012005

toward our understanding of the molecular-level identification of cellular components responsible for NP interactions with bacterial cells.
The growth curve of Serratia under normal conditions depicted the lag, log, stationary and death phase (data not shown). It is believed that the bacterial growth was not affected by the addition of the IONPs has been confirmed through the incubation of the IONPs with different surface moieties on agar growth plates.

4. Conclusions

Bacteria are vital contributors and single cell model organism for evaluating the impact of nanomaterials in the environment. Herein, we report a very simple, easy and reliable method for the selective detection of bacteria of interest from any sample. The specificity of antibodies was combined with magnetic properties of the nanoparticles enhanced the detection capabilities of bacteria detection. TEM studies reveled that no NPs were internalized by the bacterial strain. IONPs functionalized with the antibody reveled the highest binding toward the bacteria. These results demonstrate the importance of a thorough understanding of the IONPs targeting.

Acknowledgments

The authors acknowledge the financial support provided by the Deanship of the Scientific Research-Yarmouk University (Grant Number 35/2015 and Grant Number 5/2015).

References

[1] Gandhi M, Chikindas ML. Listeria: A foodborne pathogen that knows how to survive. International journal of food microbiology. 2007;113(1):1-15.
[2] Murphy C, Carroll C, Jordan KN. Environmental survival mechanisms of the foodborne pathogen Campylobacter jejuni. Journal of applied microbiology. 2006;100(4):623-32.
[3] Velusamy V, Arshak K, Korostynska O, Oliwa K, Adley C. An overview of foodborne pathogen detection: in the perspective of biosensors. Biotechnology advances. 2010;28(2):232-54.
[4] Wang Y, Ye Z, Si C, Ying Y. Subtractive Inhibition Assay for the Detection of E. coli O157:H7 Using Surface Plasmon Resonance. Sensors (Basel, Switzerland). 2011;11(3):2728-39.
[5] Wang Y, Ye Z, Ying Y. New trends in impedimetric biosensors for the detection of foodborne pathogenic bacteria. Sensors. 2012;12(3):3449-71.
[6] de Boer E, Beumer RR. Methodology for detection and typing of foodborne microorganisms. International journal of food microbiology. 1999;50(1-2):119-30.
[7] Gracias KS, McKillip JL. A review of conventional detection and enumeration methods for pathogenic bacteria in food. Canadian journal of microbiology. 2004;50(11):883-90.
[8] Yang L, Bashir R. Electrical/electrochemical impedance for rapid detection of foodborne pathogenic bacteria. Biotechnology advances. 2008;26(2):135-50.
[9] Medina C, Santos-Martinez MJ, Radomski A, Corrigan OI, Radomski MW. Nanoparticles: pharmacological and toxicological significance. British journal of pharmacology. 2007;150(5):552-8.
[10] Ag D, Bongartz R, Dogan LE, Seleci M, Walter JG, Demirkol DO, et al. Biofunctional quantum dots as fluorescence probe for cell-specific targeting. Colloids and surfaces B, Biointerfaces. 2014;114:96-103.
[11] Ionov M, Gardikis K, Wrobel D, Hatziantoniou S, Mourelatou H, Majoral JP, et al. Interaction of cationic phosphorus dendrimers (CPD) with charged and neutral lipid membranes. Colloids and surfaces B, Biointerfaces. 2011;82(1):8-12.
[12] Sapsford KE, Bradburne C, Delehanty JB, Medintz IL. Sensors for detecting biological agents. Materials Today. 2008;11(3):38-49.

[13] Hilger I, Hiergeist R, Hergt R, Winnefeld K, Schubert H, Kaiser WA. Thermal ablation of tumors using magnetic nanoparticles: an in vivo feasibility study. Investigative radiology. 2002;37(10):580-6.
[14] Li YJ, Dong M, Kong FM, Zhou JP. Folate-decorated anticancer drug and magnetic nanoparticles encapsulated polymeric carrier for liver cancer therapeutics. International journal of pharmaceutics. 2015;489(1-2):83-90.
[15] Kocbek P, Kralj S, Kreft ME, Kristl J. Targeting intracellular compartments by magnetic polymeric nanoparticles. European journal of pharmaceutical sciences : official journal of the European Federation for Pharmaceutical Sciences. 2013;50(1):130-8.
[16] Marcus M, Karni M, Baranes K, Levy I, Alon N, Margel S, et al. Iron oxide nanoparticles for neuronal cell applications: uptake study and magnetic manipulations. Journal of nanobiotechnology. 2016;14(1):37.
[17] Yoo MK, Park IY, Kim IY, Park IK, Kwon JS, Jeong HJ, et al. Superparamagnetic iron oxide nanoparticles coated with mannan for macrophage targeting. Journal of nanoscience and nanotechnology. 2008;8(10):5196-202.
[18] Yu MK, Jeong YY, Park J, Park S, Kim JW, Min JJ, et al. Drug-loaded superparamagnetic iron oxide nanoparticles for combined cancer imaging and therapy in vivo. Angewandte Chemie. 2008;47(29):5362-5.
[19] Kamat M, El-Boubbou K, Zhu DC, Lansdell T, Lu X, Li W, et al. Hyaluronic acid immobilized magnetic nanoparticles for active targeting and imaging of macrophages. Bioconjugate chemistry. 2010;21(11):2128-35.
[20] Artemov D, Mori N, Okollie B, Bhujwalla ZM. MR molecular imaging of the Her-2/neu receptor in breast cancer cells using targeted iron oxide nanoparticles. Magnetic resonance in medicine. 2003;49(3):403-8.
[21] Zhou J, Leuschner C, Kumar C, Hormes JF, Soboyejo WO. Sub-cellular accumulation of magnetic nanoparticles in breast tumors and metastases. Biomaterials. 2006;27(9):2001-8.
[22] Taylor E, Webster TJ. Reducing infections through nanotechnology and nanoparticles. International journal of nanomedicine. 2011;6:1463-73.
[23] Eliaz RE, Szoka FC, Jr. Liposome-encapsulated doxorubicin targeted to CD44: a strategy to kill CD44-overexpressing tumor cells. Cancer research. 2001;61(6):2592-601.
[24] Wang J, Tian S, Petros RA, Napier ME, Desimone JM. The complex role of multivalency in nanoparticles targeting the transferrin receptor for cancer therapies. Journal of the American Chemical Society. 2010;132(32):11306-13.
[25] Scheld WM, Valone JA, Sande MA. Bacterial adherence in the pathogenesis of endocarditis. Interaction of bacterial dextran, platelets, and fibrin. The Journal of clinical investigation. 1978;61(5):1394-404.
[26] Sethi S. Bacterial infection and the pathogenesis of COPD. Chest. 2000;117(5 Suppl 1):286S-91S.
[27] Li D, Feng Y, Zhou L, Ye Z, Wang J, Ying Y, et al. Label-free capacitive immunosensor based on quartz crystal Au electrode for rapid and sensitive detection of Escherichia coli O157:H7. Analytica chimica acta. 2011;687(1):89-96.
[28] Yang L, Li Y, Erf GF. Interdigitated Array microelectrode-based electrochemical impedance immunosensor for detection of Escherichia coli O157:H7. Analytical chemistry. 2004;76(4):1107-13.
[29] Chai R, Yuan R, Chai Y, Ou C, Cao S, Li X. Amperometric immunosensors based on layer-by-layer assembly of gold nanoparticles and methylene blue on thiourea modified glassy carbon electrode for determination of human chorionic gonadotrophin. Talanta. 2008;74(5):1330-6.
[30] Huang J, Yang G, Meng W, Wu L, Zhu A, Jiao X. An electrochemical impedimetric immunosensor for label-free detection of Campylobacter jejuni in diarrhea patients' stool based on O-carboxymethylchitosan surface modified Fe3O4 nanoparticles. Biosensors & bioelectronics. 2010;25(5):1204-11.
[31] Rosi NL, Mirkin CA. Nanostructures in Biodiagnostics. Chemical Reviews. 2005;105(4):1547-62.

[32] Alivisatos AP, Gu W, Larabell C. Quantum Dots as Cellular Probes. Annual Review of Biomedical Engineering. 2005;7(1):55-76.
[33] Perez JM, Josephson L, O'Loughlin T, Hogemann D, Weissleder R. Magnetic relaxation switches capable of sensing molecular interactions. Nature biotechnology. 2002;20(8):816-20.
[34] Koh I, Josephson L. Magnetic nanoparticle sensors. Sensors. 2009;9(10):8130-45.
[35] Sleigh JD. Antibiotic resistance in Serratia marcescens. British medical journal. 1983;287(6406):1651-3.
[36] Mahlen SD. Serratia Infections: from Military Experiments to Current Practice. Clinical Microbiology Reviews. 2011;24(4):755-91.
[37] Yoon HJ, Choi JY, Park YS, Kim CO, Kim JM, Yong DE, et al. Outbreaks of Serratia marcescens bacteriuria in a neurosurgical intensive care unit of a tertiary care teaching hospital: a clinical, epidemiologic, and laboratory perspective. American journal of infection control. 2005;33(10):595-601.
[38] Crivaro V, Bagattini M, Salza MF, Raimondi F, Rossano F, Triassi M, et al. Risk factors for extended-spectrum beta-lactamase-producing Serratia marcescens and Klebsiella pneumoniae acquisition in a neonatal intensive care unit. The Journal of hospital infection. 2007;67(2):135-41.
[39] Mascolo M, Pei Y, Ring T. Room Temperature Co-Precipitation Synthesis of Magnetite Nanoparticles in a Large pH Window with Different Bases. Materials. 2013;6(12):5549.
[40] Nahrendorf M, Waterman P, Thurber G, Groves K, Rajopadhye M, Panizzi P, et al. Hybrid in vivo FMT-CT imaging of protease activity in atherosclerosis with customized nanosensors. Arteriosclerosis, thrombosis, and vascular biology. 2009;29(10):1444-51.
[41] Heap GA, Weedon MN, Bewshea CM, Singh A, Chen M, Satchwell JB, et al. HLA-DQA1-HLA-DRB1 variants confer susceptibility to pancreatitis induced by thiopurine immunosuppressants. Nature genetics. 2014;46(10):1131-4.
[42] Fritsche U, Weisz H. [Principles of multiplication in analytical chemistry. Definition and fundamentals]. Mikrochimica acta. 1974(4):701-10.
[43] Bradford MM. A rapid and sensitive method for the quantitation of microgram quantities of protein utilizing the principle of protein-dye binding. Analytical biochemistry. 1976;72:248-54.
[44] Aljabali AA, Barclay JE, Lomonossoff GP, Evans DJ. Virus templated metallic nanoparticles. Nanoscale. 2010;2(12):2596-600.
[45] Aljabali AA, Lomonossoff GP, Evans DJ. CPMV-polyelectrolyte-templated gold nanoparticles. Biomacromolecules. 2011;12(7):2723-8.
[46] Basheva ES, Kralchevsky PA, Danov KD, Ananthapadmanabhan KP, Lips A. The colloid structural forces as a tool for particle characterization and control of dispersion stability. Physical chemistry chemical physics : PCCP. 2007;9(38):5183-98.
[47] Zhou J, Beattie DA, Ralston J, Sedev R. Colloid stability of thymine-functionalized gold nanoparticles. Langmuir. 2007;23(24):12096-103.
[48] Schrand AM, Schlager JJ, Dai L, Hussain SM. Preparation of cells for assessing ultrastructural localization of nanoparticles with transmission electron microscopy. Nature protocols. 2010;5(4):744-57.
[49] Klein ND, Hurley KR, Feng ZV, Haynes CL. Dark field transmission electron microscopy as a tool for identifying inorganic nanoparticles in biological matrices. Analytical chemistry. 2015;87(8):4356-62.
[50] Feng ZV, Gunsolus IL, Qiu TA, Hurley KR, Nyberg LH, Frew H, et al. Impacts of gold nanoparticle charge and ligand type on surface binding and toxicity to Gram-negative and Gram-positive bacteria. Chemical Science. 2015;6(9):5186-96.
[51] Matias VR, Beveridge TJ. Cryo-electron microscopy reveals native polymeric cell wall structure in Bacillus subtilis 168 and the existence of a periplasmic space. Molecular microbiology. 2005;56(1):240-51.

IOP Conf. Series: Materials Science and Engineering **305** (2018) 012006 doi:10.1088/1757-899X/305/1/012006

The prediction of a new high-pressure phase of hafnia using *first-principles* computations

Y. Al-Khatatbeh[1*], K. Tarawneh[1] and B. Hamad[2,3]

[1] *Department of Basic Sciences, Princess Sumaya University for Technology, Amman 11941, Jordan*
[2] *Physics Department, The University of Jordan, Amman 11942, Jordan*
[3] *Physics Department, University of Arkansas, Fayetteville, AR 72701, USA*

y.alkhatatbeh@psut.edu.jo

Abstract. Using density functional theory (DFT) calculations, we predicted a new high-pressure phase of hafnia (HfO_2). We found the hexagonal phase (Ni_2In-type structure; space group: $P6_3/mmc$) to be the stable phase at ultrahigh pressures greater than ~386 GPa. Our findings are consistent with recent calculations performed on the similar dioxide ZrO_2 [M. Durandurdu, J. Solid State Chem. **230**, 233 (2015)] where this phase has been claimed to be the most stable at pressures greater than 380 GPa. The Birch-Murnaghan equation of state (BM-EOS) of the new phase shows that this phase is more compressible and less dense than Fe_2P-type phase. Additionally, the hardness calculations using a scaling model confirmed that our newly predicted phase has a similar hardness compared to the other HfO_2 phases, indicating that none of the HfO_2 phases can be considered to be superhard.

1. Introduction

Hafnia (HfO_2) is an important dioxide that has been involved in many industrial and technological applications. It has important optical properties [1–4], and it is also considered as a good candidate to replace silicon dioxide (SiO_2) in electronic devices due to its high dielectric constant and thermal stability [5–8]. Additionally, the structural and mechanical properties of HfO_2 have received an extensive attention, where the high-pressure/temperature behavior of HfO_2 is considered as one of the important research directions in studying this dioxide both experimentally [9–18] and theoretically [9], [19–23].

Experimental observations [9, 11, 24] and *first-principles* computations [9, 20, 23] agree that the well-known high-pressure phase transition sequence for HfO_2 is as follows: MI (Baddeleyite) ➔ OI (Orthorhombic I) ➔ OII (Orthorhombic II or cotunnite). However, recently, a theoretical and experimental evidence for a new high-pressure post-OII phase (called Fe_2P phase) (Fig. 1) for HfO_2 has been claimed [25], in good agreement with the recently reported results in the similar dioxides ZrO_2 [26] and TiO_2 [27]. Experimentally, Fe_2P phase is found to be stable at extreme pressure-temperature (*P-T*) conditions that exceeds (175 GPa, 3000 K) for ZrO_2 [26], (210 GPa, 4000 K) for TiO_2 [27], and (105 GPa, 2000 K) for HfO_2 [25]. On the other hand, the theoretical stability of Fe_2P phase is predicted to be at pressures greater than 143 GPa (for ZrO_2) [26], 161 GPa (for TiO_2) [27], and 120-140 GPa (for HfO_2) [25]. Consequently, the updated experimental and theoretical high-pressure phase transition sequence in these three dioxides becomes as follows: MI ➔ OI ➔ OII ➔ Fe_2P.

IOP Conf. Series: Materials Science and Engineering **305** (2018) 012006 doi:10.1088/1757-899X/305/1/012006

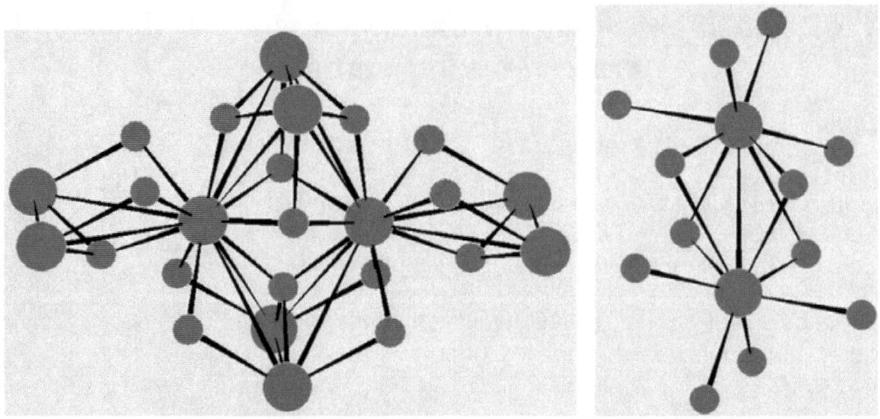

Phase: Fe$_2$P-type	Phase: Ni$_2$In-type
Crystal structure: Hexagonal	Crystal structure: Hexagonal
Space group: $P\bar{6}2m$	Space group: $P6_3/mmc$
Coordination number: 9	Coordination number: 11

Figure 1: Crystal structures and summary of the structural properties of HfO_2 phases. The larger spheres represent the Hf atom, while the smaller spheres represent the O atom.

Recently, a new phase (called Ni_2In) (Fig. 1) has been theoretically tested for ZrO_2 [28], and it has been predicted to be stable at very high pressures greater than 380 GPa. In fact, it should be noted that the transition to Ni_2In phase has been claimed to occur directly from OII phase and not from Fe_2P phase, which is not consistent with the recent calculations and measurements performed on HfO_2 [25], ZrO_2 [26] and TiO_2 [27]. This finding proves that Fe_2P-type phase is a post-OII phase. However, we should also note that the enthalpy differences between OII and Fe_2P phases are very small [25–28], and, therefore, comparable transition pressures across OII ➔ Ni_2In versus Fe_2P ➔ Ni_2In are likely.

To the best of our knowledge, the high-pressure phase stability of Ni_2In has not been previously tested for HfO_2. In this study, we use DFT calculations to test the stability and investigate the EOS of Ni_2In phase at high pressures for HfO_2 in order to better understand the phase diagram of this important dioxide.

2. Computational Method

DFT calculations are used to investigate the phase transformation and the EOSs of HfO_2 [29]. The projector-augmented wave (PAW) formalism [30, 31] was used to treat the interactions between the atoms having a core radii of 2.600 Bohr for hafnium Hf (with the valence configuration of $5p^6 6s^2 5d^2$) and 1.520 Bohr for oxygen O (with the valence configuration of $2s^2 2p^4$). The generalized gradient approximation (GGA) [32] was used to treat the electronic exchange and correlation effects. We performed our calculations using VASP software package [33–36] with an energy cutoff of 600 eV and Γ-centered *k*-point meshes [37]. For both phases, total energies and pressures were converged to better than ~0.1 meV/atom and ~0.1 GPa, respectively. The Brillouin zone integration was performed using the following *k*-point meshes for the HfO_2 phases: 6x6x10 for Fe_2P and 10x10x8 for Ni_2In. For a fixed volume, all internal degrees of freedom and unit-cell parameters of the structure were optimized simultaneously during the geometry optimizations. The ground-state energy for each phase was determined for 13-16 volumes. Up to highest pressure achieved in our calculations, both HfO_2 phases remain insulators.

3. Results and Discussion

3.1. Birch-Murnaghan equation of state determination

The applied pressure P in the third-order BM-EOS [38] is expressed as follows:

$$P(V)=\frac{3}{2}K_0\left[\left(\frac{V}{V_0}\right)^{-\frac{7}{3}}-\left(\frac{V}{V_0}\right)^{-\frac{7}{3}}\right]\left[1+\frac{3}{4}(K_0'-4)\left(\left(\frac{V}{V_0}\right)^{-\frac{2}{3}}-1\right)\right] \quad \text{Eqn. (1)}$$

where V is the volume at pressure P, V_0 is the zero-pressure volume, K_0 is the zero-pressure Bulk modulus, and K_0' is the first pressure derivative of the bulk modulus at zero pressure. From the following thermodynamic relationship: $P(V)=-\frac{\partial E}{\partial V}$, one can obtain an expression for the total energy E in the third-order form of BM-EOS as follows:

$$E(V)=\frac{9K_0V_0}{2}\left[\frac{1}{2}\left(\left(\frac{V}{V_0}\right)^{-\frac{2}{3}}-1\right)\right]^2\left[1+(K_0'-4)\frac{1}{2}\left(\left(\frac{V}{V_0}\right)^{-\frac{2}{3}}-1\right)\right]+E_0 \quad \text{Eqn. (2)}$$

where E_0 is the total energy at zero pressure. The second-order BM-EOS is obtained by fixing K_0' to 4 in Eqns. 1 and 2 to get the following second-order expressions:

$$P(V)=\frac{3}{2}K_0\left[\left(\frac{V}{V_0}\right)^{-\frac{7}{3}}-\left(\frac{V}{V_0}\right)^{-\frac{7}{3}}\right] \quad \text{Eqn. (3)}$$

$$E(V)=\frac{9K_0V_0}{4}\left(\left(\frac{V}{V_0}\right)^{-\frac{2}{3}}-1\right)^2+E_0 \quad \text{Eqn. (4)}$$

The EOS parameters for each phase were obtained by fitting the total energy as a function of volume (Eqns. 2 and 4) to a second- and a third-order BM-EOS [38] (Table I).

The calculated EOS parameters for Fe_2P and Ni_2In phases of HfO_2 are summarized in Table I. For comparison, we listed values from recent work on Fe_2P-HfO_2 [25], which are in excellent agreement with our calculations. To the best of our knowledge, the Ni_2In phase has not been previously tested for the HfO_2 case, but it has been tested for the similar dioxide ZrO_2 [28]. Table I shows that our calculated EOS parameters for Ni_2In-HfO_2 agree well with its analog Ni_2In-ZrO_2.

Futhermore, one can see an appreciable decrease in the bulk modulus upon the transformation from Fe_2P to Ni_2In phase. The calculated second-order BM-EOS shows that the Bulk modulus deceases by ~24%, indicating that Ni_2In is much more compressible than Fe_2P. However, this decrease is not unexpected as this transformation between the two phases is accociated with an increase in the zero-pressure volume (V_0) of ~2.6% and an increase in the coordination number (CN) from 9 (for Fe_2P) to 11 (for Ni_2In). We should mention that the incease in the CN is not nessecarly asscociated with V_0 increase (and thus K_0 decrease). For instance, for HfO_2 and other similar dioxides ZrO_2 [39] and TiO_2 [40], the transformation from OI phase to OII phase is asscociated with a CN incease from 7 to 9 and a V_0 decrease (and thus K_0 increase).

Table I: Theoretical EOSs of Fe_2P and Ni_2In phases of HfO_2. For comparison, we list other theoretical results for HfO_2 [25] and ZrO_2 [28]. 1σ uncertainties are given in parentheses.

Phase	Equation of state			Reference
	V_0 (Å^3)	K_0 (GPa)	K_0'	
Fe_2P	29.69 (0.03)	288 (2)	4 (fixed)	Ref. [25]
	29.74 (0.01)	270 (1)	4.32 (0.02)	Ref. [25]
	29.73 (0.02)	282 (2)	4 (fixed)	This work

	29.74 (0.01)	268 (1)	4.39 (0.02)	This work
Ni_2In	29.21	239	3.86	Ref. [28], for ZrO_2
	30.49 (0.14)	213 (6)	4 (fixed)	This work
	31.18 (0.05)	151 (2)	5.29 (0.07)	This work

To better understand the high compressibility of Ni_2In phase, we have looked at the variation of the unit-cell parameters with pressure (Fig. 2). Figure 2 clearly shows the dramatic decrease in *c*/*a* ratio up to ~50 GPa, after which it continues to decrease with increasing pressure but with a lower rate. We can also conclude that the unit-cell parameter *c* is much more compressible than the parameter *a*; therefore, the high compressibility in K_0 is likely due to the high compressibility along the unit-cell parameter *c*.

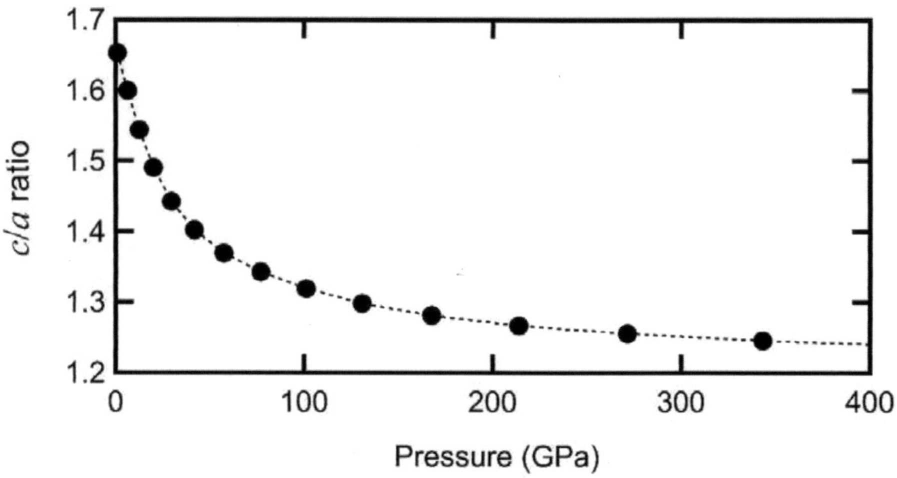

Figure 2: *c*/*a* ratio for Ni_2In-ZrO_2 phase as a function of pressure.

3.2. High-pressure phase transition

Our enthalpy calculations for the two phases (Fig. 3) show that the transition pressure from Fe_2P to Ni_2In occurs at ~387 GPa and 386 GPa using the second- and third-order BM-EOS, respectively. The transition at such ultrahigh pressure is supported by the result recently obtained for the OII ➔ Ni_2In transition in ZrO_2 [28] at 380 GPa, where the enthalpy differences are very small between OII and Fe_2P [28] and, thus, the Fe_2P ➔ Ni_2In for ZrO_2 should occur at a pressure close to 380 GPa. Therefore, our calculated transition pressure across Fe_2P ➔ Ni_2In in HfO_2 is very reasonable when compared to its analog dioxide ZrO_2.

IOP Conf. Series: Materials Science and Engineering **305** (2018) 012006 doi:10.1088/1757-899X/305/1/012006

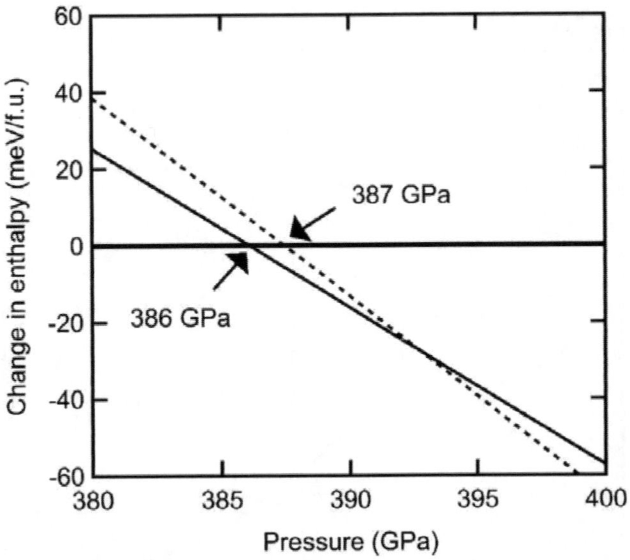

Figure 3: Change in enthalpy with respect to Fe_2P phase versus pressure of one HfO_2 formula unit. The transition pressure from Fe_2P to Ni_2In is ~387 GPa and ~386 GPa using a second- (dashed curve) and third- (solid curve) order BM-EOS [38], respectively.

In our recent theoretical study on HfO_2 [25], we found that the OII ➔ Fe_2P transition occurs at ~120-140 GPa. Consequently, the updated upper high-pressure phase transition sequence for HfO_2 is: OII ➔ Fe_2P ➔ Ni_2In. However, it is important to emphasize that the Ni_2In phase has not been observed yet experimentally for either HfO_2 or ZrO_2 in contrast to OII and Fe_2P phases, likely due to the extreme pressures required to stabilize it.

3.3. Density and volume collapse across transition

Based on our discussion in section 3.1, Ni_2In phase is found to be less dense than Fe_2P phase at zero pressure and, therefore, is more compressible. However, and due to the high compressibility of Ni_2In-HfO_2, the calculated volumes for Ni_2In are smaller than those of Fe_2P at high pressures (Fig. 4). At the transition pressure (~386 GPa) across Fe_2P ➔ Ni_2In transition, our calculated volume collapse is ~4.4% (~3.8%) using the second- (third-) order BM-EOS (Fig. 4). This relatively large volume change is expected as this transition is associated with an increase in the CN from 9 to 11. This result compares well with the volume collapse that has been experimentally observed or calculated across OI ➔ OII transition in HfO_2 and similar dioxides ZrO_2 [39] and TiO_2 [40], noting than this transition is associated with an increase in the CN from 7 (for OI) to 9 (for OII). However, a very small volume change across OII ➔ Fe_2P transition has been reported for these dioxides [25–27] as the CN remains unchanged for this transition.

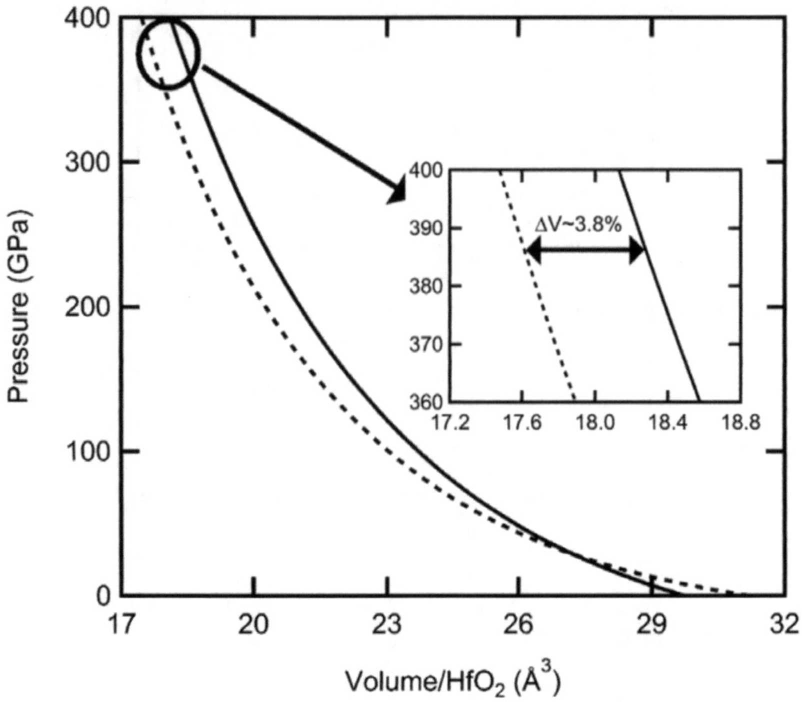

Figure 4: Pressure versus volume of one HfO_2 formula unit. The solid line represents the Fe_2P phase, whereas the dashed line represents the Ni_2In phase. The volume collapse at the transition pressure between the two phases is shown in the inset.

3.4. Hardness estimations

In this section, we discuss the mechanical hardness of our newly predicted Ni_2In phase in comparison to the other high-pressure phases of HfO_2 and the other similar dioxides that have been expected to have high hardness due to their low compressibility (see Ref. [41]). In fact, it has been recently concluded that this assumption is not valid as none of these phases for these dioxides is qualified to be an ultrahard material. The assumption of Ni_2In as being a superhard is refuted by using a scaling model [42].
Based on this model, the hardness is proportional to the average number of bonds per atom, but it is inversely proportional to average atomic volume, coordination number, and bond length [42]. Additionally, the hardness depends on R_i which represents the characteristic length scale of the charge density distribution about each atom. It has been found that R_i is nearly independent of phase[9, 39, 42, 43], where the R_i values using GGA computations [9] are 1.78 Å and 1.07 Å for Hf and O atoms, respectively.
Using this model, we estimated the hardness of Ni_2In-HfO_2 phase to ~9 GPa, which is ~10% smaller than our recent estimations [25] for the hardness of either OII-HfO_2 (9.8 GPa) or Fe_2P-HfO_2 (10 GPa). This decrease in hardness is expected due to the CN increase and V_0 increase in Ni_2In compared to OII and Fe_2P. Additionally, we should note that the material is called superhard if its hardness exceeds 40 GPa. Consequently, since the estimated hardness for the high-pressure phases of HfO_2 is ~9-10 GPa, our general conclusion is that none of these phases is qualified to be superhard.

4. Conclusions

In summary, we tested the high-pressure stability of Ni_2In-type phase for HfO_2 with respect to Fe_2P phase. Our DFT-GGA calculations predicted the transition pressure across Fe_2P ➔ Ni_2In to occur at ~386 GPa with a large volume collapse across the transition of ~4%. Additionally, our computed EOS for the newly predicted phase shows that it is obviously more compressible than Fe_2P phase. Finally, the hardness estimation for Ni_2In-HfO_2 is consistent with recent studies that have concluded that all known HfO_2 polymorphs are not superhard.

Acknowledgements
This work was completed utilizing the Holland Computing Center of the University of Nebraska.

References

[1] S. S. Lin and H. R. Li, "The optical properties of hydrophilic Hf-doped HfO_2 nanoceramic films," *Ceram. Int.*, vol. 39, no. 7, pp. 7677–7683, 2013.
[2] J. M. Khoshman and M. E. Kordesch, "Optical properties of a-HfO_2 thin films," *Surf. Coatings Technol.*, vol. 201, no. 6, pp. 3530–3535, 2006.
[3] J. Aarik, H. Mändar, M. Kirm, and L. Pung, "Optical characterization of HfO_2 thin films grown by atomic layer deposition," *Thin Solid Films*, vol. 466, no. 1–2, pp. 41–47, 2004.
[4] S.-S. Lin and C.-S. Liao, "The hydrophobicity and optical properties of the HfO_2-deposited glass," *Ceram. Int.*, vol. 39, no. 1, pp. 353–358, 2013.
[5] M.-H. Cho *et al.*, "Thermal stability and structural characteristics of HfO_2 films on Si (100) grown by atomic-layer deposition," *Appl. Phys. Lett.*, vol. 81, no. 3, p. 472, 2002.
[6] G. D. Wilk, R. M. Wallace, and J. M. Anthony, "High-κ gate dielectrics: Current status and materials properties considerations," *J. Appl. Phys.*, vol. 89, no. 10, pp. 5243–5275, 2001.
[7] V. V. Afanas'ev, A. Stesmans, F. Chen, X. Shi, and S. A. Campbell, "Internal photoemission of electrons and holes from (100)Si into HfO_2," *Appl. Phys. Lett.*, vol. 81, no. 6, pp. 1053–1055, 2002.
[8] J. Robertson and R. M. Wallace, "High-K materials and metal gates for CMOS applications," *Mater. Sci. Eng. R Reports*, vol. 88, pp. 1–41, 2015.
[9] Y. Al-Khatatbeh, K. K. M. Lee, and B. Kiefer, "Phase diagram up to 105 GPa and mechanical strength of HfO_2," *Phys. Rev. B*, vol. 82, p. 144106, 2010.
[10] J. Haines *et al.*, "Characterization of the cotunnite-type phases of zirconia and hafnia by neutron diffraction and Raman spectroscopy," *J. Am. Ceram. Soc.*, vol. 80, no. 7, pp. 1910–1914, 1997.
[11] S. Desgreniers and K. Lagarec, "High-density ZrO2 and HfO_2: Crystalline structures and equation of state," *Phys. Rev. B*, vol. 59, no. 13, pp. 8467–8472, 1999.
[12] D. M. Adams, S. Leonard, D. R. Russell, and R. J. Cernik, "X-ray diffraction study of hafnia under high pressure using synchrotron radiation," *J. Phys. Chem. Solids*, vol. 52, no. 9, pp. 1181–1186, 1991.
[13] J. M. Leger, J. Haines, and B. Blanzat, "Materials potentially harder than diamond: Quenchable high-pressure phases of transition metal dioxides," *J. Mater. Sci. Lett.*, vol. 13, pp. 1688–1690, 1994.
[14] A. Jayaraman, "Diamond anvil cell and high-pressure physical investigations," *Rev. Mod. Phys.*, vol. 55, no. 1, pp. 65–108, 1983.
[15] O. Ohtaka, T. Yamanaka, S. Kume, and F. Izumi, "Synthesis and x-ray structural analysis by the Rietveld method of orthorhombic hafnia," *J. Ceram. Soc. Japan*, vol. 99, pp. 826–827, 1991.
[16] O. Ohtaka, T. Yamanaka, S. Kume, N. Hara, and H. Asano, "Structural analysis of orthorhombic hafnia by neutron powder diffraction," *J. Am. Ceram. Soc.*, vol. 78, no. 1, pp. 233–237, 1995.
[17] H. Arashi, T. Yagi, S. Akimoto, and Y. Kudoh, "New high-pressure phase of ZrO_2 above 35 GPa," *Phys. Rev. B*, vol. 41, no. 7, pp. 4309–4313, 1990.
[18] J. M. Leger, P. E. Tomaszewski, A. Atouf, and A. S. Pereira, "Pressure-induced phase transtions and volume changes in HfO_2 up to 50 GPa," *Phys. Rev. B*, vol. 48, no. 1, pp. 93–98, 1993.
[19] R. Terki, G. Bertrand, H. Aourag, and C. Coddet, "Cubic-totetragonal phase transition in HfO_2 from computational study," *Mater. Lett.*, vol. 62, pp. 1484–1486, 2008.
[20] J. E. Jaffe, R. A. Bachorz, and M. Gutowski, "Low-temperature polymorphs of ZrO_2 and HfO_2: A density-functional theory study," *Phys. Rev. B*, vol. 72, p. 144107, 2005.

[21] J. E. Lowther, J. K. Dewhurst, J. M. Leger, and J. Haines, "Relative stability of ZrO_2 and HfO_2 structural phases," *Phys. Rev. B*, vol. 60, no. 21, pp. 14485–14488, 1999.

[22] J. K. Dewhurst and J. E. Lowther, "Highly coordinated metal dioxides in the cotunnite structure," *Phys. Rev. B*, vol. 64, p. 14104, 2001.

[23] J. Kang, E.-C. Lee, and K. J. Chang, "First-principles study of structural phase transformation of hafnia under pressure," *Phys. Rev. B*, vol. 68, p. 54106, 2003.

[24] O. Ohtaka *et al.*, "Phase relations and volume changes of hafnia under high pressure and high temperature," *J. Am. Ceram. Soc.*, vol. 84, no. 6, pp. 1369–1373, 2001.

[25] Y. Al-Khatatbeh, K. Tarawneh, H. Taani, and K. K. M. Lee, "Unpublished data."

[26] D. Nishio-Hamane, H. Dekura, Y. Seto, and T. Yagi, "Theoretical and experimental evidence for the post-cotunnite phase transition in zirconia at high pressure," *Phys. Chem. Miner.*, vol. 42, no. 5, pp. 385–392, 2015.

[27] H. Dekura, T. Tsuchiya, Y. Kuwayama, and J. Tsuchiya, "Theoretical and Experimental Evidence for a New Post-Cotunnite Phase of Titanium Dioxide with Significant Optical Absorption," *Phys. Rev. Lett.*, vol. 107, p. 45701, 2011.

[28] M. Durandurdu, "Novel high-pressure phase of ZrO2: An ab initio prediction," *J. Solid State Chem.*, vol. 230, no. July, pp. 233–236, 2015.

[29] P. Hohenberg and W. Kohn, "Inhomogeneous electron gas," *Phys. Rev.*, vol. 136, no. 3B, pp. B864–B871, 1964.

[30] G. Kresse and D. Joubert, "From ultrasoft pseudopotentials to the projector augmented-wave method," *Phys. Rev. B*, vol. 59, no. 3, pp. 1758–1775, 1999.

[31] P. E. Blochl, "Projector augmented-wave method," *Phys. Rev. B*, vol. 50, no. 24, p. 17953, 1994.

[32] J. P. Perdew, K. Burke, and M. Ernzerhof, "Generalized gradient approximation made simple," *Phys. Rev. Lett.*, vol. 77, no. 18, pp. 3865–3868, 1996.

[33] G. Kresse and J. Furthmuller, "Efficiency of ab-initio total energy calculations for metals and semiconductors using a plane-wave basis set," *Comput. Mater. Sci.*, vol. 6, no. 1, pp. 15–50, 1996.

[34] G. Kresse and J. Furthmuller, "Efficient iterative schemes for ab initio total-energy calculations using a plane-wave basis set," *Phys. Rev. B*, vol. 54, no. 16, pp. 11169–11186, 1996.

[35] G. Kresse and J. Hafner, "Norm-conserving and ultrasoft pseudopotentials for first-row and transition elements," *J. Phys. Condens. Matter*, vol. 6, no. 40, p. 8245, 1994.

[36] G. Kresse and J. Hafner, "Ab initio molecular dynamics for open-shell transition metals," *Phys. Rev. B*, vol. 48, no. 17, pp. 13115–13118, 1993.

[37] H. J. Monkhorst and J. D. Pack, "Special points for Brillouin-zone integrations," *Phys. Rev. B*, vol. 13, no. 12, pp. 5188–5192, 1976.

[38] F. Birch, "Elasticity and constitution of the Earth's interior," *J. Geophys. Res.*, vol. 57, pp. 227–234, 1952.

[39] Y. Al-Khatatbeh, K. K. M. Lee, and B. Kiefer, "Phase relations and hardness trends of ZrO_2 phases at high pressure," *Phys. Rev. B*, vol. 81, p. 214102, 2010.

[40] Y. Al-Khatatbeh, K. K. M. Lee, and B. Kiefer, "High-pressure behavior of TiO_2 as determined by experiment and theory," *Phys. Rev. B*, vol. 79, p. 134114, 2009.

[41] Y. Al-Khatatbeh and K. K. M. Lee, "From superhard to hard: A review of transition metal dioxides TiO_2, ZrO_2, and HfO_2 hardness," *J. Superhard Mater.*, vol. 36, no. 4, pp. 231–245, 2014.

[42] A. Simunek and J. Vackar, "Hardness of covalent and ionic crystals: First-principle calculations," *Phys. Rev. Lett.*, vol. 96, p. 85501, 2006.

[43] A. Simunek, "How to estimate hardness of crystals on a pocket calculator," *Phys. Rev. B*, vol. 75, p. 172108, 2007.

Modeling of Lithium Niobate ($LiNbO_3$) and Aluminum Nitride (AlN) Nanowires Using Comsol Multiphysics Software: The Case of Pressure Sensor

A.A. Ahmad*[1], A. Alsaad[1], Q.M. Al-Bataineh[1], M.A. Al-Naafa[2]

[1]Department of Physical Sciences, Jordan University of Science and Technology, P. O. Box 3030, Irbid 22110, Jordan.
[2]Department of Chemical Engineering, University of Hafr Albatin Saudi Arabia

sema@just.edu.jo

Abstract In this study, Lithium niobate ($LiNbO_3$) and Aluminum nitride (AlN) nanostructures were designed and investigated using the COMSOL Multiphysics software for pressure sensing applications. The Finite Element Method (FEM) was used for solving the differential equations with various parameters such as size, length, force, etc. The variation of the total maximum displacement as a function of applied force for various NWs lengths and the variation of the voltage as a function of applied force were plotted and discussed. AlN nanowires exhibit a better piezoelectric response than $LiNbO_3$ nanowires do.

Keywords: Piezoelectricity, Nanoelectromechanical system, COMSOL Multiphysics, Finite Element Method, Piezoelectric response

IOP Conf. Series: Materials Science and Engineering **305** (2018) 012007 doi:10.1088/1757-899X/305/1/012007

1. Introduction

Piezoelectricity is a direct electrical-mechanical conversion in certain types of crystals. It has many applications since the discovery by the Curie brothers in 1880 [1]. The reverse piezoelectric effect is observed in several crystals when an applied voltage causes bending in the crystal. The Micro Electro-Mechanical Systems MEMS were some of the most interesting applications in recent years. Crystal deformation does not exceed few nanometers, but the force exerted on the crystal to induce the deformation is in the order of few mega-newtons. For example, one cubic centimeter cube of quartz generates 12 kV with a two kilo-newtons applied force but at low current [2]. Piezoelectric devices contain materials where electric charge is created due to mechanical stress. Piezoelectric devices module based on Finite Element Method (FEM) implemented in commercial software COMSOL Multiphysics is used to study devices such as Micro Electro-Mechanical Systems (MEMS), bio sensors, controllers or Micro Machined Modal Gyroscopes [3-4]. Applications such as energy harvesting have been presented in several previous works [5-8]. Energy harvesting is an evolving area of research where unexploited energy is converted into beneficial electrical power consequently offering a promising solution to the environmental and substitution issues associated with battery powered systems. The recent advancement of micro fabrication technology has led to increasing interest in developing micro-energy harvesters in the form of MEMS devices.

In the current study, we focus on Lithium niobate ($LiNbO_3$) and Aluminum nitride (AlN) which are key candidate materials in integrated optical circuits due to their photo-stability, as well as, excellent nonlinear optical properties including large piezoelectric, electro-optic, and nonlinear optical coefficients [9-11]. The single crystal of $LiNbO_3$ lacks inversion symmetry and thus exhibits ferroelectricity, Pockels effect, piezoelectric effect, photo-elasticity and nonlinear optical polarizability [12]. $LiNbO_3$ has attracted much attention due to its potential in applications such as optical waveguides, cell phones, piezoelectric sensors, optical modulators and various other linear and non-linear optical applications [13]. In addition, we investigated the compound under ambient temperature and pressure. The AlN compound crystallizes in the hexagonal wurtzite structure (w-AlN) which belongs to the *P63mc* space group. The w-AlN is a relatively wide-band-gap semiconductor that has attracted increasing interest because of its low compressibility, good thermal stability, chemical and radiation inertness, and applications such as short-wavelength light emitting diodes, optical detectors, high-pressure, high temperature, and high-frequency optoelectronic devices [14-16]. Furthermore, because of the excellent piezoelectric response of w-AlN, it is used for surface acoustic wave sensors (SAWs) deposited on silicon wafers. One application of SAWs is an RF filter which is widely used in cell phones. The w-AlN is also used to build piezoelectric micro machined ultrasound transducers, which emit and receive ultrasound and can be used for in-air range finding over distances of up to a meter [17-19]. In this study, we use COMSOL Multiphysics software (version 5.1) that can be used to account for coupled or Multiphysics phenomena [20] to investigate the piezoelectric properties of $LiNbO_3$ and AlN systems. In particular, we determine the piezoelectric potential distributions in different geometries of $LiNbO_3$ and AlN systems at the nanoscale range. Specifically, we consider the case of nanowires (NWs) designed pressure sensors.

2. Theoretical Consideration

In general case, Hooke's law can be expressed as:

$\varepsilon_{ij} = S_{ijkl}\tau_{kl}$ .... (1)

where ε_{ij} and τ_{kl} denote the strain and stress tensors, respectively.$S_{ijkl}$is the compliance, and it is fourth-rank tensor quantity. The direct consequence of the symmetry in the strain and the stress tensors is that only 36 components of the compliance tensor are independent and distinct terms, instead of 81 components of the compliance tensor. The relation between the stress and the strain can be simplified through applying the Voigt notation as the following [21]:

$\varepsilon_i = S_{ij}\tau_j$ .... (2)

Because of the symmetry in the compliance tensor for the general anisotropic linear elastic solid, there are 21 independent elastic constants. By considering the symmetry conditions found in the wurtzite crystal structure of the AlN NWs, the stress-strain relations are expressed as (the z-axis is the c-axis for wurtzite crystal):

$$\begin{pmatrix}\varepsilon_1\\\varepsilon_2\\\varepsilon_3\\\varepsilon_4\\\varepsilon_5\\\varepsilon_6\end{pmatrix}=\begin{pmatrix}S_{11}&S_{12}&S_{13}&0&0&0\\S_{12}&S_{11}&S_{13}&0&0&0\\S_{13}&S_{13}&S_{33}&0&0&0\\0&0&0&S_{44}&0&0\\0&0&0&0&S_{44}&0\\0&0&0&0&0&2(S_{11}-S_{12})\end{pmatrix}\begin{pmatrix}\tau_1\\\tau_2\\\tau_3\\\tau_4\\\tau_5\\\tau_6\end{pmatrix}$$ .. (3) is 2 in the last row superscript or factor? 2(S1-S)?

The elastic modulus (Young's modulus) in remarkable crystallographic directions can be extracted from the compliance matrix using the following equations:

$E_{[2\bar{1}\bar{1}0]} = \tau_1/\varepsilon_1 = (S_{11})^{-1}$ .... (4)

$E_{[01\bar{1}0]} = \tau_2/\varepsilon_2 = (S_{11})^{-1}$ .... (5)

$E_{[0001]} = \tau_3/\varepsilon_3 = (S_{33})^{-1}$ ....(6)

The crystals that lack inversion symmetry exhibits piezoelectric effect and produce an electric polarization in response to mechanical stress as follow:

$P_n = e_{nj}\varepsilon_j$ ....(7)

Where P_n(n = 1, 2, 3) are the components of the polarization vector that is created as a result of piezoelectric effect, ε_j(j = 1, 2, 3, 4, 5, 6) are the components of the strain tensor, e_{nj} is the tensor of the stress constants using Voigt notation. For the wurtzite structure of AlN, the matrix of the piezoelectric stress constants can be written as [22]:

$$e_{nj}=\begin{pmatrix}0&0&0&0&e_{15}&0\\0&0&0&e_{15}&0&0\\e_{31}&e_{31}&e_{33}&0&0&0\end{pmatrix}$$ .... (8)

The matrix has only three independent elements. The converse piezoelectric effect describes how an applied electric field (E_k) will create a resultant strain (ε_i) which leads to a physical deformation of the material. The converse piezoelectric effect in Voigt notation can be written as:

$$\varepsilon_i = d_{ki} E_k \qquad \text{.......................... (9)}$$

Where d_{ki} is the tensor of the piezoelectric strain constants. For the hexagonal wurtzite crystal it is given by:

$$d_{ki} = \begin{bmatrix} 0 & 0 & d_{31} \\ 0 & 0 & d_{31} \\ 0 & 0 & d_{33} \\ 0 & d_{15} & 0 \\ d_{15} & 0 & 0 \\ 0 & 0 & 0 \end{bmatrix} \qquad \text{...................... (10)}$$

The relationship between strain tensor, stress tensor, electric displacement(D_n) and electric field for a linear piezoelectric material can be given by the constitutive equations [23].

$$\varepsilon_i = S_{ij}^{E}\, \tau_j + d_{ni}\, E_n \qquad \text{.............. (11)}$$

$$D_n = d_{ni}\tau_i + \xi_{ik}^{\tau}\, E_k \qquad \text{.............. .(12)}$$

Where the indexes i, j = 1, 2, .. 6 and n, k = 1, 2, 3 refer to different directions within the material coordinate system, and ξ_{ik} is the permittivity tensor.

Piezoelectric materials can generate electrical voltage when uniaxial strain is applied. For low frequency applications at constant temperature, the direct piezoelectric effect can be described by:

$$\vec{D} = d_1 \vec{T} + \varepsilon^T \vec{E} \qquad \text{........................ (13)}$$

$$\vec{S} = s^E \vec{T} + d_2 \vec{E} \qquad \text{........................(14)}$$

$\vec{D}$: Electric displacement and d_1 and d_2 are piezoelectric charge coefficients for the direct piezoelectric effect and the converse piezoelectric effect, respectively.

$\vec{T}$ denotes the mechanical stress.

ε^T : Permittivity at constant stress.

$\vec{E}$: Electric field.

$\vec{S}$: Mechanical strain

S^E: Mechanical compliance

Equation (13) describes the electrical field conversion into mechanical stress and equation (14) depicts the mechanical stress conversion into electrical field.

3. COMSOL Multiphysics Simulation

To simulate the pressure sensors using COMSOL Multiphysics, the piezoelectric devices Multiphysics interface has to be chosen. Solid mechanics and electrostatics are combined within the piezoelectric devices Multiphysics interface with the constitutive relationships required to model the piezoelectric device.

3.1 Geometric Modeling

3D geometry is considered for the simulations. The piezoelectric device contains a thin film deposited on silicon wafers with dimensions of 200 x 200 x 10 nm and nine NWs seeded on it. Each of the NWs has a radius of 10 nm. The height of the NWs has been varied in steps of 100, 150 and 200 nm, as shown in figure 1. The geometry has been chosen because experimental combination of electrodeposition and polymeric templates followed by chemical etching provides a large variety nanowires of controlled size, geometry, composition, and surface morphology very close to our chosen geometry..

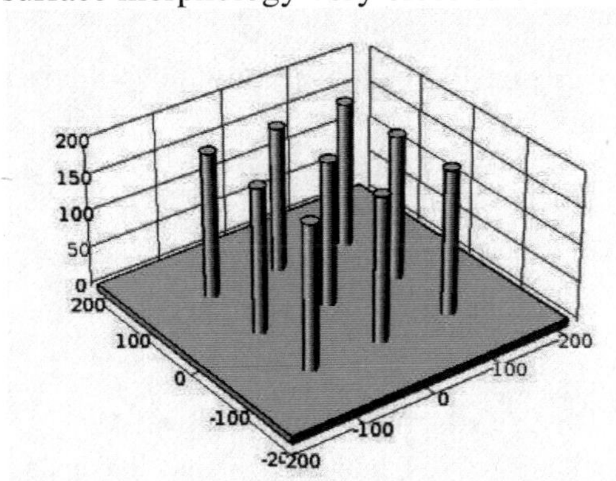

Figure 1: The geometry of piezoelectric device.

3.2 Meshing

Before starting simulation, the NW was broken into small areas; each is called a "mesh". The mesh is composed of (14809, 17725 and 21044) tetrahedral elements and (7952, 9464 and 10806) triangular elements for (100, 150 and 200) nm height of NWs, respectively. The mesh is created using the mapped mesh tool with a physics-controlled mesh and normal mesh element size. For instance, figure 2 illustrates the obtained mesh for a NW of height 200 nm.

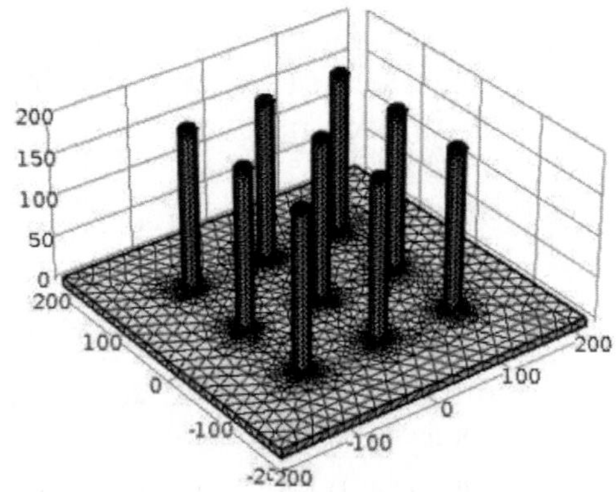

Figure 2: Piezoelectric device mesh for a NW of height 200 nm.

3.3 Sub-domain Settings

The structure of the simulated piezoelectric device consists of two compounds; the first is $LiNbO_3$ and has the following matrices and parameters:
Compliance matrix

$$S^E = \begin{pmatrix} 5.78 & -1.01 & -1.47 & -1.02 & 0 & 0 \\ -1.01 & 5.78 & -1.47 & 1.02 & 0 & 0 \\ -1.47 & -1.47 & 5.02 & 0 & 0 & 0 \\ -1.02 & 1.02 & 0 & 17 & 0 & 0 \\ 0 & 0 & 0 & 0 & 17 & -2.04 \\ 0 & 0 & 0 & 0 & -2.04 & 13.6 \end{pmatrix} * 10^{-12}\, m^2/N$$

piezoelectric Coupling

$$d = \begin{pmatrix} 0 & 0 & 0 & 0 & 68 & -42 \\ -21 & 21 & 0 & 68 & 0 & 0 \\ -1 & -1 & 6 & 0 & 0 & 0 \end{pmatrix} * 10^{-12}\ C/N$$, relative permittivity matrix

and density $\rho = 4700\, Kg/m^3$. The corresponding matrices and parameters for the AlN compound are given as:

Compliance matrix

$$S^E = \begin{pmatrix} 28.58 & 08.99 & -3.83 & 0 & 0 & 0 \\ -8.99 & 31.17 & -9.65 & 0 & 0 & 0 \\ -3.83 & -9.65 & 30.38 & 0 & 0 & 0 \\ 0 & 0 & 0 & 80 & 0 & 0 \\ 0 & 0 & 0 & 0 & 80 & 0 \\ 0 & 0 & 0 & 0 & 0 & 80 \end{pmatrix} * 10^{-13} \quad m^2 / N$$

, piezoelectric Coupling

$$d = \begin{pmatrix} 0 & 0 & -1.73 & 0 & 0 & -2.78 \\ 0 & 0 & 5.49 & 0 & -3.84 & 0 \\ -3.84 & 0 & 0 & 0 & 0 & 0 \end{pmatrix} * 10^{-12} \; C / N$$

, relative permittivity matrix

$$\varepsilon_r = \begin{pmatrix} 9.2 & 0 & 0 \\ 0 & 9.2 & 0 \\ 0 & 0 & 10.3 \end{pmatrix}$$

and density, $\rho = 3300 \; Kg / m^3$.

3.4 Boundary Settings

The base of our simulated piezoelectric device is made of a fixed substrate, on the top of which a NW is seeded. The upper end of the NW is taking to be free in response to the applied force. The base of our piezoelectric device is considered to be electrically grounded. The remaining parts of our piezoelectric device are kept at zero charge/symmetry conditions.

3.5 Simulation Results: three different kinds of forces: $F = F_0 \hat{x}$ as a lateral force, $F = F_0 \hat{z}$ have been applied as a tensile force and $F = -F_0 \hat{z}$ as a compressive force, as displayed in Figure 3.

IOP Conf. Series: Materials Science and Engineering **305** (2018) 012007 doi:10.1088/1757-899X/305/1/012007

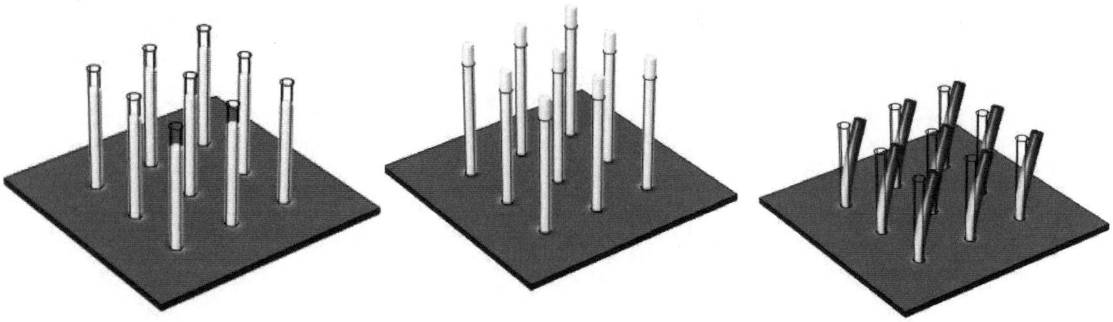

Figure 3: The three different applied forces in our study are: (a) Lateral force, $F = F_0\hat{x}$ (b) Tensile force, $F = F_0\hat{z}$ and (c) Compressive force, $F = F_0\hat{z}$.

4. Results and Discussions

In this section, we describe and interpret our results on the piezoelectric output potential and deformation (displacement) behavior of the ($LiNbO_3$) and (AlN) NWs as functions of different applied forces. We study three different heights of $LiNbO_3$ and AlN nanowires, namely 100 nm, 150 nm and 200 nm.

4.1 Nanowires of height (100) nm

Figure 4 shows the effect of the force F_x applied at the top free end perpendicular to the left side of $LiNbO_3$ and AlN NWs for a NW height of 100 nm. A maximum total displacement of 2.0983 nm for $LiNbO_3$ and 1.2705 nm for AlN at (F_x = 90 nN) was observed at the bottom of the NW from the left side. However, maximum piezoelectric potentials of 44.5 mV and 50.3 mV were observed at the left edge surface of the NWs for $LiNbO_3$ and AlN NWs, respectively.

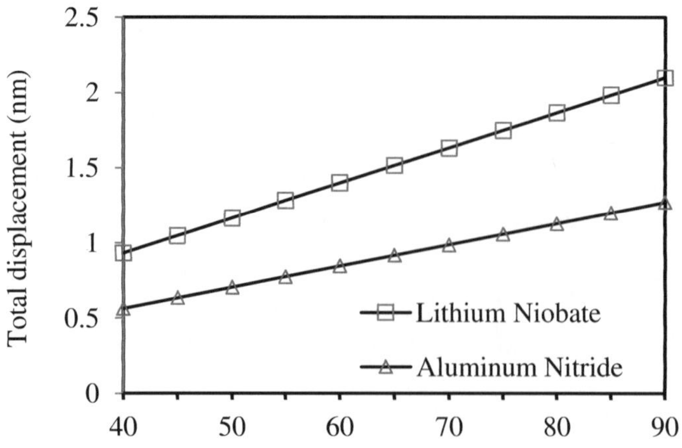

Figure 4: Total displacement (nm) of $LiNbO_3$ and AlN with boundary load $F = F_x$ (nN) for 100 nm height NWs.

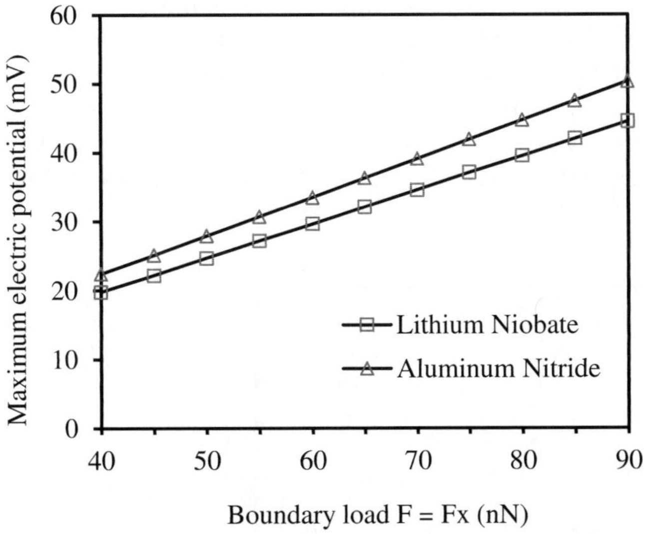

Figure 5: Maximum electric potential (mV) of $LiNbO_3$ and AlN with boundary load $F = F_x$ (nN) for 100 nm height NWs.

In addition, figures 4 and 5 clearly indicate that for the range of 40 nN to 90 nN applied force, both total displacement and electric potential exhibit a linear relationship as functions of applied load. In addition, figures 4 and 5 indicate that, for the same boundary load, $LiNbO_3$ NWs undergo total displacement larger than the one exhibited by AlN NWs. On the other hand, AlN NWs undergo maximum electric potential larger the one exhibited by $LiNbO_3$ NWs. This remarkable result indicates that AlN piezoelectric response is better than that of $LiNbO_3$ system. Furthermore, we found that, when the force F_z is applied compressively or tensely over the upper surface of $LiNbO_3$ and AlN NWs, the total displacement as a function of the boundary load remains linear as clearly displayed in figure 6.

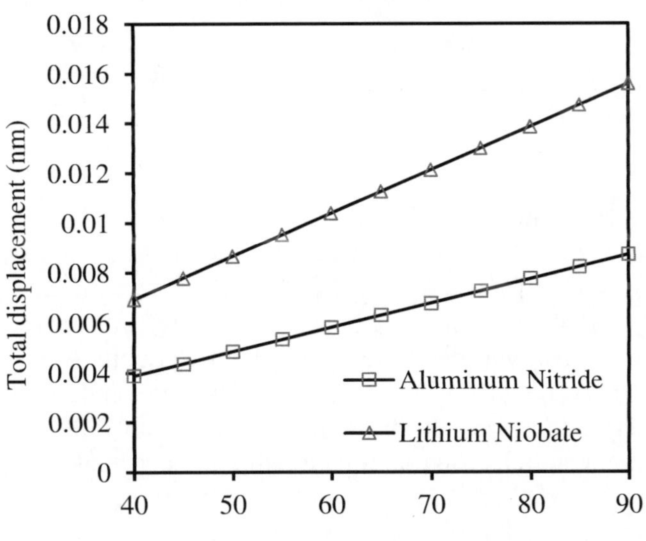

Figure 6: Total displacement (nm) of $LiNbO_3$ and AlN with boundary load $F=F_z$ (nN) for 100 nm height NWs.

Moreover, we studied the dependence of piezoelectric potential on the applied force F_z in the range of 40 nN to 90 nN for both NWs. Figure 7 shows piezoelectric potential as a function of applied force. The generated piezoelectric potentials exhibit positive values for tension forces and negative values for compressive forces for both NWs.
For instance, figure 7 shows that when F_z = 90 nN was applied, maximum piezoelectric potentials of ±73 mV and ±187 mV were generated at the top surface of $LiNbO_3$ and AlN NWs, respectively.

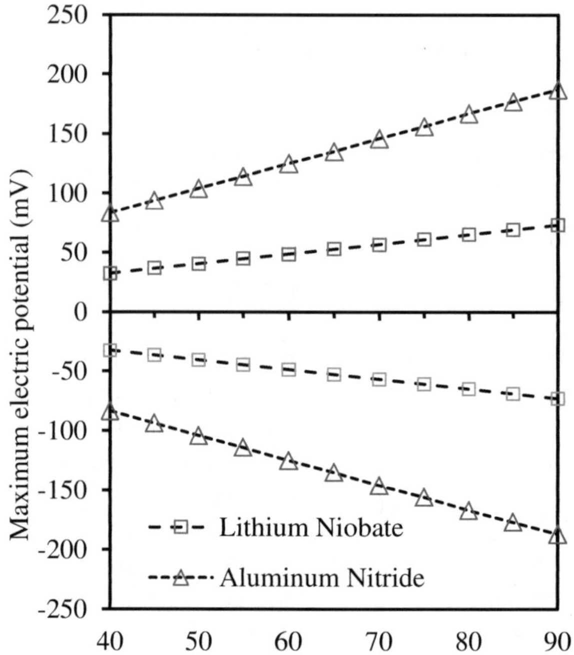

Figure 7: Maximum electric potential (mV) of $LiNbO_3$ and AlN with boundary load F = F_z (nN) for 100 nm height NWs (Positive sign for tension force and negative for compressive force).

4.2 Nanowires of height (150) nm

Figure 8 shows the effect of the force F_x applied at the top free end perpendicular to the left side of $LiNbO_3$ and AlN NWs for a NW height of 150 nm. A maximum total displacement of 7.0705 nm for $LiNbO_3$ and 4.2807 nm for AlN at (F_x = 90 nN) was observed at the bottom of the NW from the left side. However, maximum piezoelectric potentials of 72.5 mV and 84.7 mV were observed for $LiNbO_3$ and AlN NWs, respectively, at the left edge surface of the NWs.

In addition, Figures 8 and 9 clearly indicate that for the range of 40 nN to 90 nN applied force, both total displacement and electric potential exhibit a linear relationship.

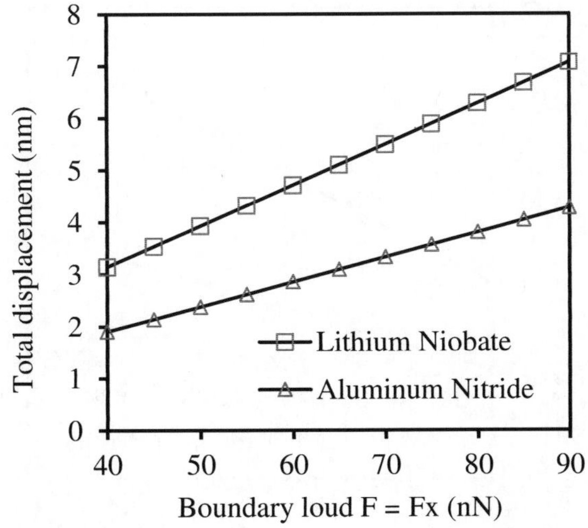

Figure 8: Total displacement (nm) of $LiNbO_3$ and AlN with boundary load $F = F_x$ (nN) for 150 nm height NWs.

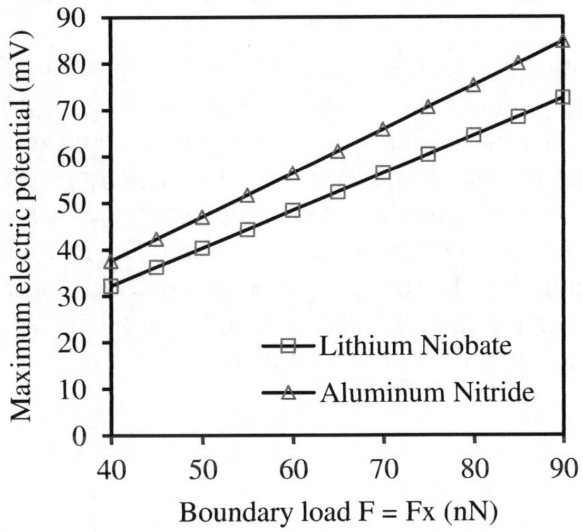

Figure 9: Maximum electric potential (mV) of $LiNbO_3$ and AlN with boundary load $F=F_x$ (nN) for 150 nm height NWs.

We have found that when the force F_z is applied compressively or tensely over the upper surface of $LiNbO_3$ and AlN NWs, the total displacement as a function of the boundary load remains linear as clearly displayed in figure 10.

IOP Conf. Series: Materials Science and Engineering **305** (2018) 012007 doi:10.1088/1757-899X/305/1/012007

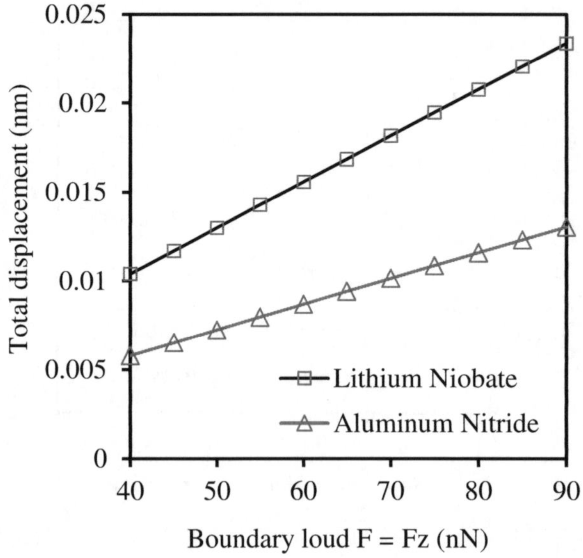

Figure 10: Total displacement (nm) of $LiNbO_3$ and AlN with boundary load $F=F_z$ (nN) for 150 nm height NWs.

Moreover, we studied the dependence of piezoelectric potential on the applied force F_z in the range 40 nN to 90 nN for both NWs. Figure 10 shows piezoelectric potential as a function of applied force. In addition, figure 11 shows that the generated piezoelectric potentials exhibit positive values for tension forces and negative values for compressive forces for both NWs. For instance, when $F_z = 90$ nN was applied, maximum piezoelectric potentials of ±109 mV and ±283 mV were generated at the top surface of $LiNbO_3$ and AlN NWs, respectively. Furthermore, figure 11 indicates that the generated piezoelectric potential exhibits a positive slope linear relationship with the applied tension boundary load and an equal but negative slope linear relationship with the applied compressive load.

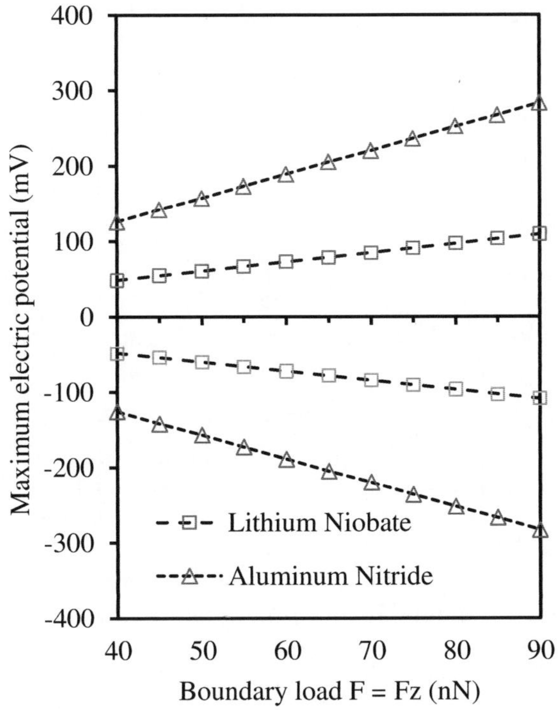

Figure 11: Maximum electric potential (mV) of $LiNbO_3$ and AlN with boundary load $F=F_z$ (nN) for 150 nm height NWs (Positive sign for tension force and negative for compressive force).

4.3 Nanowires of height (200) nm

Figure 12 shows the effect of the force F_x applied at the top free end perpendicular to the left side of $LiNbO_3$ and AlN NWs for a NW height of 200 nm. A maximum total displacement of 16.791 nm for $LiNbO_3$ and 10.164 nm for AlN at (F_x = 90 nN) was observed at the bottom of the NW from the left side. However, maximum piezoelectric potentials of 100 mV and 118 mV were observed for $LiNbO_3$ and AlN NWs, respectively, at the left edge surface of the NWs.

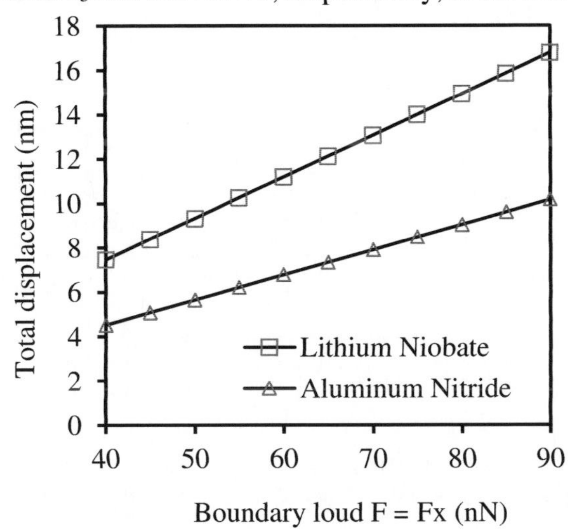

Figure 12: Total displacement (nm) of $LiNbO_3$ and AlN with boundary load $F=F_x$ (nN) for 200 nm height NWs.

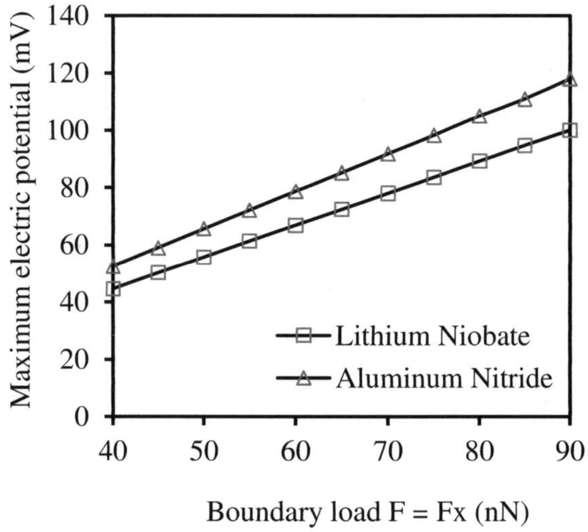

Figure 13: Maximum electric potential (mV) of $LiNbO_3$ and AlN with boundary load $F=F_x$ (nN) for 200 nm height NWs.

In addition, Figures 12 and 13 clearly indicate that for the range of 40 nN to 90 nN applied force, both total displacement and electric potential exhibit a linear relationship.

We have found that when the force F_z is applied compressively or tensely over the upper surface of $LiNbO_3$ and AlN NWs, the total displacement as a function of the boundary load remains linear as clearly shown in figure 14.

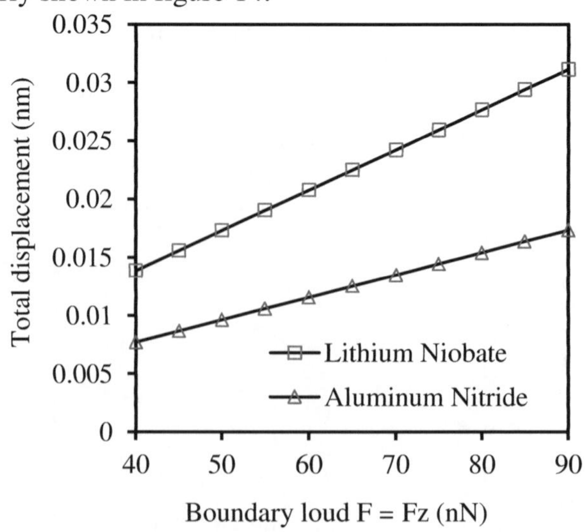

Figure 14: Total displacement (nm) of $LiNbO_3$ and AlN with boundary load $F=F_z$ (nN) for 200 nm height NWs.

Moreover, we studied the dependence of piezoelectric potential on the applied force F_z in the range 40 nN to 90 nN for both NWs. Figure 15 shows piezoelectric potential as a function of applied force.

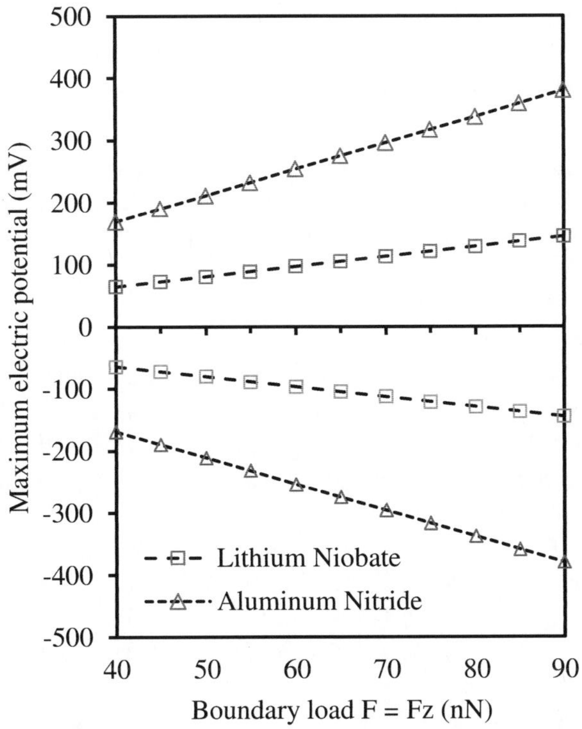

Figure 15: Maximum electric potential (mV) of $LiNbO_3$ and AlN with boundary load $F=F_z$ (nN) for 200 nm height NWs (Positive sign for tension force and negative for compressive force).

The generated piezoelectric potentials exhibit positive values for tension forces and negative values for compressive forces for both NWs. For instance, when F_z = 90 nN was applied, maximum piezoelectric potentials of ±145 mV and ±380 mV were generated at the top surface of $LiNbO_3$ and AlN NWs, respectively.

5. Conclusion

In summary, Lithium niobate ($LiNbO_3$) and Aluminum nitride (AlN) nanowires with rod geometry were investigated using COMSOL Multiphysics based on the Finite Element Method (FEM). We found that $LiNbO_3$ nanowires undergo total displacement higher than the corresponding displacement exhibited by AlN wires under the same boundary loads. However, AlN nanowires piezoelectric response is higher than the corresponding response of $LiNbO_3$ nanowires as indicated by the higher piezoelectric potentials exhibited by AlN nanowires as compared with those of $LiNbO_3$ nanowires. This remarkable finding indicates that AlN nanowires exhibit a better piezoelectric response than $LiNbO_3$ under the same boundary loads. Furthermore, we found that in both nanowires, the tension and compressive loads generate higher total displacements and piezoelectric potentials than the corresponding total displacements and potentials produced by lateral loads. One of the main findings of this work is that, total displacement and generated piezoelectric potential in both nanowires depend mainly on the height of the nanowires used in the rod geometry. We found that the 200 nm height nanowire exhibit optimized total displacement and piezoelectric response when subjected to tension, compressive or lateral boundary loads for AlN and $LiNbO_3$ nanowires. Based on our findings, we find that the piezoelectric response of both nanowires is very sensitive to the height of the nanowire in rod geometry particularly for AlN nanowires. Therefore, we recommend using the 200 nm height AlN nanowire in rod geometry for fabrication of piezoelectric and pressuring sensor devices.

Acknowledgments

The first three authors would like to acknowledge the technical and financial support provided by the faculty of scientific research at Jordan University of Science and Technology (JUST). The fourth author would like to acknowledge the support of the University of Hafr Albatin in Saudi Arabia.

References

[1] Karabalin RB, Matheny MH, Feng XL, Defay E, Le Rhun G, Marcoux C, Hentz S, Andreucci P, Roukes ML, *Piezoelectric nanoelectromechanical resonators based on aluminum nitride thin films.* Applied Physics Letters, 2009 **95(10)** 103111.

[2] Design M, *Sensor Sense: Piezoelectric Force Sensors.* Penton Media, 2008.

[3] Ayazi F, *The HARPSS process for fabrication of precision MEMS inertial sensors.* Mechatronics, 2002 **12(9)** 1185-1199.

[4] Wu X, Chen W, Zhang W, Lu Y, Cui F, Zhao X, editors. *Modeling analysis of piezoelectric micromachined modal gyroscope (PMMG).* in *Nano/Micro Engineered and Molecular Systems, 2009. NEMS 2009 4th IEEE International Conference on.* 2009. IEEE.

[5] Beeby SP, Tudor MJ, and White N, *Energy harvesting vibration sources for microsystems applications.* Measurement science and technology, 2006 **17(12)** R175.

[6] Reddy VM and Kumar GVS, Multiphysics, *Design and analysis of microcantilevers with various shapes using comsol multiphysics software.* 2013 3(3) 681-684 .

[7] Jeon YB, Sood R, Jeong JH, Kim SG., *MEMS power generator with transverse mode thin film PZT.* Sensors and Actuators A: Physical, 2005 **122(1)** 16-22.

[8] Chhabra P, Sharma A, editors. *Design, Simulation and Fabrication of Piezoelectric Energy Harvester for an Application in Tire Pressure Monitoring System (TPMS).* in *1st Winter Workshop on "Engineering at Nanoscale: From Materials to Bio-Sensors" IIT Indore.* 10. Dec. 2012.

[9] Yun BK, Park YK, Lee M, Lee N, Jo W, Lee S, et al., *Lead-free LiNbO3 nanowire-based nanocomposite for piezoelectric power generation.* Nanoscale research letters, 2014 **9(1)** 1-7.

[10] Santulli AC, Zhou H, Berweger S, Raschke MB, Sutter E, Wong SS, *Synthesis of single-crystalline one-dimensional LiNbO 3 nanowires.* CrystEngComm, 2010 **12(10)** 2675-2678.

[11] He JH, Yang RS, Chueh YL, Chou LJ, Chen LJ, Wang ZL, *Aligned AlN Nanorods with Multi-tipped Surfaces—Growth, Field-Emission, and Cathodoluminescence Properties.* Advanced Materials, 2006 **18(5)** 650-654.

[12] Volk T. and Wöhlecke M, *Lithium niobate: defects, photorefraction and ferroelectric switching.* Vol. 115. 2008: Springer Science & Business Media.

[13] Grigas A. and Kaskel S, *Synthesis of LiNbO3 nanoparticles in a mesoporous matrix.* Beilstein journal of nanotechnology, 2011 **2(1)** 28-33.

[14] Strite S. and Morkoç H, *GaN, AlN, and InN: a review.* Journal of Vacuum Science & Technology B, 1992 **10(4)** 1237-1266.

IOP Conf. Series: Materials Science and Engineering **305** (2018) 012007 doi:10.1088/1757-899X/305/1/012007

[15] Stampfl C. and Van de Walle C, *Density-functional calculations for III-V nitrides using the local-density approximation and the generalized gradient approximation.* Physical Review B, 1999 **59(8)** 5521.

[16] Berger, L.I., *Semiconductor materials*. 1996: CRC press.

[17] Przybyla RJ, Tang H-Y, Shelton SE, Horsley DA, Boser BE, editors. *12.1 3D ultrasonic gesture recognition*. in *Solid-State Circuits Conference Digest of Technical Papers (ISSCC), 2014 IEEE International*. 2014. IEEE.

[18] Ahmadi A, Hadipour NL, Kamfiroozi M, Bagheri Z, T*heoretical study of aluminum nitride nanotubes for chemical sensing of formaldehyde.* Sensors and Actuators B: Chemical, 2012 **161(1)** 1025-1029.

[19] Peyghan AA, Omidvar A, Hadipour NL, Bagheri Z, Kamfiroozi M, *Can aluminum nitride nanotubes detect the toxic NH 3 molecules?* Physica E: Low-dimensional Systems and Nanostructures, 2012 **44(7)** 1357-1360.

[20] https://www.comsol.com/.

[21] Dewit R, "Elastic constants and thermal expansion averages of a nontextured polycrystal", Journal of mechanics of materials and structures, 2008 **3(2)** 195-211.

[22] Gao, Y. and Z.L. Wang, *Electrostatic potential in a bent piezoelectric nanowire. The fundamental theory of nanogenerator and nanopiezotronics.* Nano letters, 2007 **7(8)** 2499-2505.

[23] Moheimani SR. and Fleming AJ, *Piezoelectric transducers for vibration control and damping*. 2006: Springer Science & Business Media.

IOP Conf. Series: Materials Science and Engineering **305** (2018) 012008 doi:10.1088/1757-899X/305/1/012008

The effect of cyclodextrin on both the agglomeration and the in vitro characteristics of drug loaded and targeted silica nanoparticles

Areen M. Khattabi* and Diala A. Alqdeimat

Department of Pharmaceutical Sciences and Pharmaceutics, Applied Science Private University, Amman, Jordan

a_khtabi@asu.edu.jo

Abstract One of the problems in the use of nanoparticles (NPs) as carriers in drug delivery systems is their agglomeration which mainly appears due to their high surface energy. This results in formation of NPs with different sizes leading to differences in their distribution and bioavailability. The surface coating of NPs with certain compounds can be used to prevent or minimize this problem. In this study, the effect of cyclodextrin (CD) on the agglomeration state and hence on the in vitro characteristics of drug loaded and targeted silica NPs was investigated. A sample of NPs was loaded with anticancer agents, then modified with a long polymer, carboxymethyl-β-cyclodextrin (CM-β-CD) and folic acid (FA), respectively. Another sample was modified similarly but without CD. The surface modification was characterized using fourier transform infrared spectroscopy (FT-IR). The polydispersity (PD) was measured using dynamic light scattering (DLS) and was found to be smaller for CD modified NPs. The results of the in vitro drug release showed that the release rate from both samples exhibited similar pattern for the first 5 hours, however the rate was faster from CD modified NPs after 24 hours. The in vitro cell viability assay confirmed that CD modified NPs were about 30% more toxic to HeLa cells. These findings suggest that CD has a clear effect in minimizing the agglomeration of such modified silica NPs, accelerating their drug release rate and enhancing their targeting effect.

Key words: Cyclodextrin; agglomeration; targeting; in vitro

1. Introduction

Nanoparticles are particulate dispersions or solid particles with a size in the range of 10-1000 nm [1]. Due to their unique properties, they have gained a great research interest in the area of drug delivery systems. Their ultra small size makes their taking up by cells much easier than larger molecules [2], [3]. They have a large surface area to volume ratio and their surface can be easily modified with different functional groups [4], [5]. More importantly, they can be used to improve the pharmacokinetic and pharmacodynamic properties of different types of drug molecules that are encapsulated into NPs and forming targeted delivery systems of both imaging and anticancer agents [6]. Furthermore, They have been used in vivo to protect the drug molecules in the systemic circulation, ensure delivery of the drug to the site of action at a controlled and sustained manner and to reduce the undesirable side effects of the drugs [6].

In spite of these advantages, using NPs as carriers in drug delivery systems have certain limitations. For instance, their small size and large surface area can lead to their agglomeration, making their physical handling , processing, synthesis, storage and transportation difficult in both the liquid and dry forms [1].

One of the most important characteristics of NP systems is their particle size and size distribution [7]. It has been found that these parameters determine the in vivo distribution, biological fate, toxicity, targeting ability, drug loading, drug release and stability of the NPs [1], [8]. Therefore, the problem of agglomeration and non uniform size distribution have to be overcome in order to prepare efficient drug delivery systems based on NPs.

The agglomeration state of the NPs and their size can be determined by their surface coatings [5]. Several coatings have been used to prevent or minimize the agglomeration of the NPs and keep them in colloidal state [9]. Such coatings include polymers like polyethylene glycol (PEG), poly(vinylpyrrolidone) (PVP), natural polymers like dextran, chitosan, pullulan, and surfactants like sodium oleate and dodecylamine [10].

Cyclodextrins (CDs) have also been used for surface coatings of different types of NPs [11] in order to enhance their steric hindrance and minimize their agglomeration [11], [12]. CDs are (1,4) linked macrocyclic oligosaccharides with cage-like structure having a hydrophobic interior and a hydrophilic exterior [13]. Due to these properties, they have been widely used to enhance the aqueous solubility of hydrophobic drugs [14]. In the present work, however, we intend to study the effect of CD, which conjugated to our previously investigated targeted drug delivery system [15], on the agglomeration of silica NPs and hence on their in vitro characteristics such as drug release rate and targeting ability.

2. Materials and methods

2.1. Materials

FITC-labeled propylcarboxylic acid functionalized silica NPs (particle size 200 nm, pore diameter 4 nm), Poly(propyleneglycol)bis(2-aminopropylether diamine, D4000) and carboxymethyl-B-cyclodextrin sodium salts (CM-β-CD) were purchased from Sigma Aldrich. Folic acid (purity > 98%) and Phosphate Buffered Saline (PBS, PH = 7.4) were obtained from bioworld. All other reagents and materials used for cell culture were used as received.

2.2. Preparation of drug loaded FA-CM-β-CD aminated silica NPs and FA- aminated silica NPs.

The procedures for drug loading and synthesis of these two types of modified silica NPs were presented in our previous work [15].

2.3. FITR and DLS analysis of the modified silica NPs

The FTIR spectra of the silica NPs were obtained using WQF-521 Fourier transform infrared spectrophotometer within the mid-IR range (500-4000 cm^{-1}). The mean particle size (Z-average) and PD of the NPs were measured by DLS technique using (Zetasizernano series, Malvern U.K). The NPs were diluted as suspensions (4.5 x10-3 mg/ml) in ethanol with refractive index (RI) = 1.362 for ethanol and 1.48 for silica NPs and viscosity= 1.20 cp. The suspensions were first sonicated for 5 minutes to separate the bigger aggregates. The measurements were performed in triplicates for each sample with 6 runs each time and the average was then taken.

2.4. In vitro drug release study

Certain amounts of each of drug loaded FA-CM-β-CD aminated silica NPs and FA-aminated silica NPs were mixed with 2 ml PBS (pH = 7.4) in a dialysis bag (SnakeSkin Dialysis Tubing, 22 mm x 35 feet dry diameter). Each sample was then added to 23 ml of PBS in a beaker and shaken at 37° C in water path. At different time intervals (1, 2, 4, 5, 24, 28, 30 and

48 hours), aliquots of 1.5 ml were removed, replaced with 1.5 ml of fresh PBS, to maintain the same sink conditions. The amount of released drug in the supernatant was measured via UV-Vis spectroscopy using (Uv/vis spectroscopy spuv-19). Absorbance was measured at λ_{max} specific for the anticancer agents used. The corresponding drug concentration was calculated based on the calibration curves obtained. All experiments were performed in triplicate. The percentage of cumulative drug release was then calculated and the values represent the average of three experiments for each sample.

2.5. Antiproliferative (cell viability) Assay

HeLa Cells were dispensed (100 µl/well) into 96-well tissue culture plates (flat bottom) at an optimized concentration of 15000 cells/well in complete tissue culture medium. After 24 h, the media in each well were completely removed and the attached cells were treated in triplicates with 200 µl/ of 0.05 mg/ml of two different suspensions of modified silica NPs, suspended in complete tissue culture medium. Plates were incubated for 48 h, and then cell viability was measured by using MTT (3-(4,5-Dimethylthiazol-2-yl)-2,5-diphenyltetrazolium bromide) assay. From each well, 100 µl of culture media was removed and 10 µl of thiazolyl blue tetrazolium solution was added and placed in CO_2 incubator for 3 hours. Then the reaction was stopped by adding MTT solubilization solution (100 µl/well), mixed well and incubated for another one hour. Absorbance was measured at 550 nm by microplate reader. Cell viability (% survival) was calculated and then used to get percentage of cell toxicity.

3. Results and Discussions:

3.1. Characterization of the conjugated NPs

3.1.1. FTIR Analysis

To study the effect of CD on the agglomeration and hence on the other in vitro characteristics of the silica NPs, two different modifications were prepared in this study. A sample of the commercially available propylcarboxylic acid functionalized silica NPs was first loaded with anticancer agents then modified with a long polymer (Poly(propyleneglycol)bis(2-aminopropylether diamine, D4000) , CM-β-CD and FA, respectively. The other sample was modified similarly but without CD. Thus, we have FA-CM-β-CD aminated and FA-aminated silica NPs, respectively. The FTIR spectra of the two modifications are shown in Figure 1. The first modification includes the formation of amide bonds after the addition of both the diamine and the CD, while the folic acid is embedded into CD cavity via host–gust interaction and thus the peaks of folic acid will be mainly hidden, as previously reported [12] . On the other hand, the second modification, in which the CD is absent, includes two amide bonds after both the diamine and the folic acid. In this case, the peaks specific for FA should be more clear than the first modification.

As shown in Figure 1c, the FTIR spectrum of FA-CM-β-CD aminated silica NPs has new small peaks at about 2800 – 2900 cm^{-1} compared to unmodified NPs. On the other hand, the spectrum of FA-aminated silica NPs (Figure 1d) has more and broader peaks related to FA. As shown in Figure 1b, unconjugated FA has few peaks above 3100 cm^{-1} which appeared as a broad peak in Figure 1d and one at about 2800 cm^{-1} which existed in Figure 1d at the same position. This indicated that a covalent bond was involved in FA attachment in the second modification while it was via host guest interaction in the first modification.

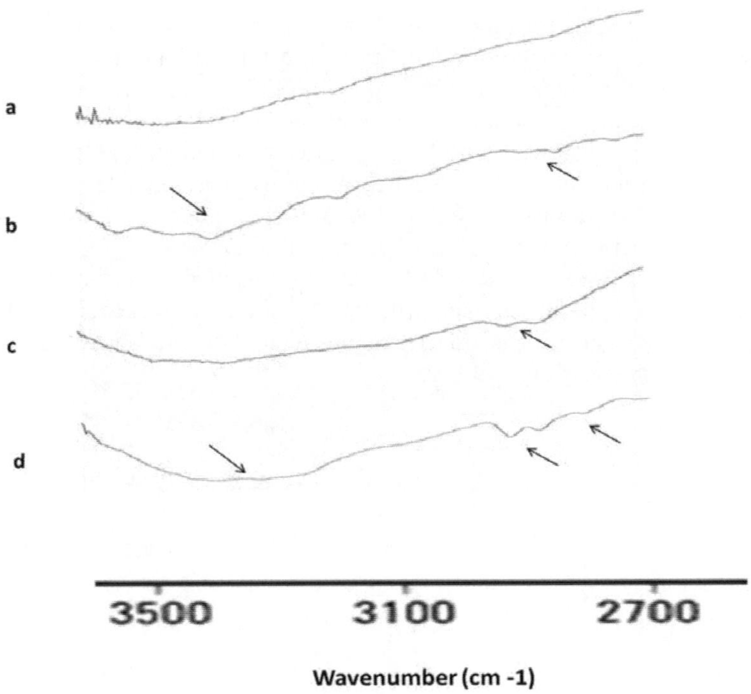

Figure 1: FTIR spectra of (a) pure (propylcarboxylic acid functionalize silica NPs, (b) unconjugated FA, (c) FA-CM-β-CD aminated silica NPs and (d) FA- aminated silica NPs.

3.1.2. DLS Analysis

The mean particle size of the two modified NPs as well as their polydispersity were measured using DLS. As shown in Table 1, the mean particle size is bigger for FA-CM-β-CD aminated silica NPs compared to FA-aminated silica NPs due to the presence of CD. However, the PD was smaller for FA-CM-β-CD aminated NPs. Since the PD measures the heterogeneity of sizes of the particles, the smaller value indicates a narrower size distribution which is most likely related to a lesser tendency of the NPs to form agglomerates [16]. These results confirmed that the presence of CD in this targeted system improves the NPs distribution and minimizes the chance of their agglomeration.

Table 1: Characterization of the modified NPs using DLS

Sample	Mean particle size (nm)	Polydispersity (PD)
FA-CM-β-CD aminated silica NPs	~ 615 nm	~ 0.4
FA-aminated silica NPs	~ 505 nm	~ 0.6

3.2. In vitro drug release study

The in vitro drug release profiles for both FA-CM-β-CD aminated and FA- aminated silica NPs were measured and the percentage of cumulative drug release was calculated. In general, as shown in Figure 2, both modifications showed an initial burst release of the drug followed by a controlled release. More importantly, they exhibited similar pattern for the first 5 hours, but a faster rate was observed for CD modified NPs after that. This is due to the fact that the particle size exert a clear effect on the various properties of nano drug delivery systems such as their release rate [17]. The absence of CD in FA-aminated silica NPs resulted in combining the NPs to create product with particle distribution shifted towards larger sizes and smaller surface area. Therefore, the actual number of NPs exposed to media will be less and the release rate of drug from such modified NPs will be slower. It also seems that the agglomeration increased with time hence the difference between both modifications was obvious after a certain length of time as shown in Figure 2.

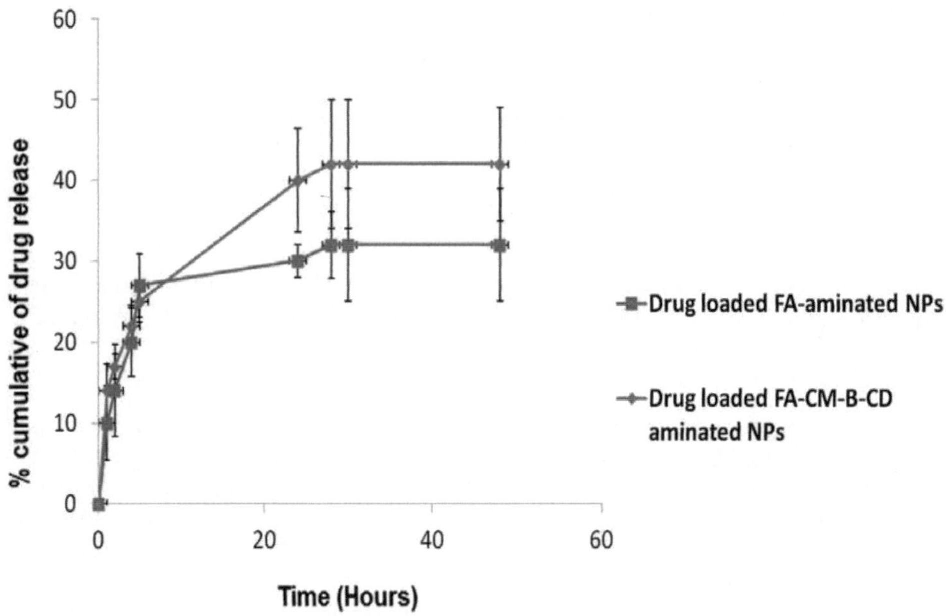

Figure 2: In vitro drug release of FA-CM-β-CD aminated silica NPs and FA- aminated silica NPs in PBS (pH 7.4).

3.3. In vitro cell viability assay

NPs (0.05 mg/ml) were for cell viability assay, two different suspensions of the two modified silica encapsulated with the same amount of drug and incubated with HeLa cells for 48 hours under the same conditions. As shown in Figure 3, FA-CM-β-CD aminated silica NPs (sample 1) were about 30% more toxic toward HeLa cells compared to FA-aminated silica NPs (sample 2).

It is well known that the size is one of the major factors determining cellular uptake efficiency of NPs and several internalization pathways depend on the size [7]. As a general rule, smaller NPs are internalized into cells easier than larger ones [7] and our results were in agreement with this fact. Since FA-aminated NPs were more susceptible to agglomeration and their particle size distribution was shifted toward larger sizes, their internalization via receptor mediated endocytosis was slower than FA-CM-β-CD aminated NPs. Thus, these two FA targeted systems have different targeting ability which was observed from the differences in their toxicity effects toward HeLa cells (Figure 3). From these findings, it was concluded that the presence of CD, in such targeted NPs, minimizes their agglomeration, accelerates their drug release rates and enhances their internalization and targeting efficiency.

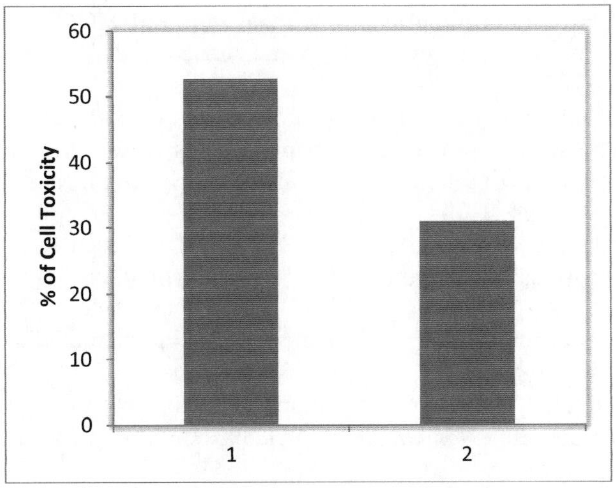

Figure 3: The percentage of cell toxicity of (1) FA-CM-β-CD aminated silica NPs and (2) FA-aminated silica NPs toward HeLa cells after incubation for 48 hours.

4. Conclusions

In this study, we investigated the effect of CD on the agglomeration and hence on the in vitro characteristics of our previously prepared TDDS which is based on silica NPs. The NPs were first loaded with anticancer agents then modified, as we previously described, with a long polymer,CM-β-CD and FA, respectively. For comparison purposes, another set of NPs were treated similarly but with the absence of CD. The results obtained showed that the polydispersity for CD modified NPs was smaller indicating a lower tendency of these NPs to form agglomerates. The results of the in vitro drug release showed that the release rate from both samples was similar for the first 5 hours, however the rate was faster from CD modified NPs after 24 hours. The in vitro cell viability assay demonstrated that CD modified NPs were more toxic to HeLa cells indicating their easier internalization. These findings suggest that CD has a clear effect in minimizing the agglomeration of such modified silica NPs, accelerating their drug release rate and enhancing their targeting effect.

Acknowledgments: The authors are grateful to the Applied Science Private University, Amman, Jordan, for the full financial support granted to this research project (Grant NO. DRGS-2015-2016-35). We also thank Dr. Wamidh H. Talib (from Department of Clinical Pharmacy and Therapeutics, Applied Science Private University, Amman, Jordan) for his assistance in performing the in vitro cell viability assay.

Conflicts of Interest: The authors declare no conflict of interest.

References

[1] V. J. Mohanraj and Y. Chen, “Nanoparticles – A Review,” *Trop J Pharm Res,* vol. 5, no. June, pp. 561–573, 2006.

[2] S. S. Suri, H. Fenniri, and B. Singh, “Nanotechnology-based drug delivery systems,” *J Occup Med Toxicol.*, vol. 2, p. 16, 2007.

[3] L. B. Peppas, “Recent advances on the use of biodegradable microparticles and nanoparticles in controlled drug-delivery,” *Int. J. Pharm.*, vol. 116, no. 1, pp. 1–9, 1995.

[4] I. Slowing, B. Trewyn, and V. Lin, “Effect of surface functionalization of MCM-41-type mesoporous silica nanoparticles on the endocytosis by human cancer cells,” *J. Am. Chem.Soc.*, vol. 128, no. 46, pp. 14792–14793, 2006.

[5] R. H. S. and M. M. Christoph Bantz, Olga Koshkina, Thomas Lang, Hans-Joachim Galla, C. James Kirkpatrick, “The surface properties of nanoparticles determine the agglomeration state and the size of the particles under physiological conditions,” *Beilstein J Nanotechnol*, vol. 15, no. 5, pp. 1774–86, 2014.

[6] S. Singh, V. K. Pandey, R. P. Tewari, and V. Agarwal, “Nanoparticle based drug delivery system : Advantages and applications,” *Indian J Sci Technol.,* vol. 4, no. 3, pp. 177–180, 2011.

[7] L. Shang, K. Nienhaus, and G. U. Nienhaus, “Engineered nanoparticles interacting with cells : size matters,” *J Nanobiotechnology*, pp. 1–11, 2014.

[8] B. G. Hoshyar N, Gray S, Han H, “The effect of nanoparticle size on in vivo pharmacokinetics and cellular interaction.,” *Nanomedicine*, vol. 11, no. 6, pp. 673–92, 2016.

[9] W. H. De Jong, “Drug delivery and nanoparticles : Applications and hazards,” *Int J Nanomedicine,* vol. 3, no. 2, pp. 133–149, 2008.

[10] G. M. Gupta AK, “Synthesis and surface engineering of iron oxide nanoparticles for biomedical applications,” *Biomaterials.*, vol. 26, no. 18, pp. 3995–4021, 2005.

[11] B. Gidwani and A. Vyas, “A Comprehensive Review on Cyclodextrin-Based Carriers for Delivery of Chemotherapeutic Cytotoxic Anticancer Drugs,” *Biomed Res. Int.*, vol. 2015, 2015.

[12] C. Su *et al.*, “Carboxymethyl-β-cyclodextrin conjugated nanoparticles facilitate therapy for folate receptor-positive tumor with the mediation of folic acid,” *Int. J. Pharm.*, vol. 474, no. 1–2, pp. 202–211, 2014.

[13] S. K. Das, R. Rajabalaya, S. David, N. Gani, J. Khanam, and A. Nanda, “Cyclodextrins-the molecular container,” *Res. J. Pharm. Biol. Chem. Sci.*, vol. 4, no. 2, pp. 1694–1720, 2013.

[14] G. Tiwari, R. Tiwari, and A. K. Rai, “Cyclodextrins in delivery systems: Applications.,” *J. Pharm. Bioallied Sci.*, vol. 2, no. 2, pp. 72–9, 2010.

[15] A. M. Khattabi, W. H. Talib and D. A. Alqdeimat, "A Targeted Drug Delivery System of Anti-Cancer Agents Based on Folic Acid-Cyclodextrin-Long Polymer Functionalized Silica Nanoparticles," *J Drug Deliv Sci Technol, (in press.)*

[16] M. D. N, R. Eskandari, H. Zolfagharian, and M. Mohammad, “Preparation and in vitro characterization of chitosan nanoparticles containing Mesobuthus eupeus scorpion venom as an antigen delivery system,” *J. Venom. Anim. Toxins incl. Trop. Dis.*, vol. 18, no. 1, pp. 44–52, 2012.

[17] S. Honary and F. Zahir, “Effect of Zeta Potential on the Properties of Nano-Drug Delivery Systems - A Review (Part 2),” *Trop J Pharm Res.,* vol. 12, no. April, pp. 265–273, 2013.

IOP Conf. Series: Materials Science and Engineering **305** (2018) 012009 doi:10.1088/1757-899X/305/1/012009

Study of the Emission Characteristics of Single-Walled CNT and Carbon Nano-Fiber Pyrograf III

Marwan S. Mousa [1], M-Ali H. Al-Akhras [2], Samer Daradkeh [1]

[1] *Surface Physics and Materials Technology Lab, Department of Physics, Mu'tah University, AL-Karak, Jordan*
[2] *Dept. of Physics, Jordan University of Science and Technology, Irbid, Jordan*

marwansmousa@yahoo.com & mmousa@mutah.edu.jo

Abstract: Field emission microscopy measurements from Single-Walled Carbon Nanotubes (SWCNTs) and Carbon Nano-Fibers Pyrograf III PR-1 (CNF) were performed. Details of the materials employed in the experiments are as follows: (a) Carbon Nano-Fibers Pyrograf III PR-1 (CNF), having an average fiber diameter that is ranging between (100-200) nm with a length of (30-100) μm. (b) Single walled Carbon Nanotubes were produced by high-pressure CO over Fe particle (HiPCO: High-Pressure Carbon Monoxide process), having an average diameter ranging between (1-4) nm with a length of (1-3) μm. The experiments were performed under vacuum pressure value of (10^{-7} mbar). The research work reported here includes the field electron emission current-voltage (I-V) characteristics and presented as Fowler-Nordheim (FN) plots and the spatial emission current distributions (electron emission images) obtained and analyzed in terms of electron source features. For both the SWCNT and the CNF a single spot pattern for the electron spatial; distributions were observed.

1. Introduction

It is known that Pyrograf-III can be uniformly dispersed in a polymer in order to provide maximum conductive network using mixing and compounding equipment [1]. Electron emission (FE) refers to the release of electrons from the surface of a material into vacuum [2], which is occurring when the electric field strength is at the scale of (10^9 $V.m^{-1}$) [3]. FE has a long history up to the derivation of the original Fowler-Nordheim (FN) equation in 1928. In early days of FE experiments, metal was used as field emitters. However, because of the active nature of materials the residual gases and ion bombardment during FE working limited the lifetime of metal field emitters, impeding the development of FE [4]. Since discovering of Carbon Nanotubes (CNTs) from 1991 to 1993 with both types Multi-Walled Carbon Nanotubes (MWCNTs) and Single Wall Carbon Nanotubes (SWCNTs) by Sumio Iijima [5] [6], it has a growing attentions due to their perfect properties that make it the best candidate to be the next generation of field emitters such as, low turn-on fields, high emission current, stable emission current, long lifetime, low energy spreading, and high brightness [7].

IOP Conf. Series: Materials Science and Engineering **305** (2018) 012009 doi:10.1088/1757-899X/305/1/012009

The history of Carbon Nano-Fibers (CNFs) goes back more than a century. In a patent published in 1889 [8], it is reported that Carbon filaments are grown from Carbon-containing gases using a metallic crucible as the catalyst [9]. Carbon nanofibers (CNFs), also known as Stacked-Cup Carbon Nanotubes (SCCNTs), is with in the class of materials termed Multi-Walled Carbon Nanotubes and are produced by the floating catalyst method. CNFs are discontinuous and highly graphitic, but there are differences in the structure as figure (1) shows.

(SCCNTs) or (CNFs) have a unique morphology in that graphene planes are canted from the fiber axis, resulting in exposed edge plane on the interior and exterior surfaces of the fiber; CNTs, on the other hand, typically resemble an assembly of concentric cylinders of graphene [10].

Fowler-Nordheim (FN) first derived a semiclassical theory of FE based on physical assumptions and mathematical approaches; the theory has been modified many times over years to describe Cold Field Emission (CFE) from bulk metal where it is known as Fowler-Nordheim (FN) equations.

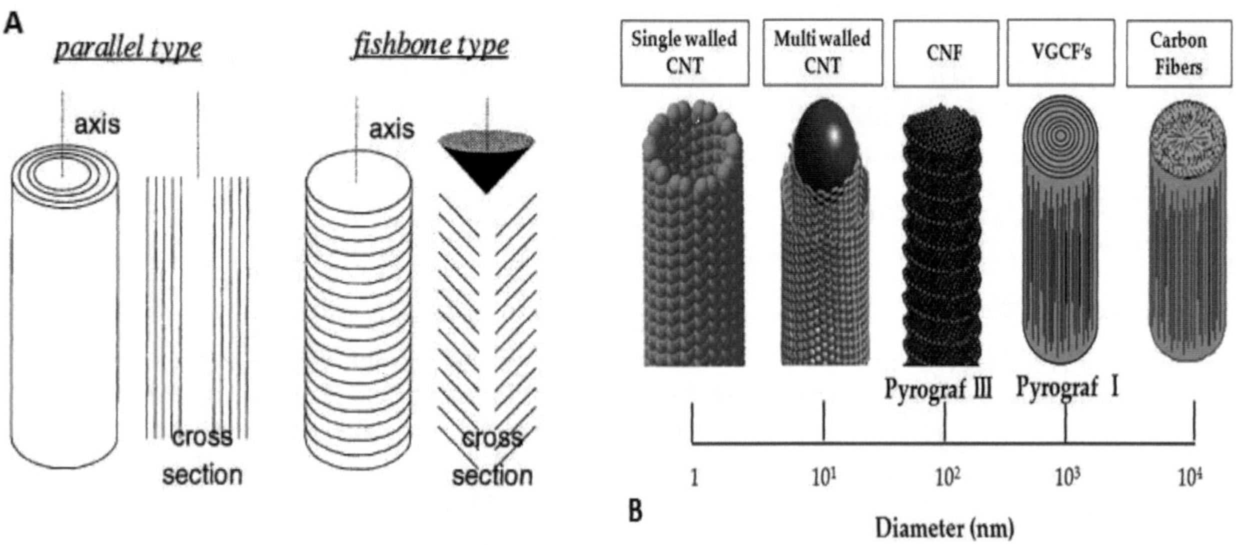

Figure 1(A) Simplified representation of the structure of parallel (left) and fishbone (right) Carbon Nanotubes. (B)Nano-Fibers sizes from Single wall Carbon Nanotubes to conventional Carbon Fibers.

The simplest of these is the so-called elementary FN-type equation

$$i = A_n\{ a\phi^{-1}(\beta_V V)^2\}\exp(-\frac{b\phi^{\frac{3}{2}}}{\beta_V V})$$

Where i is the current emission, ϕ local work function, V is the applied voltage, a and b are the first and second FN constants, as usually defined in Ref [11], β_V is the local voltage-to-surface field conversion factor at some specific reference point on the emitter surface, and A_n is the national emission area defined by considering the local emission current density at the reference point. This equation and related equations that use macroscopic field as an independent variables have often been used to describe CFE from Carbon Nanotubes (CNTs) including closed single walled (CNTs) of small apex radius, particularly in technological context [12].

2. Materials and Methods

The apparatus are used to create a prepared glass micro-point emitter where the CNTs inserted in it, so they would protrude at the tip. Figure (2) and (3) shows the produced sample.

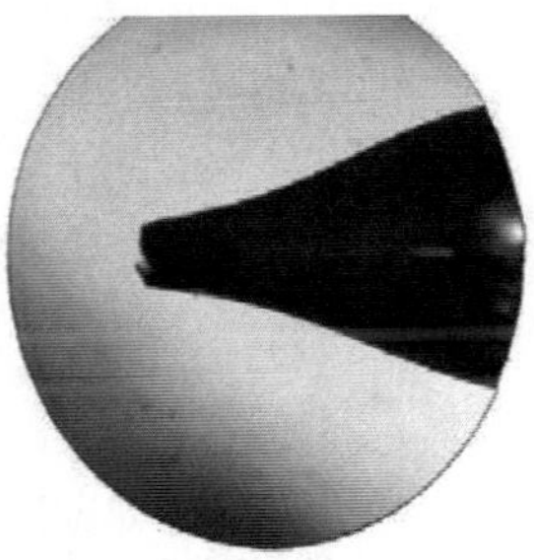

Figure 2: Optical micrograph of (SWCNT-15) tip at magnification 50x times.

Figure 3: Optical micrograph of (CNF-25) tip at magnification 50x times.

This study uses two types of CNTs, (a) Single-Walled Carbon Nanotubes (SWCNTs) were produced by high-pressure CO over Fe particles (HiPCO: High-Pressure Carbon Monoxide process), having an average diameter ranging between (1-4) nm with length (1-3) µm [13].

(b) Carbon Nano-Fibers pyrograf III PR-1 (CNFs), having average fiber diameter that is ranging between (100-200) nm, with the length of (30-100) µm.

IOP Conf. Series: Materials Science and Engineering **305** (2018) 012009 doi:10.1088/1757-899X/305/1/012009

Both tips prepared by employing a drawing technique using a glass puller, figure (4) shown the apparatus that are used to produce such tips, then the emitters are mounted in a standard Field Emission Microscopy (FEM) with an emitter-screen distance of ~10 mm [14], and current limiting resistor of (100 MΩ) is used. The system was evacuated to pressure ~10^{-8} mbar after baking at ~150°C overnight.

The FEM applied voltage V is increased slowly until a "switch-on voltage" V_{SW} is reached, at which point the emission current suddenly "switch-on", at V_{SW} the current increase rapidly from about a Nano-ampere to a much greater saturated value I_{SW}, by slowly decreasing the applied voltage to a lower value the emission current starts to be decreasing until limit reached, beyond which the emission current falls smoothly to zero as the applied voltage is decreased to the threshold value V_{TH} [15].

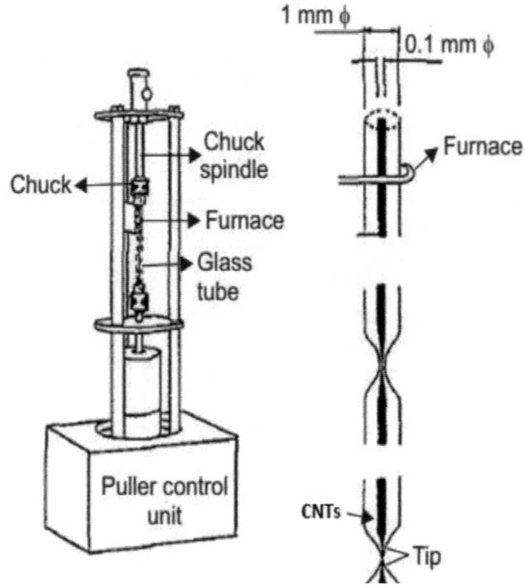

Figure 4: Schematic diagram of the glass puller.

3. Results and Discussion

This section describes the results of experiments on the emission characteristics of the Carbon Nano-Fibers and Single-Walled Carbon Nanotubes. These results include the I-V characteristics, Fowler-Nordheim (FN) plot, and emission images. Emission current measurements are made using a pico-ammeter connected between the cathode and earth.

As the applied electric field is slowly applied, the emission is only observed after a threshold field is reached, and emission suddenly jumps to a constant saturated value I_{SAT}, further increasing the applied electric field the V_{MAX} will be reached, after that the tip explode, as the voltage slowly decreasing, the emission current fall to nearly zero, reaching the threshold value V_{TH}.

Figure (2) shows an optical image of the first tip to be discussed. As we increase the applied voltage slowly, "switch-on" phenomena has been occurring at applied voltage value V_{SW} = 800 V, with emission current value I_{SW} = 4 μA, then the applied voltage was reduced, the V_{SAT} remained constant to the saturated voltage value of V_{SAT} = 400 V, with emission current value I_{SW} = 1.06 μA, further voltage reduction, the emission current falls smoothly to I_{TH} = 7 pA at applied voltage value (180 V). Figure (5) below shows the I-V behaviour during increasing and decreasing the applied voltage, and FN plot.

The experiment was repeated under the same condition and the results were as follows, the

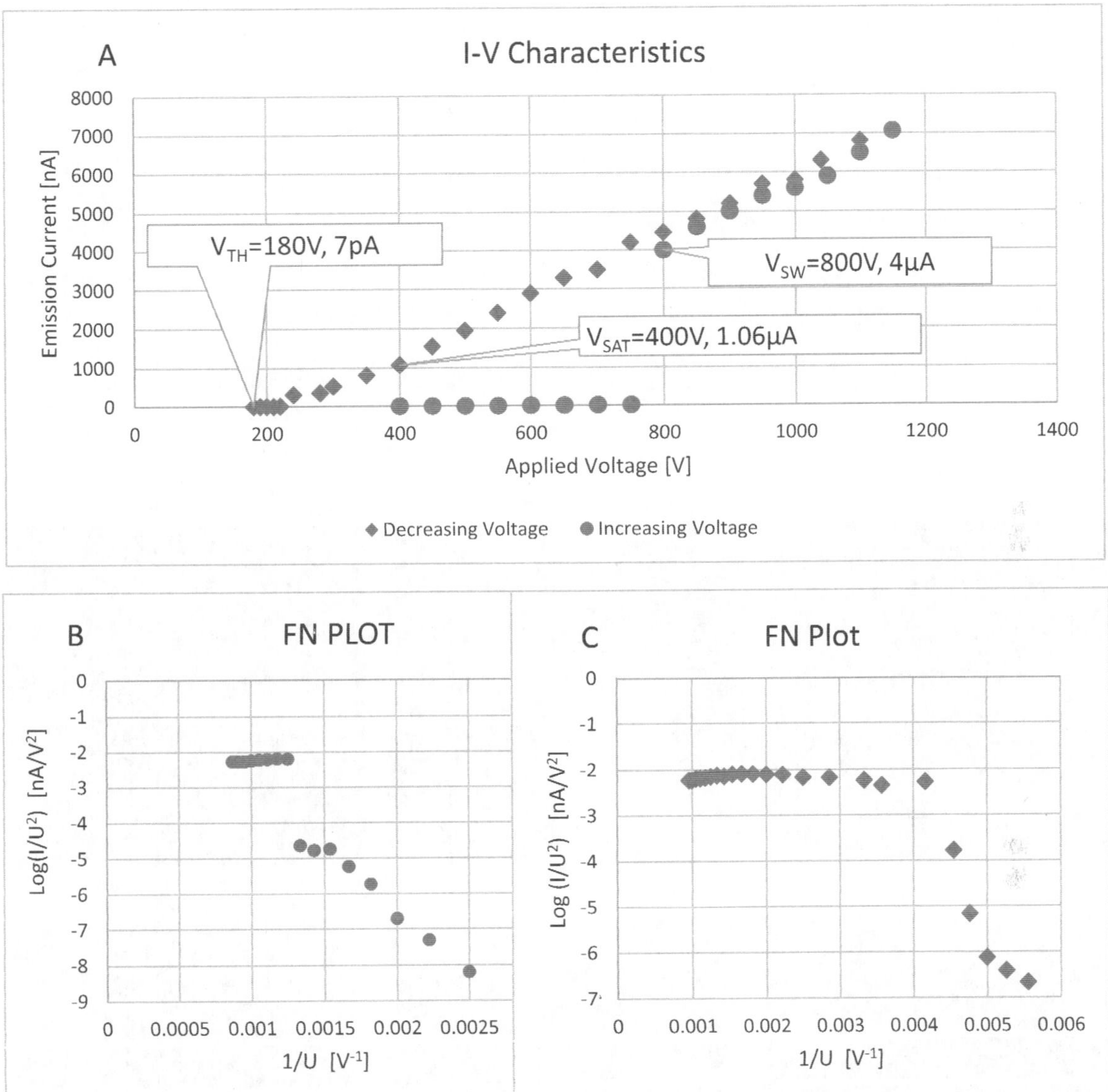

Figure 5. (A) I-V plot for increasing and decreasing voltage (B) FN plot for increasing voltage (C) FN plot for decreasing voltage.

electron emission is initiated at an applied voltage of (320 V) with emission current value (2 pA), and by continue increasing the applied voltage, a point was reached, at V_{SW} =650 V, where the emission current suddenly "switched-on" to a stable saturated value I_{SAT}, and by decreasing the applied voltage the saturated region extends down to value V_{SAT}=380 V, with emission current value (1.06 μA). Further voltage reduction causes the emission current to fall to nearly zero, reaching the threshold value V_{TH} =140 V, with emission

IOP Conf. Series: Materials Science and Engineering **305** (2018) 012009 doi:10.1088/1757-899X/305/1/012009

current value (3 pA). Figure (6) shows the I-V behavior and FN plot and, some field electron microscope images for the first cycle and second cycle.

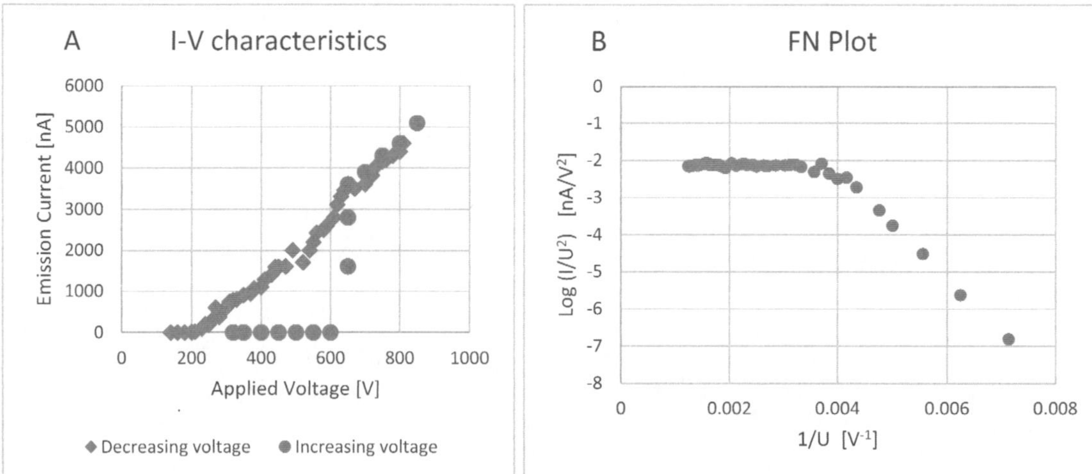

Figure 6. Second Cycle of test (A) I-V plot for decreasing and increasing voltage. (B) FN plot for decreasing voltage.

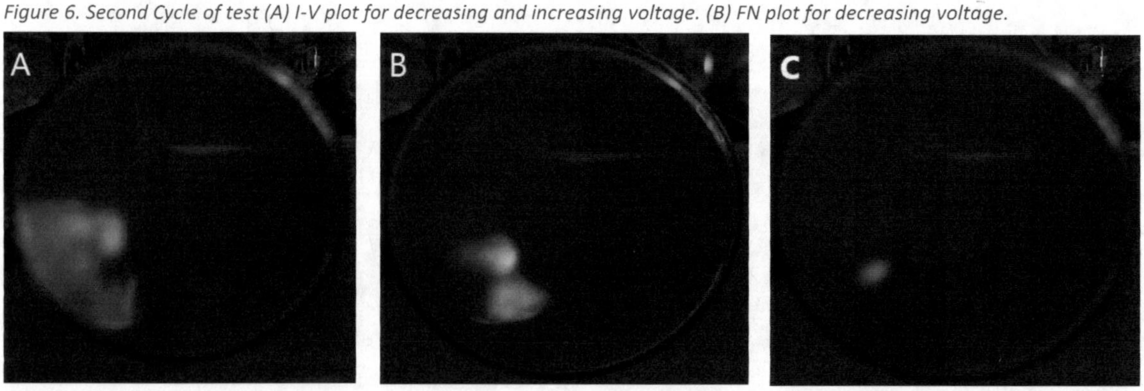

Figure 7. Emission images for first cycle at: (A) V_{SW} =800 V, 4 μA. (B) V_{SAT} =400 V, 1.06 μA. (C) V (Before switch-on) =650 V, 7.6 nA.

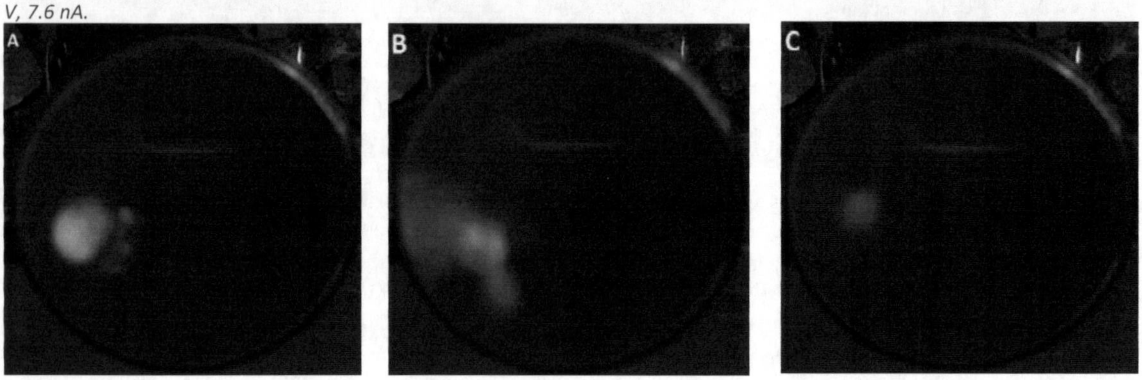

Figure 8. Emission images for second cycle at: (A) V_{SW} =650 V, 1.6 μA. (B) V_{SAT}=380 V, 1.06 μA. (C) V (Before switch-on) =590 V, 7.6 nA.

The second type of CNTs have been tested under the same conditions and the results were as follows, as the applied voltage increased slowly, the emission current initiated at applied voltage value (200 V), with emission current value (20 pA), then the emission current suddenly "switched-on" at V_{SW} =1100 V, I_{SW}=1.2 μA, then as we reduce the applied voltage, the saturation region extends down to V_{SAT} =400 V, with emission current value of (1.09 μA), further reduction, the emission current falls down to (10 pA) at applied voltage value (150 V). Figure (9) shows the I-V behavior during increasing and decreasing cycle, as well as FN plot.

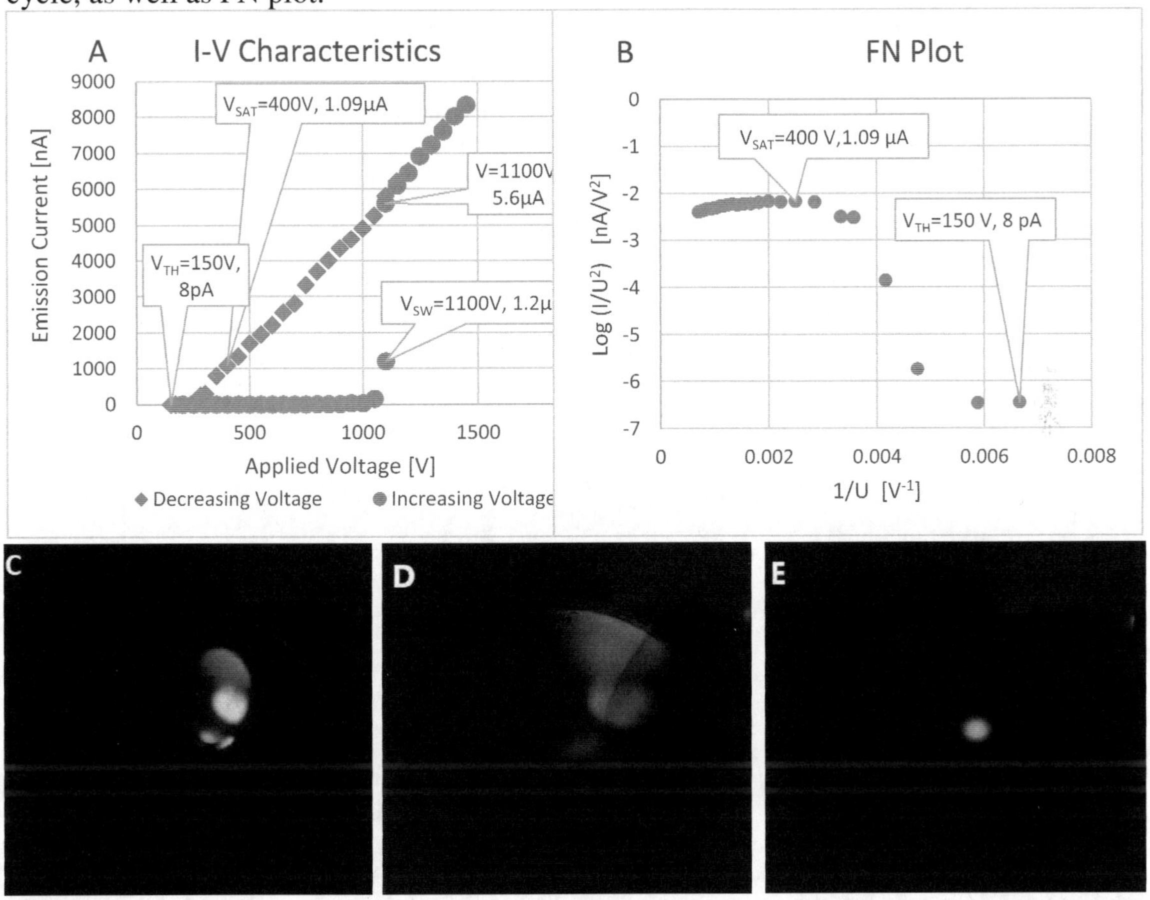

Figure 9. CNF-25 (A) I-V Plot for increasing and decreasing applied voltage. (B) FN plot for decreasing applied voltage. (C) Emission image at V_{SW}=1100 V, I_{SW}=1.2 μA. (D) Emission image at V_{SAT}=400, I_{SAT}=1.09 μA. (E) Emission image at V (Before switch-on) =850 V, 7.2 nA.

As we did for the first type of CNTs, we did the same for the second type, the test has been repeated for another cycle of increasing and decreasing the applied voltage, under the same conditions, and the results were as follows, as we increase the applied voltage the emission current start from (20 pA) at applied voltage value (350 V), then the current emission suddenly "switch-on" to (3.6 μA) at applied voltage value (750 V), we continue increasing the applied voltage until we reached (1900 V), and the emission current was (12.3 μA), we

IOP Conf. Series: Materials Science and Engineering **305** (2018) 012009 doi:10.1088/1757-899X/305/1/012009

began decreasing the applied voltage, and the saturation region extends down to (1.09 μA), at applied voltage value (350 V), further decreasing the applied voltage resulting to vanish the emission current at V_{TH}=140 V, I_{TH}=10 pA. Figure (10) shows the I-V characteristics, FN plot, and some emission images at specific applied voltage value.

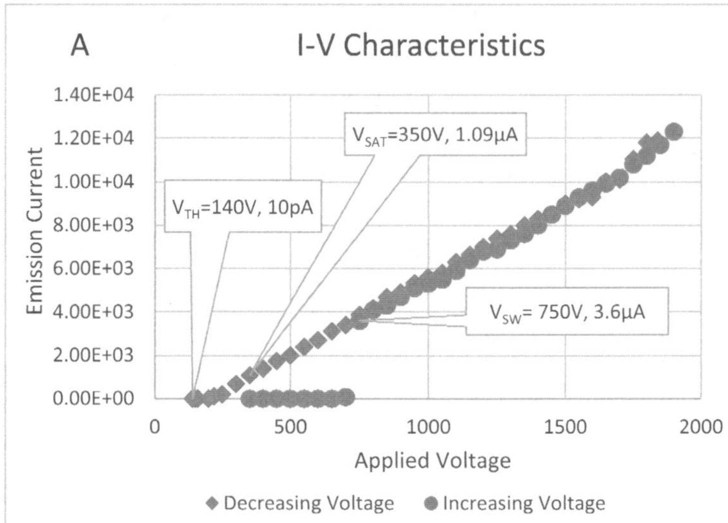

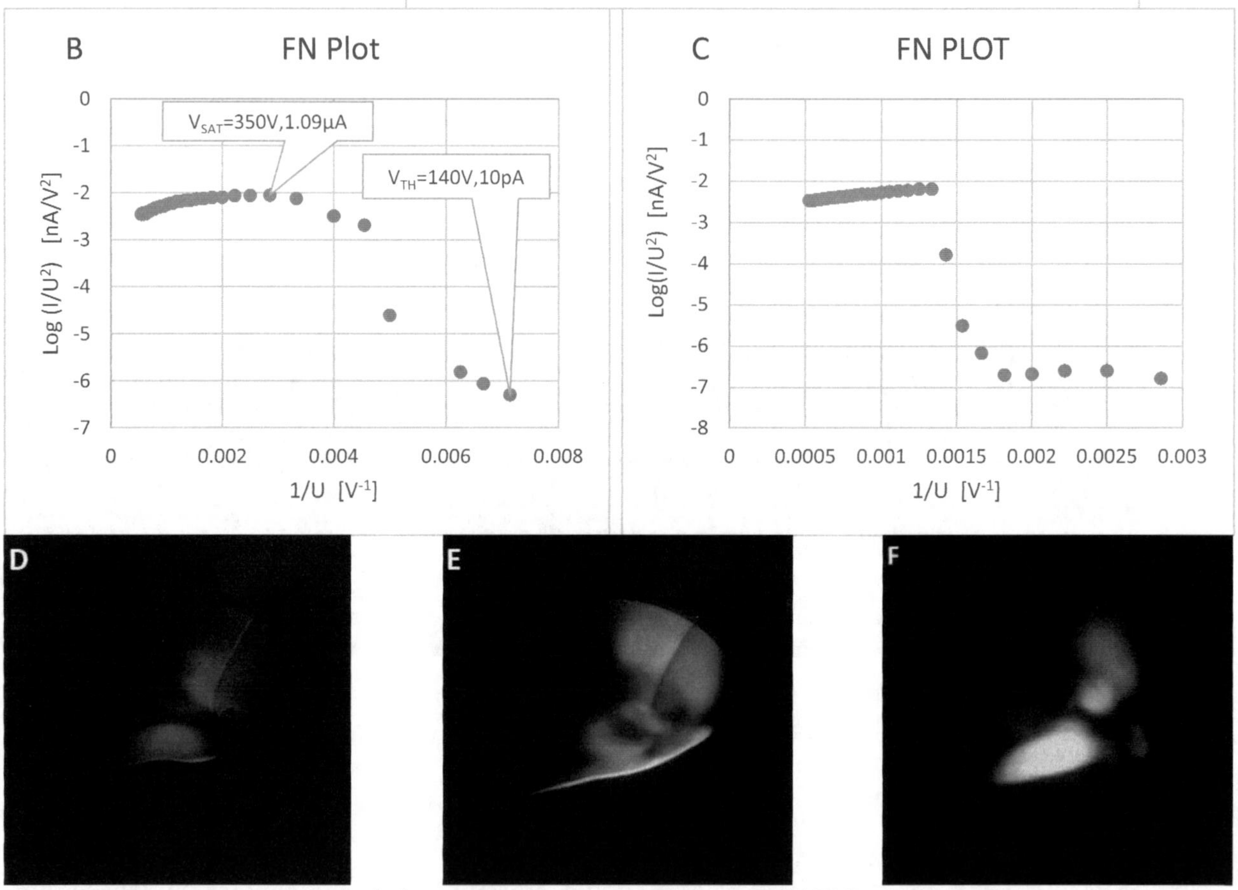

Figure 10. CNFs, second cycle: (A) I-V plot for increasing and decreasing applied voltage. (B) FN Plot for decreasing voltage. (C) FN Plot for increasing voltage. (D) Emission image at V_{SAT}=350 V,I_{SAT}= 1.09 μA. (E) Emission image at V_{SW}=750 V,I_{SW}=3.6 μA. F) Emission image at V=1700 V, I=10.1 μA.

4. Conclusions

Both types of the CNT emitters have been prepared by employing a drawing technique using glass puller, and the I-V characteristics and FN plot has been investigated, also the emission images have been studied, and it can be concluded by comparing their characteristics that SWCNTs have the lower V_{SW} than CNFs Pyrograf III PR-1 and larger emission current value at same applied voltage, the possible reason for this result is that the amount of field penetration in SWCNTs is larger than CNFs, which effects on the effective work function value. And the charge distribution on the tip apex [16].

References

[1] Pyrograf Products, Inc., an affiliate of Applied Sciences, Inc., 2011
[2] Fowler, R. H., Nordheim, L. W. (1982), Electron emission in intense electric fields. Proceedings of the Royal Society of London, series A: Mathematical, Physical and Engineering science. 119 (781): 137-181.
[3] R. V. Latham and M. S. Mousa, Hot electron emission from composite metal-insulator micro-point cathodes. J. Phys. D: Appl. Phys. 19 699.
[4] Li Y1, Sun Y, Yeow JT, Nanotube field electron emission: principles, development, and applications. Nanotechnology. 2015 Jun 19; 26(24):242001 (2015).
[5] Lijima S, Helical microtubules of graphitic carbon, Nature, 354 56-8 (1991).
[6] S. lijima, T. Ichihashi, single shell carbon Nano-tubes of 1-nm diameter, Nature 363 (6430), 603-605 (1993).
[7] Zhi Xu, X. D. Bai, E. G. Wang, and Zhong L. Wong, Field emission of individual carbon nanotube with in situ tip image and real work function, Appl. Phys. Lett. 87, 163106 (2005).
[8] T. V. Hughes and C. R. Chambers, U. S. patent 405, 480 (1889).
[9] Krijnp. DE Jong and John W. Geus, Carbon Nano fibers: Catalytic, synthesis and applications, Catal. Rev.-Sci. Eng., 42 (4), 481-510 (2000).
[10] www.pyrografproducts.com
[11] R. G. Forbes and J. H. B. Deane, Proc. R. Soc. London, Ser. A 463, 2907 (2007).
[12] R. G. Forbes, Thin-slab model for field electron emission, J. Vac. Sci. Technol. B28 (2) (2010).
[13] Carbon nanotubes. Exposure, toxicology and protective measures in the work environment. Per Gustavsson, Maria Hedmer and Jenny Rissler Report 2011:1.
[14] S. Alnawasreh, M. S. Mousa, and A. N. Al-Rabadi, Investigating the Effects of Sample Conditioning on Nano-Apex Carbon Fiber Tips for Efficient Field Electron Emission, Jo. J. Phys. 8, 95-101 (2015).

IOP Conf. Series: Materials Science and Engineering **305** (2018) 012009 doi:10.1088/1757-899X/305/1/012009

[15] M. S. Mousa, Effect of lacomit films on cold-cathode hot-electron emission, Journal de physique. C9-237 (1988).
[16] Chun-Wei Chen, Ming-Hsien Lee, S.J. Clark, field penetration induced charge redistribution effects on the field emission properties of carbon nanotubes—a first-principle study, Appl. Surf. Sci. 288 (2004) 143-150.

IOP Conf. Series: Materials Science and Engineering **305** (2018) 012010 doi:10.1088/1757-899X/305/1/012010

Investigation of composition and structure of spongy and hard bone tissue using FTIR spectroscopy, XRD and SEM

M-Ali H. Al-Akhras*, M.K. Hasan Qaseer, B. A. Albiss, M. Anwar Alebrhim, Umar S. Gezawa

Department of Physics, Bio-Medical Physics Laboratory, Jordan University of Science & Technology, P.O. Box 3030, Irbid 22110, Jordan

alakmoh@just.edu.jo

Abstract Valuable structural and chemical features can be obtained for spongy and hard bone by infrared spectroscopy and X-ray diffraction. A better understanding of chemical and structural differences between spongy and hard bone is a very important contributor to bone quality. Our data according to IR data showed that the collagen cross-links occurred to be higher in spongy bone, and crystallinity was lower in spongy bone. Deconvolution of the infrared band near 870 cm^{-1} reveals evidence for A2-type carbonate substitution on hydroxyapatite of spongy bone in addition to the A and B type carbonate substitution that are also found in hard bone. IR and XRD data confirmed the results of each other since full width at half maximum of 002-apatite pattern of XRD showed that the crystallinity was lower in spongy bone. The microstructure was examined by using scanning electron microscope and the result showed that the lattice of thin threads in spongy bone and is less dense than hard bone.

1. Introduction

A bone is a rigid organ that comprises part of the vertebrate skeleton. Bones performs many major functions, such as; support and protect the various organs of the body, production of red and white blood cells, storage of minerals, and endocrine regulation. Hard bone, is also known as cortical bone, is the dense outer surface of bone that forms a protective layer around the internal cavity, and it makes up nearly 80% of skeletal mass and is imperative to body structure and weight bearing because of its high resistance to bending and torsion. Spongy bone also is known as cancellous bone, is one of the two types of bone tissue found in the human body. Spongy bone is very porous and contains red bone marrow, where blood cells are made. It is weaker and easier to fracture than hard bone. Bone is a heterogeneous composite material of a mineral phase, hydroxyapatite ($Ca_{10}(PO_4)_6(OH)_2$) and an organic phase, (~90% collagen type I, ~5% noncollagenous proteins, ~2% lipids by weight) and water [1-3]. Chemical composition is an important contributor to bone quality, a term that encompasses the effects of architecture, composition and remodeling dynamics [4].The mineral component of human bone is carbonated hydroxyapatite because of the tendency of CO_3^{2-} to substitute for OH^- and PO_4^{3-}, with the existence of hydroxyl groups being possible theoretically but difficult to detect [5]. Carbonate increases the solubility of the apatite crystal [6], with levels in the crystal lattice commonly between 2 and 7.4 wt %, the variation of which is largely dependent on age [7, 8].There are many techniques to determine the composition and quantity of mineral present in bone [9].Vibrational spectroscopy is a powerful technique which has been used widely to characterize the bone matrix providing information about the bone quality and its composition on the microstructural level[10-13].

Age-dependent changes in FTIR spectral parameters of healthy and diseased human bone have earlier been summarized by Boskey and Mendelsohn [14]. Based on the quantitative analysis, it has been shown that the amide I and amide II components of proteins undergo frequency and intensity change as a result of changes in the protein secondary structure [15]. Moreover, X-ray diffraction is a useful non-destructive technique widely used for the characterization of structural properties. De Jong [16] was the first to apply x-ray diffraction on bone and establish the inorganic part of bone as hydroxyapatite. The aim of this study was to compare chemical and structural properties of hard and spongy bone. Both IR and XRD features were analyzed to confirm the findings.

2. Materials and methods

Our samples consist of bon pellets. Hard tissue was obtained from fresh lamb bones of the chest. After peeling off the soft tissue, it was incubated overnight to give a strong and crust form. Using mortar mould the bones were carefully broken into smaller fragments, the smaller fragments were carefully separated from the bone marrow to avoid interference and a coffee maker grinder was used to grind the bones into powder.

1g of PEG was dissolved into 5mL of distilled water and stirred for 1hr. 1g of the powdered bones was added and then stired again for 0.5hrs. This was then incubated in an oven at 60^0C for 12 hrs. The cavier hydraulic compressor was used to press the sample at 4metric tones to give a sample pellet of 2.2mm thickness and 2cm in diameter.

The PHWE Rontgengerat 35kev complete with a rate meter was used as the source of low energy photon. The crystal post was removed and the samples were placed in front of the detector. The X-ray source, sample and detector are aligned in a straight line. The beam kept normal to the surface of the sample to prevent the beam deflection. The readings from digital counter versus thickness was recorded.

The SEM images carried out at the center of nano technology available at our school.

3. Result and discussion

The aim of this study was to compare chemical and structural properties of spongy and hard bone using IR spectroscopy, XRD and SEM. In particular, crystallinity, collagen cross-linking, and carbonate substitution were analyzed. Figure1 shows IR spectra (400 - 2000 cm^{-1}) of hard and spongy bone, The IR spectral region from 800–1200 cm^{-1} is dominated by mineral bands whereas the region from 1200–1800 cm^{-1} contains mainly collagen bands. Carbonate bands are found between 860 and 880 cm^{-1} with a maximum at 877 cm^{-1}. The IR assignments are included in table 1.

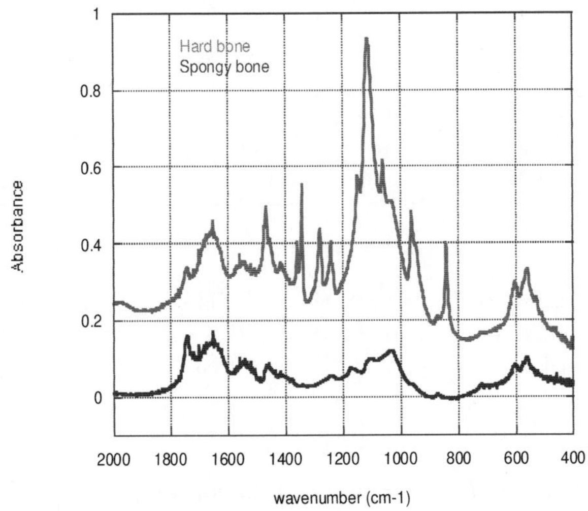

Figure 1 : IR spectra from 600 to 2000 cm^{-1} of hard (red) and spongy bone (blue).

IR band	Assignment
877	ν_2 of CO_3^{2-}
960	ν_1 of PO_4^{3-}
1003, 1013	HPO_4^{2-}
1032, 1046, 1087	ν_3 of PO_4^{3-}
1238	Collagen (amide III)
1412, 1456	Type B CO_3^{2-}
1545	Type A CO_3^{2-}, amide II
1660	Collagen(amide I)

Table 1: Band assignment in Raman spectra of hard and spongy bone

3.1 Curve fitting

Originlab was used to analyze the acquired IR spectra data to reveal the subbands which have been used in this finding. Curve-fitting process of ν_1 ν_3 phosphate, amide I envelope and the weak shoulder of carbonate CO_3^{2-} ν_1 . The Gaussian shape was considered and a linear baseline was assumed. Peak finding method based on 2nd derivative and the filtering peak depend on height peak with no smoothing.

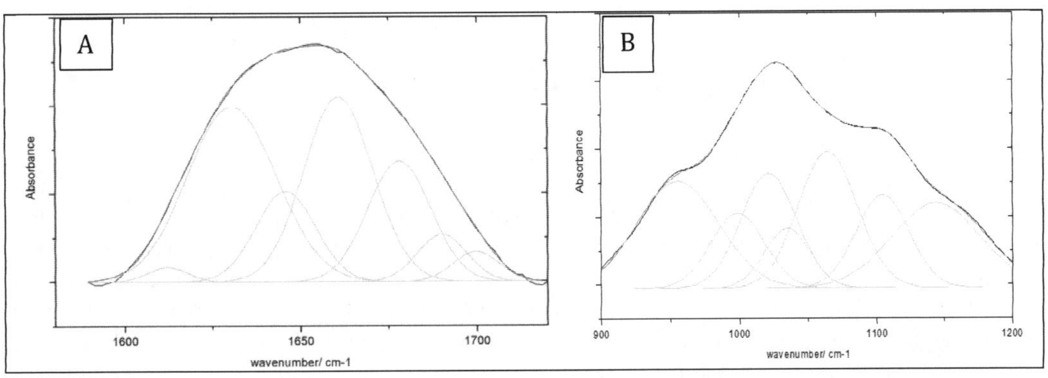

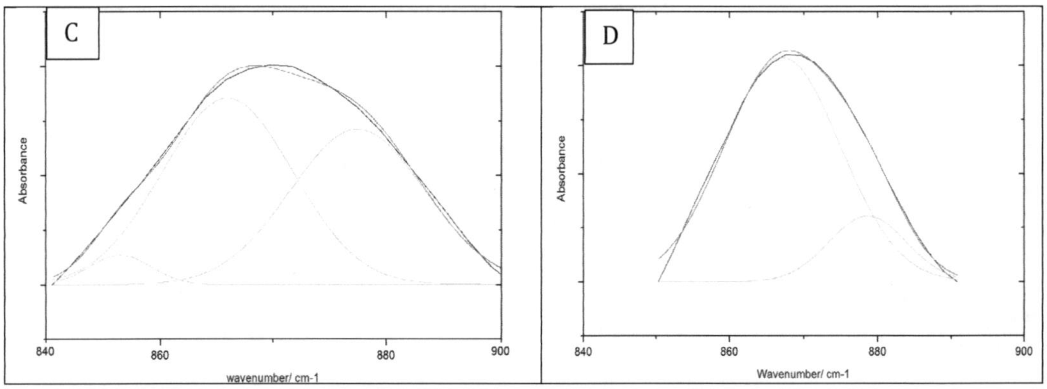

Figure2: representative of curve-fitting analysis of IR (A) amide I band, Phosphate band (B) and carbonate region near 876 cm^{-1} of spongy bone (C) and hard bone (D). The experimental spectrum (black) is overlaid with the reconstructed envelope (red) and fitted curves (green).

3.2 Collagen cross-links

The amide I band from 1590 to 1700 cm^{-1} is assigned to the peptide backbone of proteins. It is composed in IR spectra of seven major subbands that have been correlated with secondary structure motifs. They are located at 1610 cm^{-1} (aromatic rings of amino acids), 1630 cm^{-1} (β-sheets), 1645 cm^{-1} (random coils), 1660 cm^{-1} (α-helix), 1678 cm^{-1} (β-sheets), 1690 cm^{-1} (turns) and a component at 1699 cm^{-1}. Representative curve fitting results are presented in figure 2 A. Collagen cross-links provide the fibrillar collagen matrices with properties like tensile strength and viscoelasticity.Collagen crosslinking is measured as changes in the amide I envelope. The biochemical analysis of collagen model peptides showed that pyridinoline (Pyr) crosslinks resulted in a band at 1660 cm^{-1} and dehydrodihydroxylysinonorleucine (de - DHLNL) crosslinks in a band at 1690 cm^{-1} . It is known that the content of de - DHLNL crosslinks decreases with bone collagen maturity, while Pyr crosslink content increases.
The results are summarized in table 2. The collagen cross links index was found to be approximately twice higher in spongy (cancellous) bone compared to the hard (cortical) bone.

	Hard bone	Spongy bone
Collagen cross link	1.59	3.91

Table2: collagen cross-link index in bone obtained from IR spectra.

3.3 Carbonate substitutions

The carbonate ν_2 vibration of CO_3^{2-} in IR spectra at 876 cm^{-1} was subjected to curve fitting from 850 to 900 cm^{-1} and revealed differences in the carbonate substitution of spongy and cortical bone (Fig. 2 C and D). The curve-fitting was performed with two components for the cortical bone spectra, while three components were used in spongy spectra. Bands at 871 and 878 cm^{-1} are related to ν_2 of B-type and A-type CO_3^{2-}, respectively. The band at 860 cm^{-1} was assigned to ν_2 of A2-type CO_3^{2-}. The appearance of the A2 type substitution could be attributed to the fact that the hydroxyl site can accept two vacancies and the $CO_3{}^{2-}$ has consequently more degree of freedom that causes the increase of disorder. The ratios are included in table 3. The values (C/P) are highest for spongy bone. The higher carbonate content in spongy bone is responsible for a higher crystal disorder and is consistent with lower crystallinity according to Table 3.

IOP Conf. Series: Materials Science and Engineering **305** (2018) 012010 doi:10.1088/1757-899X/305/1/012010

	Hard bone	C/P ratio	Spongy bone	C/P ratio
A type %	16.4	0.076	13.2	0.086
A2 type %	-	-	11.2	0.073
B type %	83.6	0.263	75.6	0.303

Table 3: Relative amount of A, A2 and B carbonate substitution in hard and spongy bone obtained by curve fitting of IR band near 870 cm^{-1}. The carbonates to phosphate ratios using the subbands of carbonate profile at 870 cm^{-1} to phosphate at 960 cm^{-1}, were included.

3.4 Crystallinity

The phosphate IR band was fitted under the area from 900 to 1200 cm^{-1} with seven components near 960, 1020, 1030, 1055 and 1110 cm^{-1}. The band at 960 cm^{-1} is associated to the $v1$ symmetric stretching vibration of PO_4^{3-}, 1020 cm^{-1} is assigned to $HPO_4{}^{2-}$ in crystalline apatite, 1030 and 1055 cm^{-1} are related to the $v3$ vibration of $PO_4{}^{3-}$, and 1070 and 1110 cm^{-1} to $v1$ $CO_3{}^{2-}$ at the substitution site B and A, respectively [17]. A representative fitting result was included in Fig. 2 B. The ratio of the subbands 1030 cm^{-1} to 1020 cm^{-1} is widely used in the literature to study the mineral crystallinity and has been correlated with the crystal size and perfection as determined by X-ray diffraction line broadening [18]. Crystallinity is a metric related to mineral maturity, and is a measure of mineral crystallite size, mineral maturity, and the amount of substitution into the apatitic lattice. Crystallinity increases when crystals are larger and more perfect (i.e., less substitution), and is directly proportional to the inverse width of the 002 reflection (c - axis reflection) in the powder X - ray diffraction pattern of bone mineral as shown in figure 3. The result summarized on table 4, it is clearly shown that the hard bone is more crystallized than the spongy bone, the IR data and the XRD data confirmed each other.

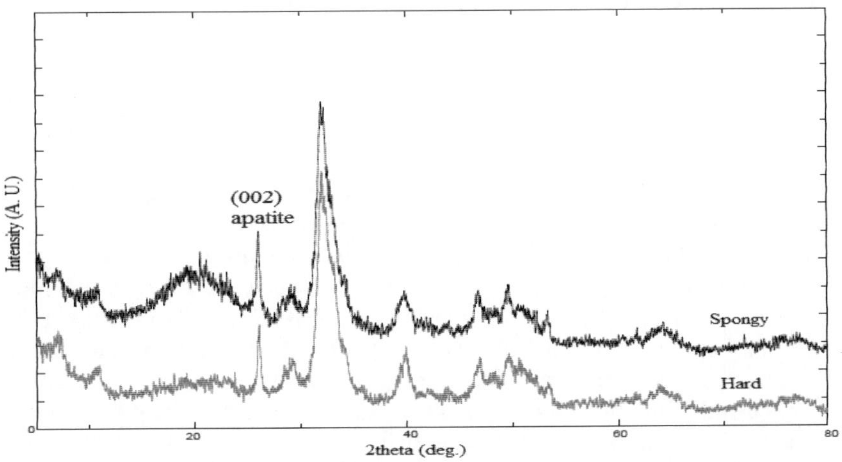

Figure 3: XRD patterns of spongy (black) and hard bone (red).

IOP Conf. Series: Materials Science and Engineering **305** (2018) 012010 doi:10.1088/1757-899X/305/1/012010

	IR ratio: I_{1030} / I_{1020}	XRD: FWHM of (002) pattern
Hard bone	0.523	1.11
Spongy bone	0.202	1.76

Table 4: crystallinity in hard and spongy bone obtained from IR and XRD. IR data were calculated from the ratio of subbands 1030 cm^{-1} and 1020 cm^{-1}. The XRD data were calculated from the width at half maximum of (002) apatite pattern.

3.5 SEM

The bone microstructures were examined by using a scanning electron microscope (SEM). The hard bone, is made up many rod-like units called osteons or Haversian systems which run longitudinally within the bone. Haversian systems have a central Haversian canal which carries blood and lymphatic vessels and nerve branches.
In figure 4, osteons or Haversian systems can be seen. The central dark circle is the Haversian canal, the canal is surrounded by rings a concentric lamellae of calcified bone matrix. Spongy bone consists of a lattice of thin threads of bone called trabeculae and is less dense than hard bone.

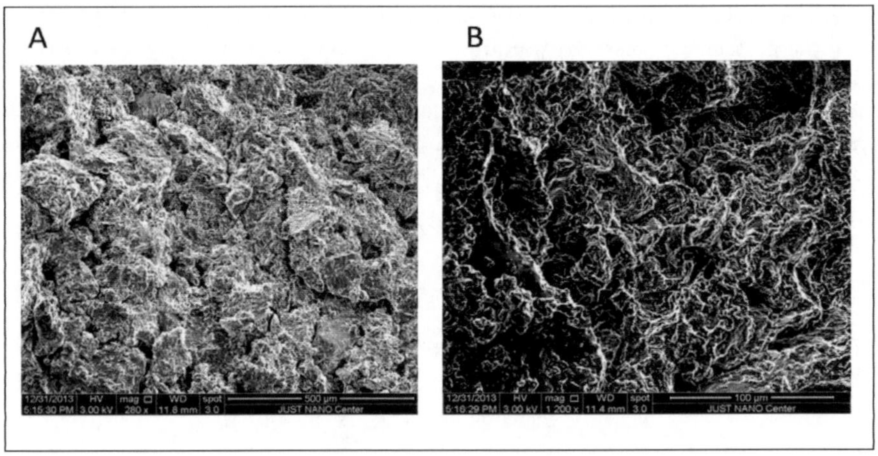

Figure 4: SEM micrographs of bone tissue: (A) spongy and (B) hard bone.

4. Conclusion

On this investigation we focused on study of structural and chemical features of hard and spongy bone. FTIR, XRD and SEM were used. IR data showed that the collagen cross-links occurred to be higher in spongy bone, and crystallinity was lower in spongy bone. Deconvolution of the infrared band near 870 cm-[1] reveals evidence for A2-type carbonate substitution on hydroxyapatite of spongy bone in addition to the A and B type carbonate substitution that are also found in hard bone. IR and XRD data confirmed the results of each other. . The microstructure was examined by using scanning electron microscope and the result showed that the lattice of thin threads in spongy bone and is less dense than hard bone. These results probably give more information beside the published scientific articles in the literature for more understanding the nature of bone structure to develop next generation of biomaterials and clinical application.

Acknowledgements: This work was supported by a grant from the Deanship of Scientific Research at Jordan University of Science and Technology (JUST) No. (173/2013). The Authors would like to thank Miss Duaa Al-Khalili for her kind help and support.

References

[1] Boskey A. Mineralization of bones and teeth. Elements Mag 2007;3: 385-392
[2] Young MF. Bone matrix proteins: their function, regulation, and relationship to osteoporosis. Osteoporos Int 2003; 14:S35-S42
[3] Boskey A. Bone composition: relationship to bone fragility and antiosteoporotic drug effects. Int Bone & Mineral Society 2013.
[4] Reiner Salzer and Heinz W. Siesler. Infrared and Raman Spectroscopic Imaging; 2nd edition 2013.
[5] Rey, C., et al., Hydroxyl groups in bone mineral. Bone, 1995. **16**(5): p. 583-6.
[6] Arends, J. and W.L. Jongebloed, Apatite single crystals. Formation, dissolution and influence of CO_3^{2-} *ions*. Journal of the Royal Netherlands Chemical Society, 1981. **100**(1): p. 3-9.
[7] Liao, S., et al., The preparation and characteristics of a carbonated hydroxyapatite /collagen composite at room temperature. Journal of Biomedical Materials Research Part B: Applied Biomaterials, 2005. 74B(2): p. 817-821.
[8] Meneghini, C., et al., Rietveld Refinement on X-Ray Diffraction Patterns of Bioapatite in Human Fetal Bones. Biophys. J. 2003. **84**(3): p. 2021-2029.
[9] Curry JD. Bones Structure and Mechanics. Princeton: Princeton University Press, 2002.
[10] Figueiredo MM. et al., Infrared Spectroscopy – Life and Biomedical Sciences. Ch 18, 2012; P. 315.
[11] Penel G. et al. composition of bone and apatitic biomaterials as reveald by intravital Raman microspectroscopy. Bone. 2005; 36(5): 893-901
[12] Kozielsky M. et al., Determination of composition and structure of spongy bone tissue human head of femur by Raman spectral mapping. J Mater Sci Med. 2011; 22(7): 1653-1661
[13] Arundhati Bhowmick. et al., Development of bone-like zirconium oxide nanoceramic modified chitosan based porous nanocomposites for biomedical application. Int Journal of Biological Macromolecules. 2017, 95: 348-356.
[14] Boskey A and R. Mendelsohn, J. Biomed. Opt., 2005, 10, 031102.
[15] Paschalis EP. Osteoporosis Int., 2009, 20, 1043–1047.
[16] De Jong W F 1926 Recueil des Travaux Chimiques des Pays-Bas (45) 445
[17] G. Penel, G. Leroy, C. Rey and E. Bres, MicroRaman spectral study of the PO4 and CO3 vibrational modes in synthetic and biological apatites, Calcified Tissue International **63** (1998), 475–481.
[17] S.J. Gadaleta, A. Gericke, A.L. Boskey and R. Mendelsohn, Two-dimensional infrared correlation spectroscopy of synthetic and biological apatites, Biospectroscopy **2** (1996), 353–364.

IOP Conf. Series: Materials Science and Engineering **305** (2018) 012011 doi:10.1088/1757-899X/305/1/012011

Magnetocaloric behavior of Mn rich $Ni_{46}Cu_2Mn_{43}In_{11}$ alloy

Mayukh K. Ray[1, 2], I M Obaidat[1]*, Sangam Banerjee[2]*

[1]*Department of Physics, United Arab Emirates University, Al-Ain 15551, United Arab Emirates*
[2]*Surface Physics and Material Science Division, Saha Institute of Nuclear Physics, 1/AF Bidhannagar, Kolkata-700064, India*

iobaidat@uaeu.ac.ae and sangam.banerjee@sha.ac.in

Abstract: In this work, we studied the magnetic entropy change (ΔS_M) across the martensite transformation (MT) in Mn-rich $Ni_{46}Cu_2Mn_{43}In_{11}$ alloy. This compound undergoes a MT and a magnetic phase transition around the temperatures (T_M=) 272 K and (T_{CA}=) 325 K, respectively. A large field induced shift (=0.28 K/kOe) of the MT temperatures is observed. An application of magnetic field (H =) of 50 kOe causes a large ΔS_M of 20 J/kg-K and -4.4 J/kg-K around T_M and T_{CA}, respectively. We also found that the change in magnetic field induced isothermal $\Delta S_M(H)_T$ is mainly depends on the induced austenite phase fraction by the applied magnetic field at that temperature. Possible reasons for the observed behaviours are comprehensively discussed.

Keywords: Magnetization, Huesler, magnetic entropy, phase transition.

1. Introduction

In recent times, non-stoichiometric Ni-Mn-Z (Z=Sn,In) intermetallic compounds has attracted much attention due to their potential applications in the field of magnetic refrigeration, spintronics and magnetic read head etc [1-5]. It is believed that such novelties are primarily due to the presence of a strong coupling between the structural, the magnetic and the electronic degrees of freedom [2]. It is also well-known that these types of systems are characterized by first-order martensitic transition (MT) at T_M as well as a second-order paramagnetic (PM) to ferromagnetic (FM) transition at the Curie point (T_{CA}). The T_{CA} in these systems must be higher than their T_M ($T_{CA} > T_M$) to realize the magneto-functionality around the T_M, i.e., the material should be in a magnetically well ordered state when MT takes place during cooling [6-7]. The MT in these materials also alters the magnetic state in which material may lose its ferromagnetism or attain a state having antiferromagnetic (AFM)-like correlation below the MT. Since the MT occurs between the magnetically order (austenite) to a magnetically disorder (martensite) state consequently a large change in the magnetization (ΔM) and hence a large magnetic entropy change (ΔS_M) is expected to achieve across the MT [8]. It is noteworthy that the high ΔM is one of the necessary conditions for any material to be a good magnetocaloric material (MCM). The presence of large ΔS_M produces the magnetocaloric effect (MCE) in these materials has great potential applications in the cooling technology without any harmful impact on the environment unlike the vapour compression based refrigerator. It is also well known that the MCE is a thermal response of a magnetic material against the variation of an external magnetic field and usually expressed in terms of ΔS_M in an isothermal process, or adiabatic temperature change (ΔT_{ad}). As we already mentioned earlier that a large ΔS_M can be achieved across the MT therefore the researchers paid significant efforts in designing Ni-Mn based MCM with a strong magneto-structural coupling at MT. The studies also focused on the improvement of another vital MCE parameter, called the refrigeration capacity (RC) or the relative cooling power (RCP), which quantifies the amount of heat transferred from cold to hot sink through the ideal refrigeration cycle [9]. In order to achieve a large RCP, it is important to widen the MT temperature range as this determines the working temperature range of a MCM. Upto now, the expensive rare-earth metal Gd, which experiences a second-order magnetic transition near room temperature, is regarded as a benchmark refrigerant material in magnetic refrigerator prototype [10, 11]. Since these rare earth systems are costly, researchers are considering Ni-Mn based intermetallic compounds as an alternative to its rare earth counterpart. But the problems in Ni-Mn based materials are, the T_M in these systems is much lower than the room temperature (RT) for most of the Ni-Mn based martensite compounds and also the narrow MT width makes them low cooling efficient materials. In search for a better MCM with improved magneto-functionalities for more useful applications within the family of Ni-Mn we have synthesized the $Ni_{44}Cu_2Mn_{43}In_{11}$ compound and studied its magnetic and magnetocaloric properties.

2. Experimental details

The ascast sample was prepared by using tri-arc melting furnace from pure metals (99.995%, Alfa Aesar) in argon atmosphere. The sample was then annealed at 1173 K for 24 hrs in vacuum followed by quenching in normal water for homogenization. The crystal structure was determined using room temperature powder X-Ray diffraction (XRD) (Bruker D8 advance diffractometer). The nominal compositions of the annealed alloy was confirmed by energy dispersive X-Ray analysis (EDX) attached with a field emission scanning electron microscope (FESEM) within the error of ±0.1 at.%. Calorimetric measurements were done using Differential Sacnning Calorimetry (DSC). The magnetic measurements were carried out using Physical Property Measurement System (PPMS, Quantum Design) in the temperature range T=5-380 K and the field range H=0-90 kOe.

3. Result and discussions

The temperature dependent dc susceptibility χ(T) measurements were conducted using the zero field cooled warming (ZFCW) and the field cooled cooling (FCC) protocols and the results are shown in Fig.1. In the measurement of ZFCW χ(T), we first cooled the sample under zero magnetic field down to T= 5 K and then applying a magnetic field (H =) 100 Oe we recorded the χ (T) during warming the sample upto 395 K whereas the FCC χ(T) was measured in subsequent cooling of the sample. The χ(T) data show multiple features depending on the change in magnetic phases due to the thermal variations. At T= 330 K, the sample shows a sharp upturn with the decreasing temperature, which corresponds to the paramagnetic (PM) to ferromagnetic (FM) transition within the austenite phase, called Curie temperature of austenite phase (T_{CA}). On further cooling, χ (T) starts to decrease and a sharp fall is observed around T_{ms}. The fall in χ (T) is possibly due to the loss of ferromagnetism and the sample presumably becomes PM or AFM below the MT [12, 13]. On the other hand χ(T) starts to increase around T_{as} during warming the material and a thermal hysteresis between the χ_{ZFCW} (T) and χ_{FCC} (T) curve appears in the temperature range of 245-310 K. Such hysteresis signifies the first order nature of MT. All the characteristic transition temperatures, i.e., the austenite start and finish temperatures (T_{as} and T_{af}) and the martensite start and finish temperatures (T_{ms} and T_{mf}) are labelled in Fig.1. Now below the MT, the compound continues to be in a low magnetic phase and eventually shows a spin glass type magnetic ground state below T=65 K. To validate the transformations characteristics we have carried out differential scanning calorimetry (DSC) in the absence of any magnetic field. The thermodynamic heat profiles pertaining to endothermic and exothermic peak are obtained during the heating and cooling cycles, respectively (inset of Fig.1). The exothermic and endothermic transformation peaks represent forward (austenite phase to martensite phase trasition) and reverse (martensite phase to austenite phase transition) MT, respectively. The values of the characteristic MT temperatures obtained from the magnetic measurement are in good agreement with our calorimetric measurements. Small discrepancy between the results of magnetic and calorimentric measurement can be ascribed to the influence of some factors such as the different thermal contacts of the samples with the thermocouples in the DSC and PPMS instruments and the two tangent procedures used for the determination of the characteristic temperatures etc. After the baseline correction of our calorimetric data we have quantified the transition entropy change using the following equation:

$$S(T) = \int_{T_1}^{T_2} \frac{1}{T}\left(\frac{dQ}{dT}\right) dT \dots\dots\dots\dots (1)$$

where T_1 and T_2 denote the temperatures well below and above the phase transition, respectively and $\left(\frac{dQ}{dT}\right)$ is the heat capacity. Using this equation we have numerically calculated the transition entropy change (ΔS_{tr}) with the value of 17.7 J/kg-K. It is also important to mention that in the measurement of $\chi(T)$, $\chi(T)$ starts to drop below the T_{CA} and the thermal hysteresis even exists in temperature range T_{ms}/T_{af} to T_{CA} but we did not observed any trace of this hysteric region in our calorimetric and resistivity measurements.

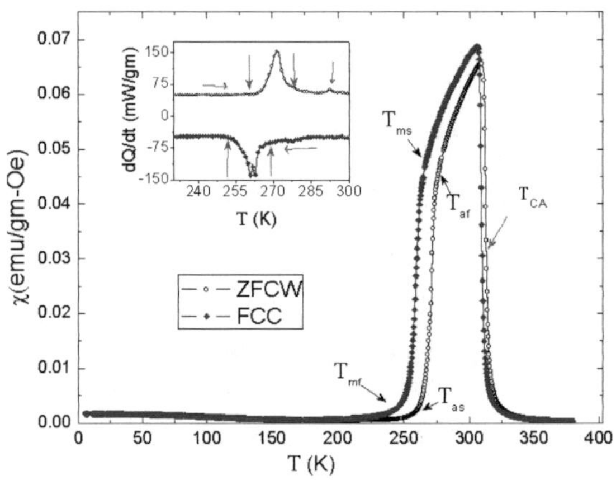

Figure 1. The XRD Temperature variation of dc susceptibility at H=100 Oe and the inset shows the Differential Scanning Calorimetry (DSC) in heating and cooling run.

We have further measured the thermo-magnetization M(T) cuves using field cooled cooling (FCC) and field cooled warming (FCW) protocols in the temperature range T=150-395 K under different applied fields (see the Fig.2). It can be seen from the Fig.2 that the magnetization difference ($\Delta M = M|_{Tmf} - M|_{Tms}$) between the austenite and martensite phases keeps on increasing with the increase of the magnetic field. Also the M(T) below T_{mf} remains nearly temperature independent for all the applied fields which signifies the low magnetic behaviour of this phase. It is also evident from the Fig.2 that the characteristic martensite transformation temperatures shift towards the lower temperatures side with the application of magnetic field. The change in T_{ms} with magnetic field H, i.e., $\frac{dTms}{dH}$ is found to be 0.28 K/kOe. This observed value of $\frac{dTms}{dH}$ has good agreement with the one we obtained using the Clausius-Clapeyaron (C.C) equation 2. If we consider the FCC M(T) curve for H=70 kOe then the ΔM comes out to be 55 emu/g and from our DSC measurement we found the ΔS_{tr} is 17.7 J/kg-K.

$$\frac{dTms}{dH} = -\frac{\Delta M}{\Delta Str} \dots\dots\dots\dots\dots\dots \quad (2)$$

Therefore, if we use these values in C.C equation, $\frac{dTms}{dH}$ comes out to be 0.31 K/kOe which is very close to that obtained from direct determination of T_{ms} shift. The shift in T_{ms} is easy to understand by considering the change in the free energy during the MT. We know that the reverse martensitic transformation (RMT) is motivated by the chemical free energy difference between the martensite and austenite phases and the elastic energy stored in the martensite phase. Therefore, one can express the Gibbs free energy difference between the martensite and austenite phases during the RMT (martensite to austenite transformation) as

$$\Delta G^{M\to A} = \Delta G_{ch}^{M\to A} + \Delta G_{el}^{M\to A} + \Delta E_{irr}^{M\to A} \dots\dots\dots\dots\dots\dots\dots\dots (3)$$

where $\Delta G^{M\to A}$ represent the total Gibbs free energy dieerence between the two phases and it should be < 0 to trigger the RMT [14, 15]. While $\Delta G_{ch}^{M\to A}$ represent the chemical free energy difference, $\Delta G_{el}^{M\to A}$ is the elastic energy stored in the martensite phase and $\Delta E_{irr}^{M\to A}$ is the dissipation energy generated by defects in the system. Now when we apply a magnetic field, an additional and magnetic contribution to the free energy comes into play a crucial role in the total Gibbs free energy equation. Then the modified Gibbs free energy difference can be rewritten as follows [16]

$$\Delta G^{M\to A} = \Delta G_{ch}^{M\to A} + \Delta G_{el}^{M\to A} + \Delta E_{irr}^{M\to A} + \Delta G_{mag}^{M\to A} \dots\dots\dots (4)$$

where $\Delta G_{mag}^{M\to A}$ is the Zeeman energy difference between the martensite and austenite phases and can written as

$$\Delta G_{mag}^{M\to A} = -HM_S^A - (-HM_S^M) = H(M_S^M - M_S^A) \dots\dots\dots\dots\dots\dots\dots\dots\dots (5)$$

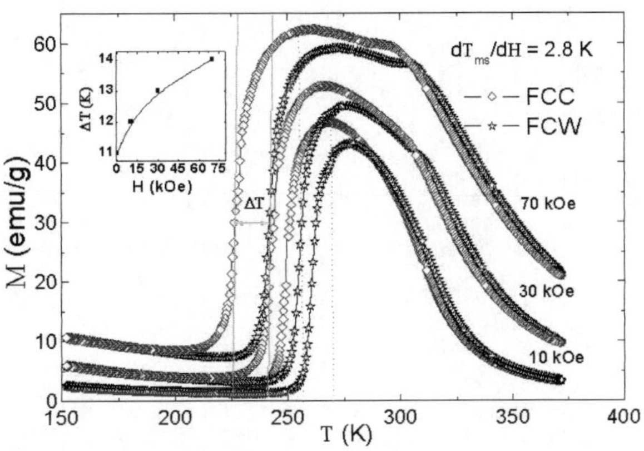

Figure 2. Temperature variation of magnetization under different applied fields and inset shows the variation of ΔT with applied field.

Here M_S^A and M_S^M are the saturation magnetizations of the austenite and martensite phases, respectively. In our studied compound the martensite phase is weak-magnetic, and the austenite is a strong FM (M_S^A>M_S^M), which results in $\Delta G_{mag}^{M\to A}$< 0. Therefore, the magnetic field favours the RMT, leading to a decrease in transformation temperatures. Using the same ideas one can also explain the increase of the thermal hysteresis width (ΔT) across the MT with measuring H (see the inset of Fig.2). For a given M(T) curve (say H=70 kOe), the ΔM is little higher for FMT (austenite to martensite) than that for RMT. Therefore, the shifts of T_{ms} and T_{mf} to the lower temperature side are larger than the shifts of T_{as} and T_{af} which eventually leads to the increase in ΔT with magnetic fields. Similar hysteresis loops were obtained at 50 K and 100 K.

To find out the effect of magnetic fields on the ΔS_M we have conducted the isothermal M-H measurements in the temperature range of T=220-275 K. up to a magnetic field of H= 90 kOe.

From the M-H curves we have calculated the magnetic field induced ΔS_M using the following equation:

$$\Delta S_M(T_{av})_{\Delta H} = \frac{\Delta H}{2\Delta T}(\Delta M_1 + 2\sum_{k=1}^{n-1}\Delta M_k + \Delta M_n) \ldots\ldots\ldots\ldots\ldots (6)$$

where T_{av} and ΔT represents the average temperature and temperature difference between two ZFC M-H curves, respectively. ΔH is the changing step in the applied field and ΔM_k is the difference in magnetization at the field k×H between two isothermal curves, where k is the step number of applied field, k=1 and k = n are the step correspond to H=0 and the last step of maximum field, respectively. ΔS_M versus H at several temperatures is shown in Fig.3. From the Fig.3 we can see that when the system is far below of its MT region i.e. below 240 K, only a small part of martensite phase is transformed into the austenite by the available magnetic field and thus the induced ΔS_M remains low and unsaturated. In this region even a magnetic field as large as 90 kOe is not enough to produce a complete field induced martensite to austenite transformation. Now as we move towards the MT region, more fractions of the martensite phase is now start to convert into the austenite phase by the available field and $\Delta S_M(H)_T$ starts to increase up to T=250 K. Whereas for the temperatures just below of martensite temperature T_M ($=\frac{1}{2}(T_{ms} + T_{af})$), a field of a lower values than H=90 kOe is sufficient to produce saturated ΔS_M. With the increase of temperatures, the austenite phase fraction increases and ΔS_M starts to the decrease again due to the decrease of the induced austenite phase fraction. So, one can conclude from Fig.3 that the induced ΔS_M is proportional to the induced austenite phase fraction.

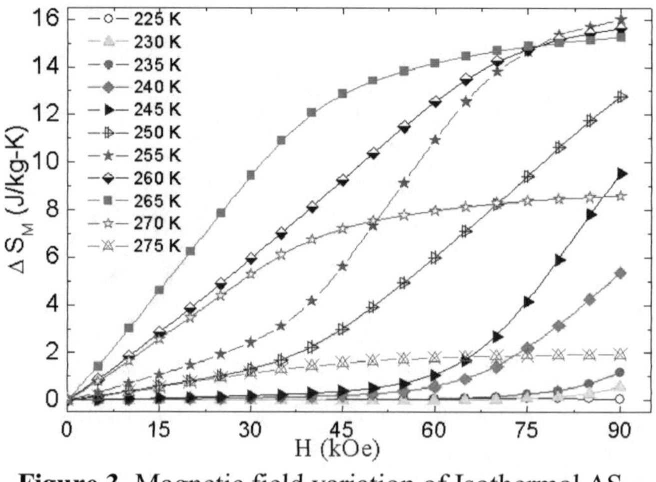

Figure 3. Magnetic field variation of Isothermal ΔS_M.

Apart from the magnetic entropy change ΔS_M, the relative cooling power (RCP) of a magnetic material is another important parameter as it measure the efficiency of the material to be used as a refrigerant in magnetic refrigerator. The RCP is defined as: RCP=$\Delta S_M(\max)\times\delta T_{FWHM}$, where δT_{FWHM} is the full-width-at-half-maximum of the $\Delta S_M(T)$ curve. There are several approaches to determine RCP, though all of them target the range of working temperature [17]. Therefore there is further need to correlate MT with ΔS_M as well as with the working temperature range. Amorphous materials with second order transition favour large RCP and low magnetic entropy change opposite to the crystalline materials with first order transition [18]. The key concern is to develop magnetic refrigerant with high magnetic entropy change without compromising the working temperature range. Since this system shows two different types of transition i.e., MT (at

T_M) and magnetic transition (at T_{CA}), it can show both the inverse and conventional MCE. Both, the conventional magnetocaloric effect (negative ΔS_M) and the inverse magnetocaloric effect (positive ΔS_M) can be used in the refrigeration cycling.

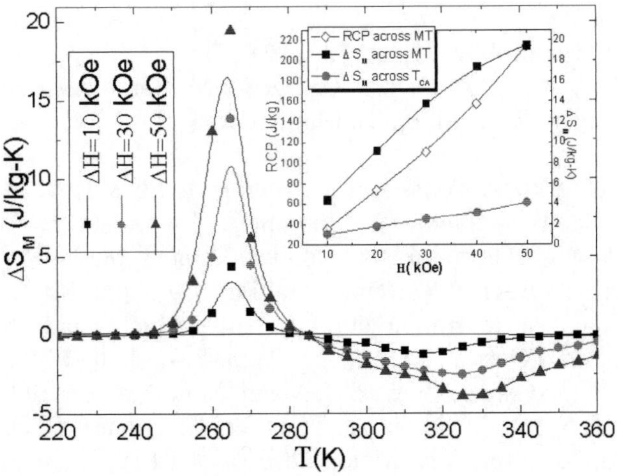

Figure 4. Temperature variation of ΔS_M at different applied field and inset shows the field variation of ΔS_M and the RCP.

To quantify the $\Delta S_M(T)_H$ and the RCP of our material we obtained a series of magnetization isotherms M-H in the temperature range of 220-360 K with an interval of 3K and calculated the ΔS_M using the Maxwell relation. The temperature dependence of ΔS_M for this compound up to H=5T are presented in Fig.4. With a magnetic field change of 50 kOe, a positive ΔS_M with a maximum value of 20 J/kg K and a negative ΔS_M with a maximum value of -4.4 J/kg-K are observed around the T_M and T_{CA}, respectively. The positive and negative values of $\Delta S_M(T)_H$ indicate the existence of inverse and direct MCE in this system. Plots of the magnetic field dependence of $\Delta S_M(T)_H$ across the T_M and T_{CA} and also the RCP across the MT are shown in the inset of Fig.4. A strong field dependence of both the ΔS_M and RCP across the MT is observed. We have obtained a RCP value of 215 J/kg across the MT for the field change of 50 kOe which is comparatively very high considering Ni-Mn based systems. Moreover, in contrary to what was expected in this kind of material, the temperature of the negative peak shifts towards higher temperatures than T_{CA}. The ΔS_M has the characteristic shape of a first and second-order phase transitions both expanded over a large temperature range.

Conclusions:

In summary, the magneto-structural martensite transitions and magnetic entropy change of Mn-rich $Ni_{44}Cu_2Mn_{43}In_{11}$ were studied. A strong field dependence of ΔS_M and RCP across the MT is observed. From the field induced behaviour of ΔS_M it is evident that the magnetic field induced austenites phase contributing more in the total ΔS_M. The observed large RCP in this system can be useful for practical applications as magnetic refrigerant material.

Acknowledgement

This work was financially supported by the UAEU Program for Advanced Research (UPAR) under the grant no. 31S169.

IOP Conf. Series: Materials Science and Engineering **305** (2018) 012011 doi:10.1088/1757-899X/305/1/012011

References:

[1] T. Krenke, M. Acet, E.F.Wassermann, X. Moya, L. Manosa, A. Planes, Phys. Rev. B **72**, 01441 ((2005).
[2] T. Krenke, E. Duman, M. Acet, E.F. Wassermann, X. Moya, L. Maosa, A. Planes, Nat. Mater. **4**, 450 (2005).
[3] D.C. Dunand, P. Mllner, Adv. Mater. **23**, 216 (2011).
[4] V. Chernenko, E. V. Lvov, A. Cesari, Kosogor, J.M. Barandiarn, Metals **3**, 237 (2013).
[5] M. K. Ray, K. Bagani, R. K. Singh, B. Majumdar, S. Banerjee, J. Appl. Phys. 114, 123904 (2013).
[6] J. Enkovaara, O. Heczko, A. Ayuela, R.M. Nieminen, Phys. Rev. B **67**, 212405 (2003).
[7] P.J. Brown, A.P. Gandy, K. Ishida, R. Kainuma, T. Kanomata, K.U. Neumann, K. Oikawa, B. Ouladdia, K.R.A. Ziebeck, J. Phys.: Condens. Matter **18**, 2249 (2006).
[8] M. K. Ray B. Maji, K. Bagani, S. Banerjee, J. Phys. D: Appl. Phys., **47**, 385001 (2014).
[9] C.F. Sanchez-Valdes, P.J. Ibarra-Gaytan, J.L. Sanchez Llamazares, M. Avalos- Borja, Pablo Alvarez-Alonso, Pedro Gorria, J.A. Blanco, Appl. Phys. Lett. **104**, 212401 (2014).
[10] V.K. Pecharsky, K.A. Gschneidner Jr., J. Magn. Magn. Mater. **200**, 44 (1999).
[11] S. Singh, L. Caron, S.W. DSouza, T. Fichtner, G. Porcari, S. Fabbrici, C. Shekhar, S. Chadov, M. Solzi, C. Felser, Adv. Mater. **28**, 3321 (2016).
[12] R.Y. Umetsu, R. Kainuma, Y. Amako, Y. Taniguchi, T. Kanomata, K. Fukushima, A. Fujita, K. Oikawa, K. Ishida, Appl. Phys. Lett. **93**, 042509 (2008).
[13] V.V. Khovaylo, T. Kanomata, T. Tanaka, M. Nakashima, Y. Amako, R. Kainuma, R. Y. Umetsu, H. Morito, H. Miki, Phys. Rev. B **80**, 144409 (2009).
[14] N. Hatcher et al., Phys. Rev. B **80**, 144203 (2009).
[15] X. Q. Wang et al., Phys. Rev. B **78**, 092103 (2008).
[16] R.C. O'Handley Modern Magnetic Materials: Principles and Applications (New York:Wiley).
[17] M.E. Wood, W.H. Potter, Cryogenics **25**, 667 (1985).
[18] P. Gorria, J.L. Sachez Llamazares, P. Alavarez, M.J. Perez, J. Sanchez Macros, J. A. Balnco, J. Phys. D: Appl. Phys. **49**, 192003 (2008).

Size-dependent magnetic anisotropy of PEG coated Fe_3O_4 nanoparticles; comparing two magnetization methods

C Nayek[1], K Manna[2], A A Imam[1], A Y Alqasrawi[3] and I M Obaidat[1]*

[1]*Department of Physics, United Arab Emirates University, Al-Ain 15551, United Arab Emirates*
[2] *Max-Planck-Institute for Chemical Physics of Solids, Nöthnitzer Straße-40, 01187 Dresden, Germany*
[3] *Department of Mechanical Engineering, United Arab Emirates University, Al-Ain 15551, United Arab Emirates*

iobaidat@uaeu.ac.ae

Abstract: Understanding the size dependent magnetic anisotropy of iron oxide nanoparticles is essential for the successful application of these nanoparticles in several technological and medical fields. PEG-coated iron oxide (Fe_3O_4) nanoparticles with core diameters of 12 nm, 15 nm, and 16 nm were synthesized by the usual co-precipitation method. The morphology and structure of the nanoparticles were investigated using transmission electron microscopy (TEM), high resolution transmission electron microscopy (HRTEM), selected area electron diffraction (SAED), and X-ray diffraction (XRD). Magnetic measurements were conducted using a SQUID. The effective magnetic anisotropy was calculated using two methods from the magnetization measurements. In the first method the zero-field-cooled magnetization versus temperature measurements were used at several applied magnetic fields. In the second method we used the temperature-dependent coercivity curves obtained from the zero-field-cooled magnetization versus magnetic field hysteresis loops. The role of the applied magnetic field on the effective magnetic anisotropy, calculated form the zero-field-cooled magnetization versus temperature measurements, was revealed. The size dependence of the effective magnetic anisotropy constant K_{eff} obtained by the two methods are compared and discussed.

Keywords: Magnetization, Nanoparticles, Magnetic anisotropy, Coercivity.

1. Introduction

In magnetic nanoparticles (MNPs), magnetic anisotropy employs a direct impact on the magnetization, remanence, reversal and relaxation. Magnetic anisotropy is therefore a crucial parameter in the preparation of magnetic nanoparticles designed for high-density data storage applications and for medical applications [1–8]. For this reason, a controlled modulation of the magnetic anisotropy for a given size of crystals is one of the most efficient ways which improves and tunes their magnetic properties. The anisotropy of the magnetic material is utilized to control the critical applied magnetic field required to switch the magnetization of the element between its two magnetic stable states, thus creating a binary barcode. The suspension of MNPs in biomedical application may be delivered to the site of application by intravenously or by direct localized injection method. These methods require agglomeration-free MNPs which cannot inhibit their own distribution and can be achieved either by reducing the size of the NP or by modifying their surface chemistry. The modification of the surface of nanoparticles is done by using various biocompatible and biodegradable high molecular weight polymers [9,10]. The coating also can provide the colloidal stability of the particles against aggregation and gravitational settling by reducing the inter-particle dipole-dipole interactions [11-16]. The most commonly available coatings are; derivatives of dextran [17,18], polyvinyl alcohol (PVA), Polyethylene glycol (PEG) [19], and some nonpolymeric coatings *etc.* [20].

As mentioned, the MNPs have attracted a lot of interest in the field of biomedical applications, [21–24] such in magnetically induced hyperthermia (MIH), [25,26] magnetic resonance imaging, [27] targeting drug delivery, [28] and radioisotope-labeled targets [29]. The MIH is one of the most promising techniques for cancer treatment [30] where localized MNPs are heated by an applied alternating magnetic field. The heated MNPs kill only tumor cells and thus reduce injuries to the surrounding healthy tissues. In addition, enhancing the particles' magnetic anisotropy results an increase in the relaxation time and allows for the use of lower frequencies of the alternating magnetic field. Because of their low toxicity, high saturation magnetic moment at room temperature, and highest ordering temperatures among spinel ferrites [31], both the magnetite (Fe_3O_4) and maghemite (γ-Fe_2O_3) MNPs are considered very suitable MNPs for the use in MIH. Both types of MNPs were approved by the food and drug administration (FDA) as the biocompatible magnetic materials for MIH in humans [32]. Increasing heating efficiency of these iron oxide nanoparticles will allow for the reduction of the amount of nanoparticles delivered to the targeted organ and thus will minimize negative side effects. Improving the heating capability of these MNPs can be achieved by controlling the particle size and some special magnetic properties. Possible ways to increase this heating rates are by increasing the magnetic anisotropy of the MNPs. Pure Fe_3O_4 MNPs have some physical limits of magnetically induced heat generation [33]. In the last few years, the magnetic anisotropies of MNPs were tuned by mutual coupling of magnetic hard and/or soft components for enhancing the heating power to therapeutic efficacy [26]. Hence, there is a direct relation between magnetic anisotropy and the magnetic induction performance.

Using magnetic nanoparticles in magnetic recording depends on their magnetic thermal stability which directly correlated with the magnetic energy barrier of the nanoparticles [34-35]. The magnetic energy barrier depends on both the size of the nanoparticle and on its magnetic anisotropy. Hence, the knowledge of the size -dependent magnetic anisotropy in nanoparticles is essential in their applications in data storage devices.

In this paper, we have studied the role of particle size on the effective magnetic anisotropy, K_{eff} of PEG coated Fe_3O_4 nanoparticles. This was done by calculating the K_{eff} using two methods that are based on magnetic measurements.

2. Experimental details

The PEG functional iron oxide (Fe_3O_4) nanoparticles were synthesized by the usual co-precipitation method. The $FeCl_3.6H_2O$ (0.1M) and $FeCl_2.4H_2O$ (0.05M) were used as precursor materials. They were dissolved in deionized water separately and mixed at 300 K. The 25 % ammonium hydroxide solution was promptly added to the mixture and a 25% NH_4OH solution continuously mixed drop by drop until PH became 10. The final solution was then stirred for 30 minutes. The reaction time was varied from 1 h to 5 h to get nanoparticles with different sizes. Finally, the Fe_3O_4 nanoparticles were collected by magnetic field separation and washed properly. The collected precipitation was dried at 40 °C in vacuum. The PEG solutions with the proportionate concentration were prepared by dissolving PEG powder in distilled water. Particular amount of water and Fe_3O_4 nanoparticles were vigorously stirred for 15 minutes. Afterwards, the PEG-solution was mixed to the suspension and stirred for about 30 min. The mixture was then centrifuged at 6000 rpm to obtain the ferrofluid. Finally, the precipitated ferrofluid was collected and dried overnight in vacuum condition. The sample which were made with a reaction time of 1 h, 2 h, and 5 h were assigned as S1, S2, and S3, respectively. The morphology of the nanoparticles were confirmed by several microscopic tools such as high resolution transmission electron microscopy (HRTEM), and selected area electron diffraction (SAED). Structure and phase analysis was conducted by X-ray diffraction (XRD) using SHI-MADZU Lab X-XRD-6100 with CuKα (λ =1.5418 A°) diffractometer. The dc magnetic measurements were done using SQUID from quantum design. Magnetization (*M*) versus applied field (*H*) up to ±3 T was measured for all the samples at several temperatures under zero field-cooling (ZFC) condition. ZFC and field-cooled (FC) magnetization versus temperature (*M-T*) measurements were

IOP Conf. Series: Materials Science and Engineering **305** (2018) 012012 doi:10.1088/1757-899X/305/1/012012

conducted at several applied magnetic fields (20, 50, 100, 200, and 500 Oe) in the temperature range of 2 – 300 K.

3. Result and discussions

Figure 1 displays the X-ray diffraction plots of the PEG-coated magnetite (Fe_3O_4) nanoparticles. The corresponding diffraction peaks were attributed to Fe_3O_4 confirming the phase purity of the samples.

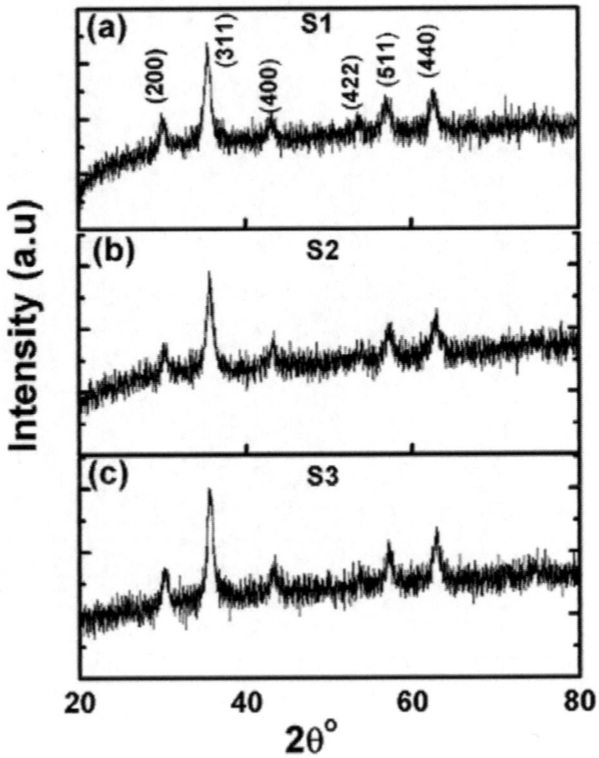

Figure 1. The XRD patterns for sample (a) S1, (b) S2, and (c) S3.

The XRD data was plotted for 2θ ranging from 20° to 70°. The characteristic diffraction peaks at $2\theta = 30.1°$, 35.4°, 43.3°, 53.6°, 57.2°, and 62.7°, were attributed to (220), (311), (400), (422), (511), and (440) Bragg reflection of Fe_3O_4. The preponderance of amorphous phase of PEG clarified that the crystalline nature of Fe_3O_4 was suppressed due to the presence of PEG. Note that the diffraction peaks of Fe_3O_4 nanoparticles were matched with cubic spinel structure. The average particle sizes were calculated from the XRD by using Scherrer formula; $D_P = \frac{0.94\lambda}{\beta \cos\theta}$, where D_P is the average crystallite size, λ is the x-ray wavelength, β is the full width of half maximum (FWHM) of the XRD line, and θ is the Brag's angle. The average particle sizes calculated were 12 nm, 15 nm, and 16 nm, respectively.

The morphology of the nanoparticles were confirmed by several microscopic tools such as transmission electron microscopy (TEM), high resolution transmission electron microscopy (HRTEM), and selected area electron diffraction (SAED). Figure 2 (a) shows the TEM image of particles in sample S1. The particles are observed to be almost spherical in shape. The histogram for sample S1 is shown in the inset of figure 2 (a) where it confirms the narrow size distribution of the nanoparticles. The average diameter of the core of the Fe_3O_4 particles is 12 nm and the average coating thickness is around 2 nm. The circular rings shown in the SAED pattern in the inset of figure 2 (b) confirms the polycrystalline nature of the sample. The TEM and HRTEM images of the other samples are typical to those shown in figure 2 but with different core sizes.

The average Fe_3O_4 particle diameters calculated from the TEM images for samples S2, and S3 are 15 nm, and 16 nm, respectively. The shell thickness is 2 nm for all samples.

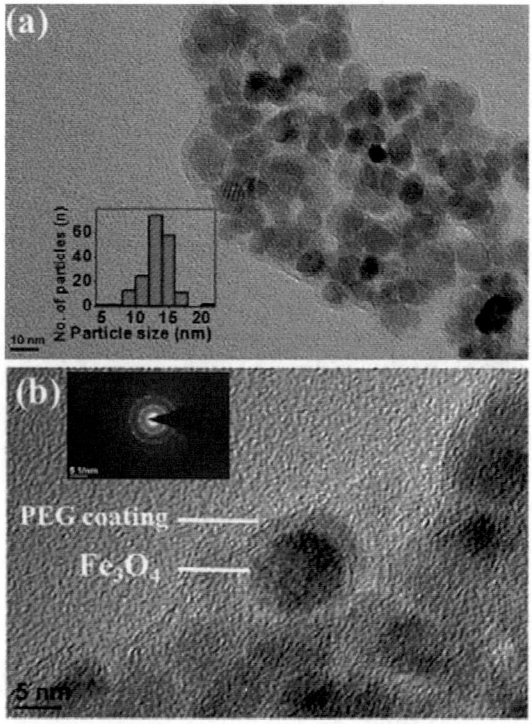

Figure 2. (a) TEM image and (b) HRTEM image of sample S1. The inset of (a) represents the size histogram for the corresponding sample. The inset of (b) is the SAED pattern confirming the polycrystalline nature of the sample.

Magnetic anisotropy is defined as the directional dependence of a material's magnetic properties. The magnetic moment of magnetically anisotropic materials will like to align with an "easy axis", which is an energetically favorable direction of spontaneous magnetization. There are different experimental methods that could be used to obtain the effective anisotropy constant, K_{eff} [36]. Some of these include obtaining K_{eff} from Mössbauer spectra, the coercive field, and the ac magnetic susceptibility. One other method to obtain K_{eff} is from the zero-field-cooled (ZFC) magnetization versus temperature measurements. In this study, K_{eff} was calculated using two methods from the magnetization measurements. In the first method, K_{eff} was calculated from the temperature-dependent coercivity curves which were obtained from ZFC magnetization versus magnetic field (*M-H*) hysteresis loops at several temperatures. In the second method, K_{eff} was calculated from the ZFC magnetization versus temperature (*M-T*) measurements at several applied magnetic fields.

In the ZFC *M-H* measurements, the temperature of the sample was cooled down from room temperature to a lowest available temperature (2 K) in the absence of any applied magnetic field. After that, a magnetic field was applied and the magnetic moment was recorded. The magnetic moment was measured while the applied magnetic field varied between -3 T and +3 T. The *M-H* hysteresis loops under ZFC measured at T = 2 K, for samples S1, S2, and S3 were shown in figure 3. Similar hysteresis loops were obtained at 50 K and 100 K.

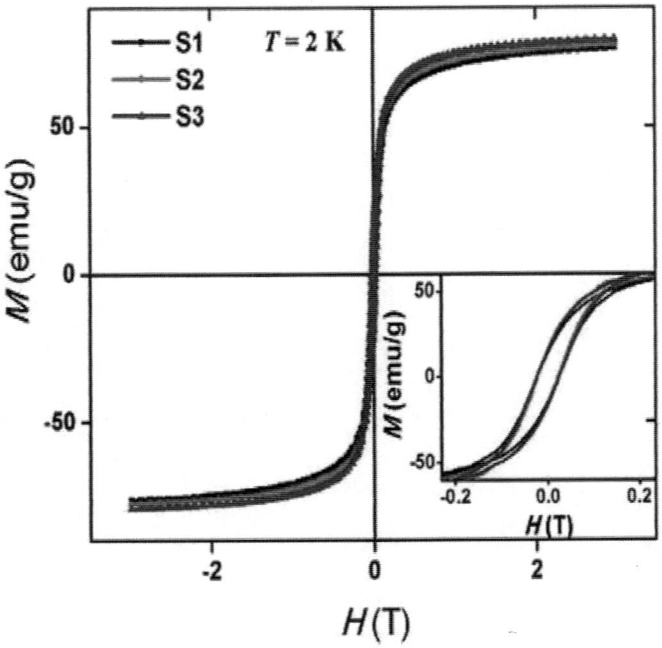

Figure 3. Magnetization versus magnetic field measured at $T = 2$ K for all samples. The inset shows an enlarged portion around the center of the loop.

It is worth mentioning that the real values of magnetization for these coated nanoparticles are larger than the reported values here. The magnetization is the magnetic moment measured by the SQUID divided by the measured mass of the sample. But this mass is composed of both the mass of the PEG coating and the mass of the iron oxide core. The PEG coating does not contribute to the measured magnetization. The reported magnetization values should include the iron oxide core only. Hence, the measured values should be multiplied by the ratio of the total mass of the particle to the mass of the core (m_{total}/m_{core}) to obtain the real magnetization values. We calculated the average sizes of the core and the shell in each sample from the HRTEM images. Then by using a density of 5 g/cm^3 for the Fe_3O_4 core and 1.13 g/cm^3 for the PEG coating, the calculated mass ratios (m_{total}/m_{core}) are calculated to be 1.13, 1.11, 1.097 for samples S1, S2, and S3, respectively. Thus to obtain the real magnetization values, the reported values should be multiplied by 1.13, 1.11, and 1.097 for samples S1, S2, and S3, respectively.

The coercive field, H_C was inferred by taking the average of the values between negative and positive field axes of the hysteresis loop where magnetization is zero ($H_C = \frac{|H_{C1}|+|H_{C2}|}{2}$). It should be noted that any experimental sample containing large numbers of NPs with various sizes, shapes, surface effects, and inter-particle interactions has K_{eff} with the superposition of all these effects. Figure 4 shows the coercive field as a function of temperature under ZFC protocol for all samples. The coercivity was obtained from the *M-H* hysteresis loops which were obtained at $T = 2$, 50 K, and 100 K. The role of the size distribution on the magnetic properties of nanocrytalline PEG-coated Fe_3O_4 is better understood through *T*-dependence of the coercive field. Generally, for an assembly of identical and non-interacting magnetic nanoparticles, the temperature dependence of coercive field can be studied on the basis of Néel relaxation and the Bean-Livingston approaches. Temperature dependence of H_C can be written as [37]:

$$H_C = 0.96\,\frac{K_{eff}}{M_S}\left[1-\left(\frac{T}{T_B}\right)^{\alpha_K}\right] \quad \text{........ (1)}$$

IOP Conf. Series: Materials Science and Engineering **305** (2018) 012012 doi:10.1088/1757-899X/305/1/012012

Here α_K is called the Kneller exponent, K_{eff} is the effective anisotropy constant for spherical particles, M_S is the saturation magnetization, T_B is the mean blocking temperature, and $H_0 = 0.96 \frac{K_{eff}}{M_S}$ .... (2).

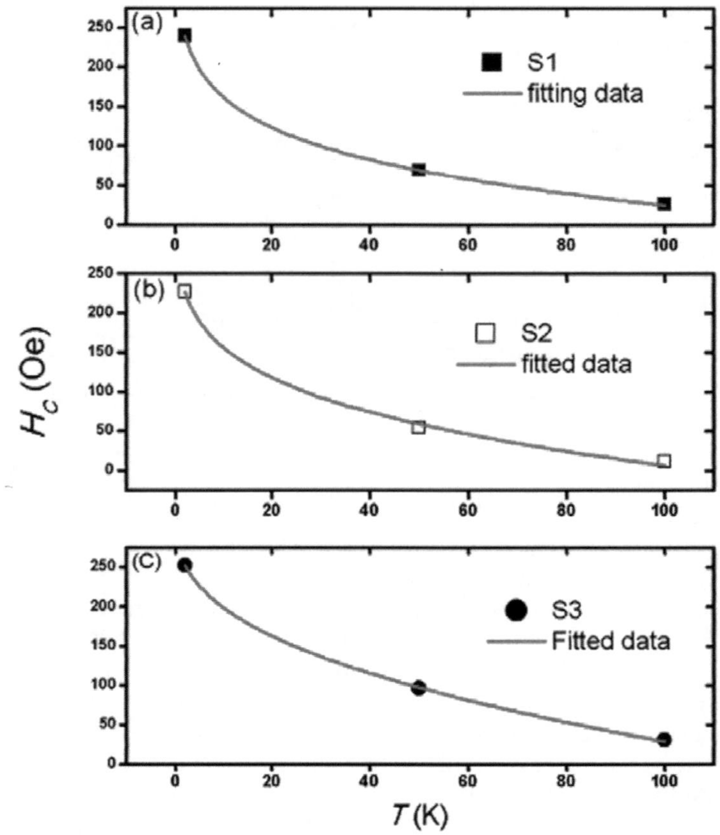

Figure 4. The temperature dependent coercivity for sample (a) S1, (b) S2, and (c) S3. The solid lines represent the fittings of the measured data to H_C that is expressed in equation 1.

For non-interacting particles with similar sizes, $\alpha_K = \frac{1}{2}$. In fact the real nanoparticles systems can present large size distribution and significant interparticle interactions once magnetic measurements are generally carried out at powder samples. The K_{eff} values were calculated by fitting the experimental data of the temperature dependence H_C as expressed in equation 1. The M_S values were taken from the magnetic hysteresis loops at the corresponding temperatures.

It should be noted that this approach is in agreement with experimental results only in the low temperature region when most of the particles are blocked. Figure 5 displays the calculated K_{eff} values as function of the size of particles.

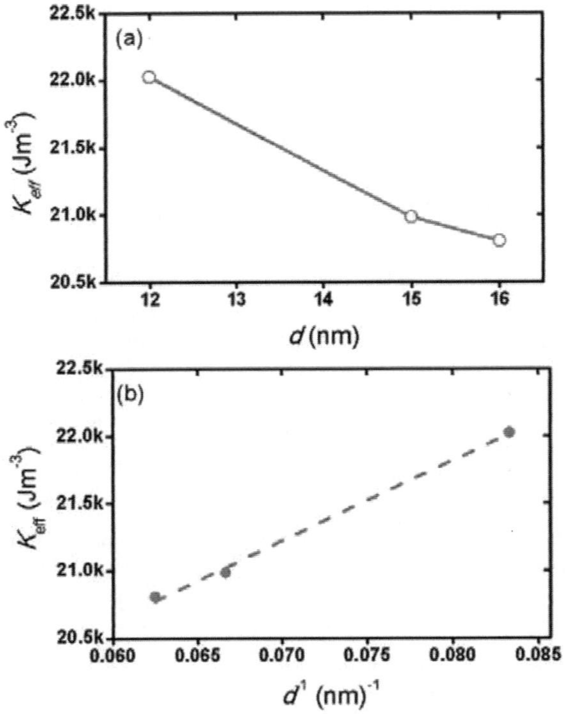

Figure 5. The effective anisotropy, K_{eff} as function of (a) the particle diameter, and (b) the inverse of the particle diameter. The solid line in (a) is a guide to eyes, and the dashed line in (b) is the best fit for the data points.

These values of the effective anisotropy are very close to the values reported for Fe_3O_4 nanoparticle in other studies [38]. These values are larger than the bulk value for Fe_3O_4 which is $K_{eff} = 1.1 \times 10^4$ J/m^3 [39]. This larger value could be attributed to the large contribution of surface spin effects which result in surface anisotropy. The nanoparticles exhibit a variation in values of K_{eff} when different molecules are chemisorbed on the particles [40, 41]. This indicates that surface anisotropy gives crucial contribution to K_{eff}. In addition, deviations from a perfect spherical shape should also have influence on these anisotropy values. The effective anisotropy can be regarded as a result of volume (core) anisotropy and surface anisotropy. It is not an easy task to separate the parameters that control the effective anisotropy in nanoparticles. However, by considering several nanoparticle samples with different core dimensions, the difference in effective anisotropy can be suggested to be due to contributions of different surface to core ratios. When using this method, it is assumed that other parameters controlling the effective anisotropy remain unchanged by changing the sample. We have analyzed K_{eff} on the basis of a model [42] in which the total energy barrier is given by:

$$\Delta E = K_{eff}V = K_V V + K_S S \dots\dots\dots\dots (3)$$

Where K_V and K_S are the volume and surface anisotropy energy constants, respectively. Considering that the particles are spherical with average diameter *d*, Eq. 3 can be written as

$$K_{eff} = K_V + \frac{6}{d} K_S \dots\dots\dots\dots (4)$$

The size dependence of K_{eff} is in agreement with Eq. 4. As can be seen in figure 5 (a), at all temperatures (2, 50, 100 K), K_{eff} increases as the size of NPs decreases. This behaviour could be attributed to the enhancement of the surface anisotropy with decreasing the size of NPs. The straight-line fits for K_{eff} versus

d^{-1} data which is shown in figure 5 (b), yield the values K_V and K_S. According to Eq.4, the slope of the straight line = 6 K_S and the intercept of the extrapolated straight line with the K_{eff} axis is K_V.

There are several factors that could influence surface magnetization and hence, the surface magnetic anisotropy. For example, the complex redistributions of the spin-polarized electronic density, which results from the reduced coordination of atoms at the surface of the nanoparticles, lead to changes in the spin-orbit energy and hence in the magnetic anisotropy energy [43]. In addition, expansion or contraction of the lattice structure [44-45], and the breaking of the crystallographic symmetry at the surface [46, 47] could also contribute towards the surface anisotropy of nanoparticles.

The second method that was used to calculate K_{eff} is based on ZFC and FC magnetization versus temperature (*M-T*) measurements. For the typical time window of SQUID or VSM measurements, the following equation will be used to calculate the effective anisotropy.

$K_{eff} < V > \approx 25\, k_B < T_B >$………. (5)

Which is obtained from:

$$< T_B > = \frac{K_{eff}<V>}{\ln(\frac{\tau_{SQUID}}{\tau_0})k_B} = \frac{K_{eff}<V>}{25k_B} \text{………(6)}$$

Here K_{eff} is the effective magnetic-anisotropy energy per unit volume, τ_{SQUID} is the characteristic measuring time of the instrument (~10 s) and τ_0 is related to the natural frequency of the gyromagnetic precession (~10^{-9}–10^{-13} s). $< V >$ is the average volume of the nanoparticles obtained either form the HRTEM images or the XRD patterns. $< T_B >$ is the average blocking temperature of the nanoparticles. The blocking temperatures was obtained directly from ZFC and FC *M-T* magnetization curves. Figure 6 displays the ZFC and FC magnetization versus temperature curves for sample S1 obtained at 200 Oe. Similar curves were obtained at several other fields for all samples.

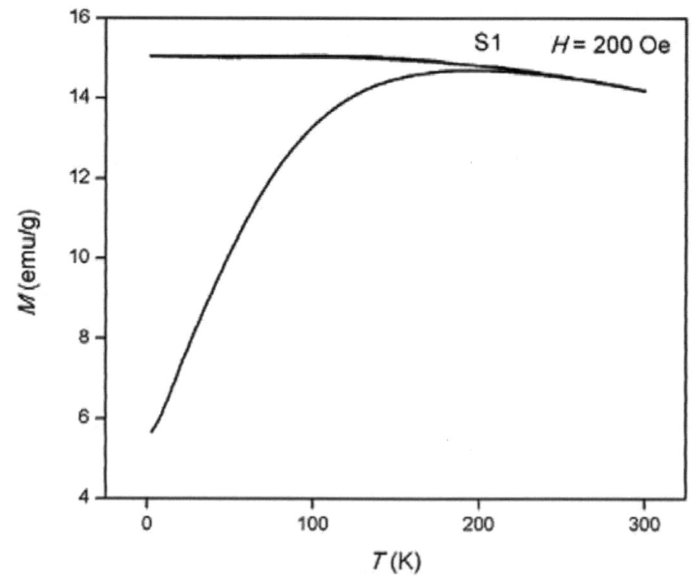

Figure 6. The ZFC and FC magnetization versus temperature curves for sample S1 obtained at the applied filed of 200 Oe.

In the ZFC M-T magnetization measurements, the sample was initially cooled from room temperature to 2 K. A small magnetic field (50-500 Oe) was applied to the sample and the magnetization was measured as the temperature was being raised to 300 K. For the FC *M-T* measurements, the magnetization was recorded again while the temperature was lowered down to 2 K under the application of the same magnetic field.

The K_{eff} values were calculated using Eq. 5 where the particle volume was calculated using the diameters obtained from the TEM images. Figure 7 displays the calculated K_{eff} values as function of the particle diameter at several applied fields. As can be seen in this figure, K_{eff} values display a similar behavior to the obtained from the first method where K_{eff} increases with the decrease of the size of the particle. However, the magnitudes of K_{eff} obtained using the second method are considerably larger. We can see that as the applied filed increase, the values of K_{eff} obtained in the second approach those obtained in the first method. The volume and surface contributions to K_{eff} were obtained using Eq. 4 at all the applied fields and the results are displayed in figure 8. Both contributions increase with decreasing the size of the particles.

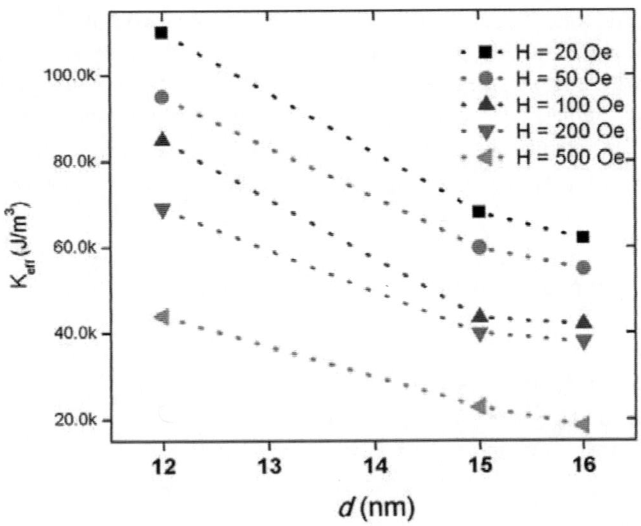

Figure 7. K_{eff} values as function of the particle diameter at several applied fields.

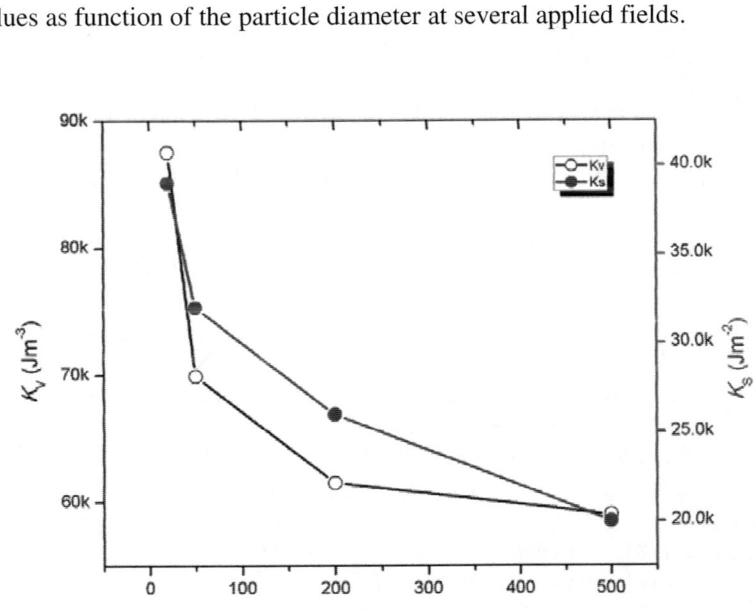

Figure 8. The volume, K_V and surface K_S contributions to K_{eff} at several applied fields.

The difference in magnitude and in the behavior of effective anisotropy with size of the nanoparticles can be understood based on the different mechanisms of two magnetization protocols. In the *M-T* ZFC measurements, the magnetic moment of each one of the nanoparticles is thermally activated and thus it always oscillate between the two easy axis directions which are the lowest two energy states of magnetization. The applied magnetic field has a small magnitude and thus less influence on the orientation of the magnetic moments of the nanoparticles. In the *M-H* ZFC measurements, magnetic moment each nanoparticle is being forced to rotate along the magnetic applied field direction.

4. Conclusions:

The morphology, structure, and effective anisotropy of PEG-coated iron oxide (Fe_3O_4) nanoparticles were investigated. The TEM, HRTEM, and SAED images show the crystalline nature of the nanoparticles with nearly spherical shape with narrow size distributions. Using magnetic measurements, the size dependence of the effective magnetic anisotropy constant K_{eff} of the nanoparticles were obtained using two dc magnetization methods. In both methods, K_{eff} was found to decrease significantly and almost linearly with increasing the size of particles. However, the magnitudes of K_{eff} were considerably different.

Acknowledgement

This work was financially supported by the SUREPLUS Program under the Grant no. 31S277.

References:

[1] *Fuji Electric Review* 2011 vol 57, pp30–62.
[2] Rosensweig R E 2002 *J. Magn. Magn. Mater.* **252** 370.
[3] Jun Y, Seo J and Cheon J 2008 *Acc. Chem. Res.* **41** 179.
[4] Frey N A, Peng S, Cheng K and Sun S 2009 *Chem. Soc. Rev.* **38** 2532.
[5] Fortin J P et al. 2007 *J. Am. Chem. Soc.* **129** 2628.
[6] Mehdaoui B et al. 2011 *Adv. Funct. Mater.* **21** 4573.
[7] Georgelin T, Bombard S, Siaugue J M and Cabuil V 2010 *Angew. Chem. Int. Ed.* **49** 8897.
[8] Sato T, Nagaoka K, Kobayashi S, Manjanna J and Murakami T 2017 *AIP Adv.* **7** 056319.
[9] Auzans E, Zins D, Blums E and Massart 1999 *R. J. Mater. Sci.* **34** 1253.
[10] Zaitsev V S, Filimonov D S Presnyakov I A, Gambino R J and Chu B 1999 *J. Colloid Interface Sci.* **212**, 49.
[11] Orive G, Ali O A, Anitua E, Pedraz J L and Emerich D F 2010 *Biochim. Biophys. Acta*, **1806** 96.
[12] Kateb B, Chiu K, Black K L, Yamamoto V, Khalsa B, Ljubimova J Y, Ding H, Patil R, Portilla-Arias J A and Modo M. et al. 2011 *NeuroImage* **54** S106.
[13] Butterworth M D, Illum L and Davis S S 2001 *Colloids Surf. A* **179** 93.
[14] Josephson L, Tung C H, Moore A and Weissleder R 1990 *Bioconjug. Chem.* **10** 186.
[15] Kim D K, Mikhaylova M, Zhang Y and Muhammed M 2003 *Chem. Mater.* **15** 1617.
[16] Liu Q and Xu Z 1995 *Langmuir* **11** 4617.
[17] Bautista M C, Bomati-Miguel O, Zhao X, Morales M P, González-Carreño T, de Alejo R P, Ruiz-Cabello J and Veintemillas-Verdaguer S 2004 *Nanotechnology* **15** S154.
[18] Berry C C, Wells S, Charles S, Aitchison G and Curtis A S G 2004 *Biomaterials* **25** 5405.
[19] Nayek C, Manna K, Bhattacharjee G, Murugavel P and Obaidat I M 2017 *Magnetochemistry* **3** 19.
[20] Portet D, Denizot B, Rump E, Lejeune J J and Jallet P 2001 *J. Colloid Interface Sci.* **238** 37.
[21] Wu W, He Q and Jiang C 2008 *Nanoscale Res. Lett.* **3** 397.
[22] Kim D H, Rozhkova E A, Ulasov I V, Bader S D, Rajh T, Lesniak M S and Novosad V 2010 *Nature Mater.* **9** 165.
[23] Lu Z and Yin Y 2012 *Chem. Soc. Rev.* **41** 6874.

[24] Granitzer P, Rumpf , Tian Y, Akkaraju G, Coffer J, Poelt P and Reissner M 2013 *Appl. Phys. Lett.* **102** 193110.
[25] Jeun M, Moon S J, Kobayashi H, Shin H Y, Tomitaka A, Kim Y J, Takemura Y, Paek S H, Park K H, Chung K W and Bae S 2010 *Appl. Phys. Lett.* **96** 202511.
[26] Lee J H, Jang J T, Choi J S, Moon S H, Noh S H, Kim J W, Kim J G, Kim I S, Park K I and Cheon J 2011 *Nat. Nanotechnol.* **6** 418.
[27] Yang H C, Liao S H, Huang K W, Chieh J J, Chen H H, Chen M J, Chen K L, Wang L M and Horng H E, 2013 *Appl. Phys. Lett.* **102** 013119.
[28] Kumar C S S R and Mohammad F 2011 *Adv. Drug Delivery Rev.* **63** 789.
[29] Shubayev V I, Pisanic T R and Jin S H 2009 *Adv. Drug Delivery Rev.* **61** 467.
[30] Ito A, Tanaka K, Honda H, Abe S, Yamaguchi H and Kobayashi T 2003 *J. Biosci. Bioeng.* **96** 364.
[31] Smit J and Wijin H P J *1959 Ferrites: Physical Properties of Ferrimagnetic Oxides in Relation to Their Technical Applications, Philips' Technical Library, Eindhoven, The Netherlands.*
[32] Tartaj P 2004 *in Encyclopedia of Nanoscience and Nanotechnology*, edited by Nalwa H S (American Scientific Publishers, Los Angeles), Vol. 6 p 823.
[33] Jeun M, Lee S, Kang J K, Tomitaka A, Kang K W, Kim Y I, Takemura Y, Chung K W, Kwak J and Bae S 2012 *Appl. Phys. Lett.* **100** 092406.
[34] Skumryev V et al 2003 *Nature* **423** 850.
[35] Sun S et al 2000 *Science* **287** 1989.
[36] Yoon S 2011 *J. Korean Phys. Soc.* **59** 3069.
[37] Mendonça E C, Jesus C B R, Folly W S D, Meneses C T, Duque J G S and Coelho A A 2012 *J. Appl. Phys.* **111**, 053917.
[38] Morup S and Topsbe H 1976 *Appl. Phys.* **11** 63.
[39] Klaus D 2010 Sattler, Edit. *Handbook of Nanophysics: Nanomedicine and Nanorobotics* vol. 23 p 11.
[40] Clausen B S, Morup S and Topsoe H 1979 *Surf. Sci.* **82**, L589.
[41] Morup S, Topsoe H and Clausen B S 19882 Phys. Scr. **25**, 713.
[42] Batlle X and Labarta A 2002 *J. Phys. D: Appl. Phys.* **35** R15.
[43] Bruno P 1989 *Phys. Rev. B* **39**, 865.
[44] Dorfbauer F et al 2007 *J. Magn. Magn. Mater.* **316** E791.
[45] Evans R et al 2007 *IEEE Trans. Magn.* **43** 3106.
[46] Jamet M et al 2004 *Phys. Rev.* B **69** 24401.
[47] Obaidat I M, Mohite V, Issa B, Tit N, Haik Y 2009 *Cryst. Res. and Technol.* **44**, 489.

Effect of laser beam on temperature distribution on artificial cylindrical shaped hard tissue bones

M-Ali H. Al-Akhras*, Mohammad-Khair Qaseer, B. A. Albiss, Umar S. Gezawa

Department of Physics, Bio-Medical Physics Laboratory, Jordan University of Science & Technology, P.O. Box 3030, Irbid 22110, Jordan

alakmoh@just.edu.jo

Abstract: Samples from fresh lamb chest bones were made in cylindrical shapes to study the time variation of temperature T as functions of the cylindrical radius and depth when its front surface exposed to a laser beam of 110Mw power and 642nm wavelength. The laser beam was directed at the center of the front surface of the horizontal cylinder. The measurements were done in vacuum and at atmospheric pressure. Our data reveal the linear variation of T with time, followed by a gradual increase before it reaches a plateau value at higher time. This sort of behavior independent of the radius or the depth where the temperature was measured. Moreover, the maximum variation occurs on the front surface where the laser beam was hitting and diminishes gradually with depth deep inside the cylinder. Data at atmospheric pressure showed less changes in temperature. The temperature distribution in bone due to laser irradiation is very important for a rational use of laser therapy as well as in the surgery to minimizes the thermal tissue damage.

IOP Conf. Series: Materials Science and Engineering **305** (2018) 012013 doi:10.1088/1757-899X/305/1/012013

1. Introduction

In recent years, laser heating treatment soldering and ablation have become some of the most investigated methods in nano-engineering, microbiology and medicine sciences [1-5]. Understanding of the underlying physics and interrelation of the processes taking place in tissues irradiated by continuous or pulsed laser allows optimization of relevant parameters in therapeutic and medical applications. In this context, Stanowski *et al.* [3] achieved a quantitative description of the IR laser-QWI process leading to an accurate knowledge of temporal and spatial temperature profiles induced by the laser solving a 3-dimensional heat diffusion. Mrochen *et al.* [6,7] investigated in the same context the influence of temporal and spatial spot sequences on an *in-vivo* surface temperature increase during laser surgery with a high-repetition-rate excimer laser.
The most interesting measurements in this field are hence temperature profiles. In fact, in performing laser heating treatments, tissues temperature rise has always been considered as the key for determining response amplitude [6-10]. Besides research interests, it is desirable to have complete knowledge of the temperature distribution during medical laser applications based on heating. Nevertheless, the complexity of the problem lies in the fact that this temperature rise depends conjointly on factors intrinsic to the body such as the thermal conductivity and diffusivity of the tissue and the energy dissipated inside it by the source.
In this study the temperature profile to the heat equation is proposed in a modeling animal hard bone tissue.

2. Experimental set up and samples preparation:

The schematic of the experimental set-up is shown in Fig. 2. The objective of the research is to study the temperature rise of the bone pellets. A 640 nm desktop diode laser (Voltron Inc.) operating at a wavelength of 640 nm was used. The objective of this research is to study the temperature rise of the bone pellets. A 640 nm desktop diode laser (Voltron Inc.) operating at a wavelength of 640 nm was used The samples are well insulated on all sides to prevent heat loss to the surroundings. The experiment is categorized into two aspects; one under atmospheric pressure and the other under vacuum. The bone pellets were mounted onto the sample holder of the cryostat. K-type thermocouples having diameter of 0.5 mm are inserted in-between the bone pellete for temperature measurements. Temperature readings of the thermocouples are recorded through a voltmeter. These thermocouples provide a measurement range from 0^0C to 480^0C with a sensitivity of $\pm 1^0$C and 0.2s of response time.
Hard tissue was obtained from fresh lamb bones of the chest. After peeling off the soft tissue, it was incubated overnight to give a strong and crust form. Using mortar mould the bones were carefully broken into smaller fragments, the smaller fragments were carefully separated from the bone marrow to avoid interference and a coffee maker grinder was used to grind the bones into powder.
1g of PEG was dissolved into 5mL of distilled water and stirred for 1hr. 1g of the powdered bones was added and then stired again for 0.5 hrs. This was then incubated in an oven at 60^0C for 12 hrs. The cavier hydraulic compressor was used to press the sample at 4 metric tones to give a sample pellet of 2.2 mm thickness and 2 cm in diameter. Figure (1)a shows the location of the thermocouples at the top surface to measure the temperature T. Figure (1)b shows the location of the thermocouples as the sample depth increased. On each surface the locations of the thermocouples as the at the above surface.

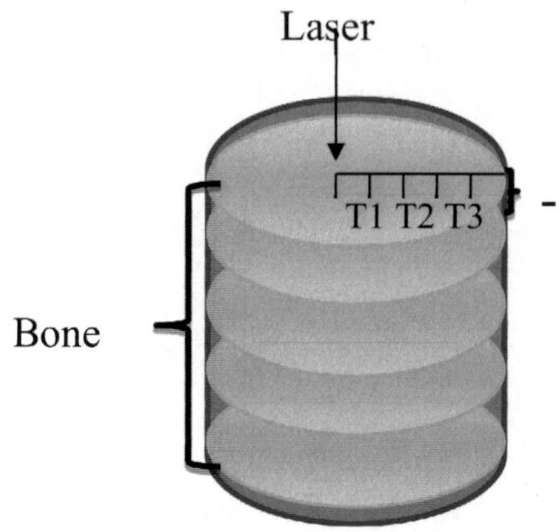

Figure (1)a Sechmatic of a sample shows the position of the thermocouples at the top surface.

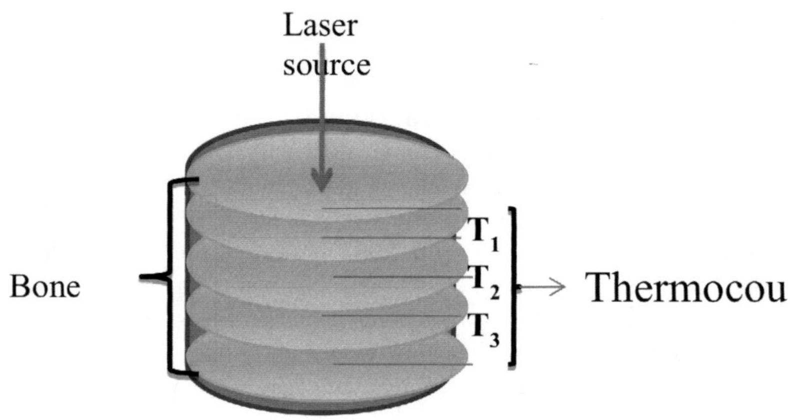

Figure (1)b Sechmatic of a sample shows the position of the thermocouples with sample depth.

3. Results and Discussion

Figure (2) shows our experimental results for the variation of the temperature as function of the distance from the center of the sample disk to the edge at different positions as the depth of the layer changes. The figure shows the maximum variation in the temperature is achieved by using the maximum power of the laser beam. Hence, in all of the following results we used the maximum power of the laser beam.

IOP Conf. Series: Materials Science and Engineering **305** (2018) 012013 doi:10.1088/1757-899X/305/1/012013

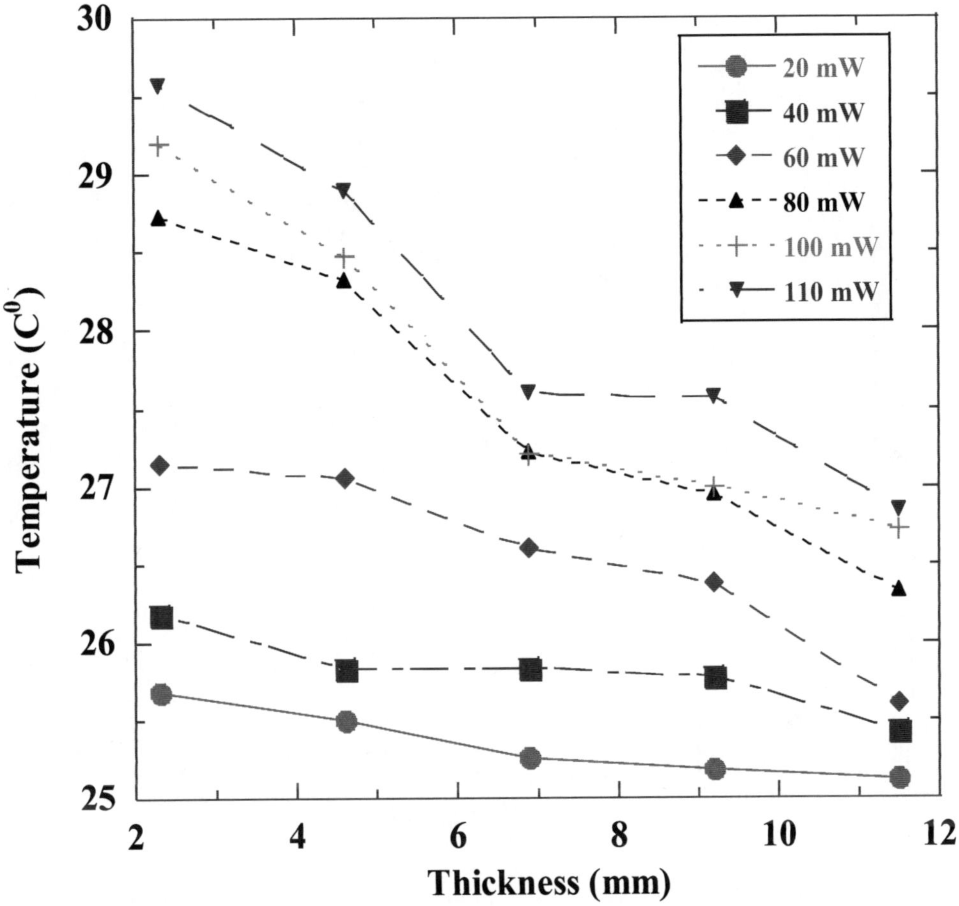

Figure (2): T versus R of the top surface of the cylinder at different powers of laser.

Our results for the time dependent variations of the temperature T on the upper surface of the cylindrical shape samples is shown in figure 3a when the sample is in vacuum and in figure 3b when the sample is in atmosphere pressure. Both figures show a fast and linear increase in T at the beginning followed by a gradual increase in T before it reaches a platue value at higher times. The figures show that in vacuum, the temperature change is higher than when the sample is exposed to the atmosphere.

IOP Conf. Series: Materials Science and Engineering **305** (2018) 012013 doi:10.1088/1757-899X/305/1/012013

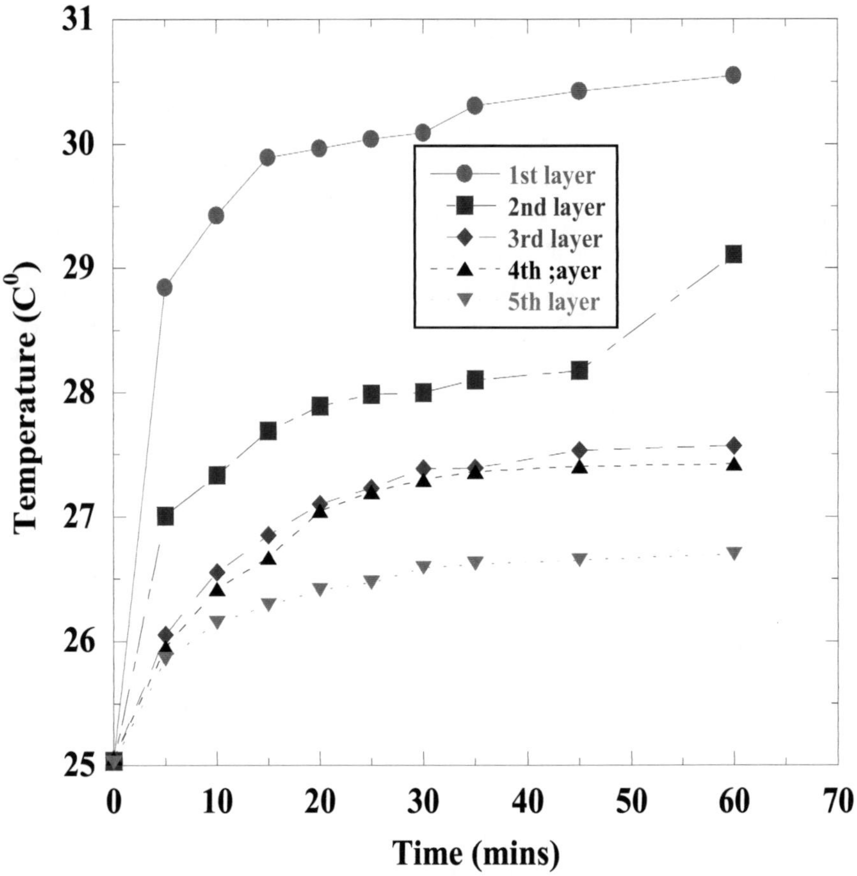

Figure (3)a Typical results of T versus time at different depths where samples in vacuum state.

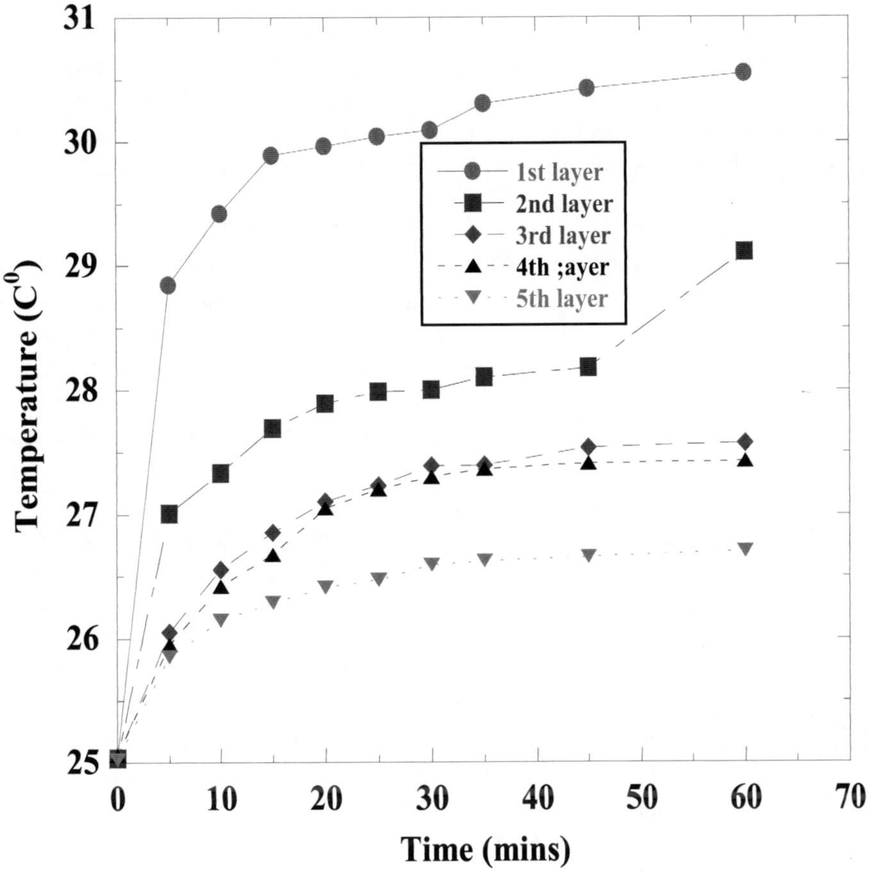

Figure (3)b Typical results of T versus time at different depths where samples in the vacuum state.

Our results of the variation of T versus R at different depth in vacuum state are shown in table (1).

Table 1: The change in temperature for different layer at different time interval.

Time	ΔT^0C (1st layer)	ΔT^0C (2nd layer)	ΔT^0C (3rd layer)	ΔT^0C (4th layer)	ΔT^0C (5th layer
10.0	4.275	1.425	0.475	0.300	1.200
15.0	4.025	1.400	0.575	0.475	0.475

20.0	3.870	1.425	0.775	0.475	0.525
25.0	3.850	1.325	0.325	0.800	0.475
60.0	4.375	1.225	0.675	0.725	0.725

The maximum variation on T exceeds 4 °C at the top surface and decrease to 1.2 °C at the top of the bottom layer. It is worth mention that each layer has 2.2 mm in thickness so the depth of the depth of the fifth layer is 8.8 mm. The platue values of T vary from 1.2 °C at the top surface to 0.7 °C the top of the bottom layer. Therefore, we conclude that the first ten minutes has the most important in any medical processes or operations for exposure to the laser beam.

4. Conclusions:

The normal incidence of laser power with 110 mW power of wave length 642 nm on cylindrical samples made from bones showed that the three regimes of T with time. For each layer there is a linear change in T with t and gradual increase to finally a platue value depends on the depth. The linear regime is the most important because in any medical processes the temperature rise takes place at the begining of the process. Our data showed more than 4 °C the rise in the temperature at the upper surface reduced to 1.2 °C as the depth increase to 8.8 mm the top surface of the fifth layer.

Acknowledgements: This work was supported by a grant from the Deanship of Scientific Research at Jordan University of Science and Technology (JUST) No. (173/2013). The Authors would like to thank Miss Duaa Al-Khalili for her kind help and support.

References

[1] M.A. Awedaa , M. Agidab , M. Dadab , O. B. Awojoyogbeb , K. Isahb , O.P. Faromikab , K. Boubaker c* , K. Deb and O.S. Ojambatib," A solution to Laser-induced Heat Equation inside a Two-layer Tissue Model Using Boubaker Polynomials Expansion Scheme", JLMN-Journal of Laser Micro/Nanoengineering Vol. 6, No. 2, 106 (2011).

[2] T. Furumoto, T.Ueda, T. Tsukamoto, N. Kobayashi, A. Hosokawa, S. Abe, Study on Laser Consolidation of Metal Powder with Yb:fiber laser - Temperature Measurement of Laser Irradiation Area -, J. Laser Micro/Nanoeng. 4 (2009) pp. 22-27

[3] R. Stanowski, O. Voznyy, J. J. Dubowski, Finite element model calculations of temperature profiles in Nd:YAG laser annealed GaAs/AlGaAs quantum well microstructures, J. Laser Micro/Nanoeng. 1 (2006) pp.17-22.

[4] N. M. Bulgakova, I. M. Burakov, Y. P. Meshcheryakov, R. Stoian, A. Rosenfeld, I. V. Hertel, Theoretical Models and Qualitative Interpretations of Fs Laser Material Processing, J. Laser Micro/Nanoeng. 2 (2007) pp.76-86.

[5] M. Strassl, V. Wieger, D. Brodoceanu, F. Beer, A. Moritz, E. Wintner, Ultra- short pulse laser ablation of biological hard tissue and biocompatibles, J. Laser Micro/Nanoeng. 3 (2008) pp. 30-40.

[6] Micheal Morchen," Effect of time sequences in scanning algorithms on the surface temperature during corneal laser surgery with high-repetition-rate excimer laser", J. of Catact & Refractive Surgery", 35, p738 – 746 (2009).

[7] M. Mrochen, U. Schelling, C. Wuellner, C. Donitzky, Effect of time sequences in scanning algorithms on the surface temperature during corneal laser surgery with high-repetition-rate excimer laser, J Cataract Refract Surg. 35 (2009) pp. 738–746
[8] M. Ben-David, R. Cantor, N. Balbul, M. Yehuda, I. Gannot, Measuring tissue heat penetration by scattered light measurements, Lasers in Surgery and Medicine 40 (2008) pp. 494-499.
[9] B. Laikhtman, A. Gourevitch, D. Donetsky, D. Westerfeld, G. Belenky, Current spread and overheating of high power laser bars, J. Appl. Phys. 95 (2004) pp/ 3880-3889
[10] P.L. Djorev, E. Borisova, L. Avramov, Interaction of the IR laser radiation with human skin – MonteCarlo simulation, Proc. SPIE (2003) pp. 403-407

Effect of Photon Radiations in Semi-Rigid Artificial Tissue Sensitized by Protoporphyrin IX Encapsulated with Silica Nanoparticles

Ghaseb N. Makhadmeh[1, 3, 4 *], Azlan Abdul Aziz[1, 3] and Khairunisak Abdul Razak[2, 3], M-Ali H. Al-Akhras[5]

[1]*School of Physics, Universiti Sains Malaysia, Penang, Malaysia.*
[2]*School of Materials and Mineral Resources Engineering, Universiti Sains Malaysia, Penang, Malaysia.*
[3] *NanoBiotechnology Research and Innovation (NanoBRI), Institute for Research in Molecular Medicine (INFORMM), Universiti Sains Malaysia, Penang, Malaysia*
[4]*Arts and Science College, Amman Arab University, Amman, Jordan*
[5]*Bio-Medical Physics Laboratory, Department of Physics, Jordan University of Science & Technology, Irbid, Jordan*

ghaseb84@yahoo.com

Abstract: This study involves the synthesis of Protoporphyrin IX (PpIX) encapsulated with Silica Nanoparticles (SiNPs) as an application for Photodynamic therapy. Semi-rigid artificial tissues with optical features similar to human tissue were used as sample materials to ascertain the efficacy of PpIX encapsulated with SiNPs. The disparity in optical characteristics (transmittance, reflectance, scattering, and absorption) of tissues treated with encapsulated PpIX and naked PpIX under light exposure (Intensity at 408 nm ~1.19 mW/cm^2) was explored. The optimal exposure times required for naked PpIX and SiNPs encapsulated PpIX to engulf Red Blood Cells (RBCs) in the artificial tissue were subsequently measured. Comparative analysis showed that the encapsulated PpIX has a 91.5 % higher efficacy than naked PpIX. The results prove the applicability of PpIX encapsulated with SiNP on artificial tissue and possible use on human tissue.

Keywords: Photodynamic Therapy, Silica Nanoparticles, Encapsulation, Protoporphyrin IX, Optical Properties, Artificial Tissue.

1. Introduction

The current advancement in Photodynamic Therapy (PDT) has offered new medical approaches to cancer treatment, such as the use of photosensitizers (PSs) under light exposure of a specific intensity [1]. Singlet oxygen $^1O_2$ produced by the light exposure engulfs the cells by means of cytotoxicity [2-4]. However, the efficiency of $^1O_2$ production decreases significantly because of the small volume of PS that reaches the target cells. This decline in PS efficacy is attributable to the defense mechanism of reticulo-endothelial system (RES) or the engulfing of PS by the macrophage system in the body. Nonetheless, macrophage normally gives no attention to biocompatible foreign objects with sizes under 100nm [5]. Thus, the inhibition effect of macrophage can be resolved using nanoparticles as a medium of PS transportation. This study proposes the use of silica nanoparticles (SiNPs) as an encapsulating medium for PS. The intrinsic features of SiNPs such as low toxicity and high biocompatibility ensures its effectiveness as an ideal encapsulation material for Drug Delivery System (DDS) [6]. In addition, it is easily synthesized with low polydispersity at low temperatures. Moreover, bio-molecular compounds have a propensity to agglutinate to the external surface of SiNPs, while its inner surface has the distinctive capability of encapsulating PS [7-9].

Protoporphyrin IX (PpIX) is an essential derivative of porphyrin [10-12]. The absorption of emitted energy light source by PpIX leads to a series of chemical reactions which generate singlet oxygen. The singlet oxygen produced is capable of engulfing the artificial tissue [13-16]. Therefore, this study evaluates the effect of PpIX encapsulated by SiNPs on Red Blood Cells (RBC's) using artificial tissue (light absorption component) as test sample [17-18]. The transmittance, reflectance and optimal exposure time properties of the test sample were determined. Variations in absorption, scattering coefficient, the relation between the depth and the transmittance and Half Value Layer (HVL) of the RBC's were also analyzed. The efficacy of encapsulated PpIX on RBCs destruction was then compared with that of naked PpIX.

2. Materials and Methods

SiNPs were prepared using a microemulsion method [19]. The first step involves mixing 0.1ml of ammonia (Sigma Aldrich Cat No: 1336, 21, 6), 200ml distilled water, and 5.5g Tween 80 (Sigma Aldrich Cat. No. P1754). The solution was constantly stirred for 15 min. The measured pH value was found to exceed 9.0. A 6 ml of 1-Butanol (Sigma Aldrich Cat No: B8, 590) was subsequently added, and the solution stirred for further 5 minutes at room temperature. The next step entails the encapsulation of 15 ml PpIX (10.04 μM as a final concentration). Following the encapsulation, the sample was wrapped in aluminum foil because of its high sensitivity to light. After 1 hour, 2 ml Triethoxyvinylsilane (TEVS) (Sigma Aldrich Cat. No. 175560) was added, and the solution was continually stirred at a temperature of 27°C for 20 hours. After carrying out the sample preparation, dialysis was performed on the solution using a dialysis membrane (by soaking the dialysis membrane in distilled water for 5 min) [18].

100 ml of artificial tissue was prepared by dissolving 2 grams of agar (No. 9405, Sigma Aldrich) in 40 ml distilled water and 40 ml of the Phosphate Buffered Saline (PBS) (1 tablet in 200 ml distilled water, No. P4417, Sigma Aldrich). The temperature of the solution was increased to 100°C for 3 min. The sample was stirred continuously during the course of heating up and cooling down. Following the drop in temperature to less than 80°C, 3gm of 5-10μm diameter silica powder was added. The solution was subsequently mixed for 5 min. A 1 ml of solution ink (Abs. equal to 0.18 at 500 nm after diluting for 100 times), 20 ml of bovine serum and 0.2 ml of intralipid were incorporated into the solution at 45°C. 1.5 ml of concentrated blood (Abs. equal to 0.90 at 500 nm after diluting for 100 times), 10.04 μM PpIX encapsulated with SiNPs, and 1 ml of penicillin were also added at 37 °C.

The resultant solution was poured into sample holders containing semi-rigid artificial tissue slices of 0.71 mm thickness. Afterward, the solution was stored overnight at temperatures below 4 °C. The next day, 5 sample slices were exposed to light (intensity at 408 nm ~1.19 mW/cm^2) provided by an arc lamp positioned 30cm away from the sample. The experiment was carried out at various exposure times (0, 10, 25, 45 and 70 min.). The transmittance (T) and reflectance (R) of the samples were measured using NIR-UV-VIS Spectrophotometer. The scattering and absorption coefficients spectrums were derived using the Kubelka-Munk theory [19]. Different parameters were correlated, and such as: thickness of the sample and transmittance; thickness of the material at which the intensity decreased by one half (Half Value Layer (HVL)), and the optimal exposure time that is required to engulf half of RBCs. A similar sequence of measurements was also applied on naked PpIX.

3. Results and Discussions

3.1. Impact of Naked and Encapsulated PpIX on the Optical Properties of An Artificial Tissue

In order to determine the efficacy of encapsulation, the effects of using encapsulated PpIX were compared with those of naked PpIX. The difference in optical properties for both treatments was examined. As indicated by the encircled cusps in Fig. 1 (a) and (b), variations in minimum point of reflectance after light exposure were observed for samples treated with encapsulated PpIX and naked PpIX, respectively. For the sample treated with encapsulated PpIX, a comparably slightly higher increase in reflectance is observed at the minimum point of 577nm (maximum peak of RBCs absorption), possibly due to enhanced light energy absorption by encapsulated PpIX. This absorption process produces more $^1O_2$ which increases the destruction of RBCs in the tissue.

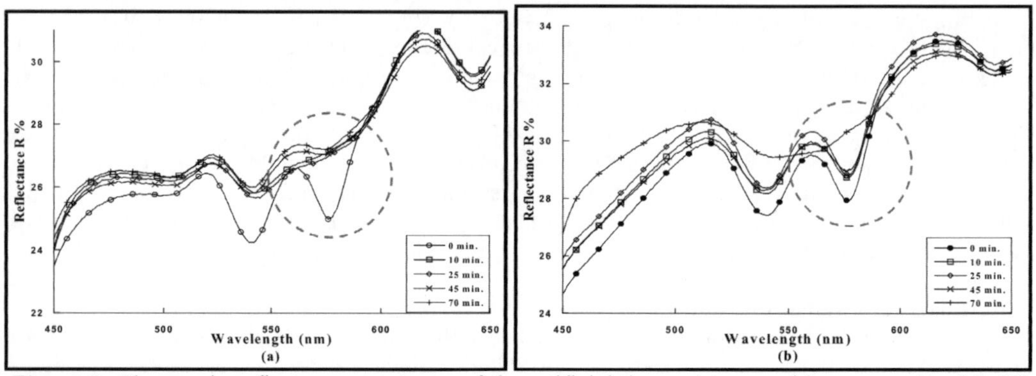

Figure 1. Changes in reflectance spectrums of the artificial tissue: (a) treated by encapsulated PpIX, (b) treated by naked PpIX.

Shifts in the spectrum are also observed in the minimum points of transmittance for both samples treated with encapsulated PpIX and naked PpIX as shown in Fig. 2 (a) and (b), respectively. The increase in minimum point of transmittance (577 nm) is attributable to the singlet oxygen produced by PpIX, which engulfs the RBCs in the sample, consequently increasing the transmittance value.

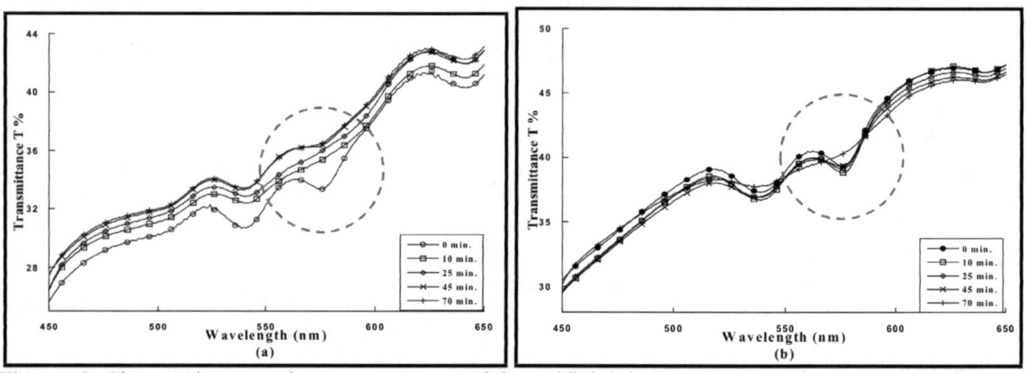

Figure 2. Changes in transmittance spectrums of the artificial tissue: (a) treated by encapsulated PpIX, (b) treated by naked PpIX.

This trend is also evident in the scattering values at 577 nm as depicted in Fig. 3 (a) and (b). This is due to the fact that the singlet oxygen generated from PpIX light energy absorption engulfs the RBCs in the sample.

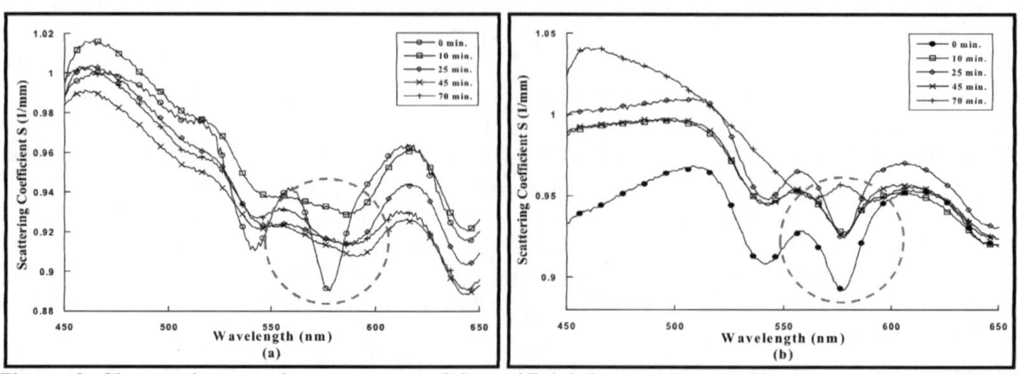

Figure 3. Changes in scattering spectrums of the artificial tissue: (a) treated by encapsulated PpIX, (b) treated by naked PpIX.

However, a contrasting observation was made for the absorption values, as shown in Fig. 4 (a) and (b). Rather, a decline in the maximum peak of absorption coefficient was noted. This is because the singlet oxygen, generated from the absorption of light energy by PpIX, engulfed the RBCs membrane, resulting in the release of hemoglobin in the sample which reduced the absorption peak.

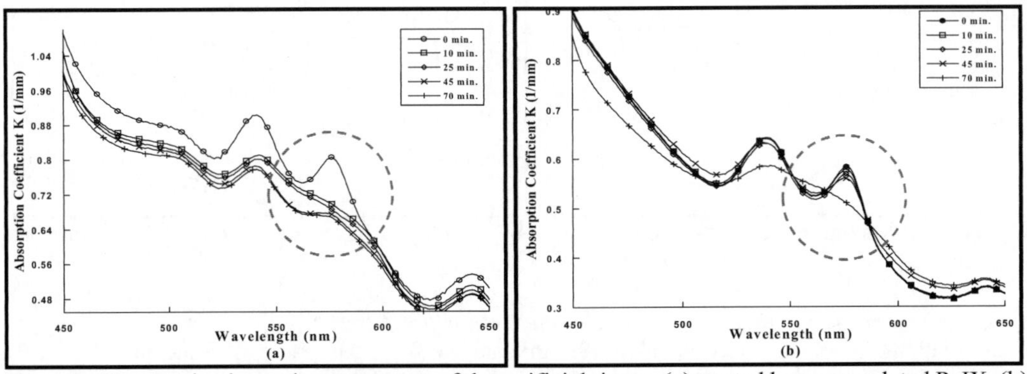

Figure 4. Changes in absorption spectrums of the artificial tissue: (a) treated by encapsulated PpIX, (b) treated by naked PpIX.

The higher decrease in the absorption coefficient of samples treated with encapsulated PpIX indicates a relatively faster rate of RBCs destruction compared to naked PpIX treatment because the agglomeration of encapsulated PpIX molecules by the SiNPs membrane enhances light energy absorbance and singlet oxygen production in contrast to the naked PpIX which is dispersed in the sample.

3.2 Optimal Exposure Time Measurement

This section entails the measurement of the optimal exposure time required for the samples (RBCs) to achieve half life time. Fig. 5 (a) depicts a decrease in the K's maximum peak of the samples treated with naked and encapsulated PpIX, following the exposure to light source. This is due to the engulfing of RBCs by $^1O_2$ produced through PpIX absorption of light energy [21]. This phenomenon is also indicated by the decrease in absorption coefficient. For the sample treated with encapsulated PpIX, an abrupt drop in K's maximum peak is observed in the first 10mins of light exposure, followed by a more subtle gradual decline, which indicates an initial rate of rapid RBC destruction. In contrast, treatment with naked PpIX show a steady decline in K's maximum peak for the first 50mins of light exposure, accompanied a faster decline, which indicates the naked PpIX takes a longer time to take effect. The relatively higher decline in the K's maximum peak of encapsulated PpIX treatment signifies RBCs are engulfed faster in the sample treated with encapsulated PpIX compared to naked PpIX.

Values of 5.78 and 68.00 min. were obtained for half life time of RBCs (fractional haemolysis of 50%) of samples treated with naked and encapsulated PpIX, respectively, as shown in Fig. 5 (b). The higher value for encapsulated PpIX is possibly attributable to the clustering and agglomeration of encapsulated PpIX molecules by SiNPs membrane, also pointed out by the increased absorbance of light energy and higher volume of singlet oxygen produced. For naked PpIX molecules, the molecules are homogeneously dispersed in the sample, which initiates a relatively slower interaction during exposure to a light source.

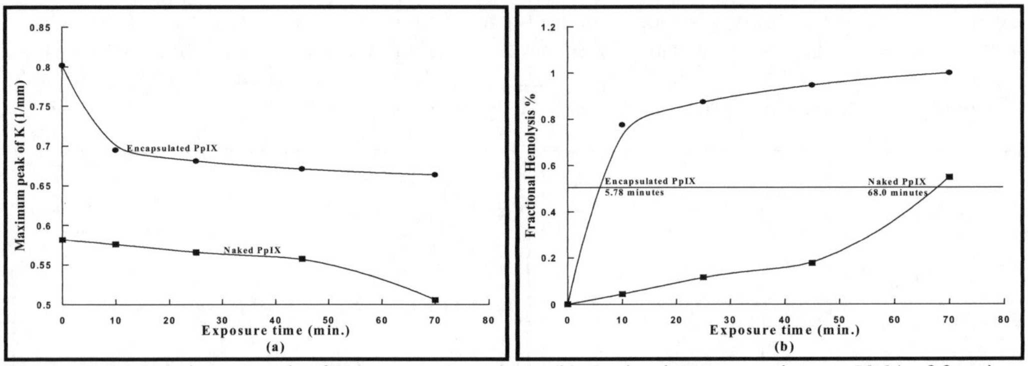

Figure 5. (a) Maximum peak of K vs. exposure time, (b) Optimal exposure time at 50 % of fractional haemolysis.

3.3 Correlation of Transmittance with the Thickness of Artificial Tissue

The variations in transmittance at 577nm as thickness (D) of the artificial tissue increases were measured for five samples, as shown in Fig. 6 (a). The Half Value Layer (HVL) of the samples was determined to be 0.45 and 0.53 mm for samples treated with encapsulated and naked PpIX respectively. The equation denoting the relationship between Transmission (T) and thickness (D) of the samples was derived from their Logarithm relation as shown in Fig. 6 (b).

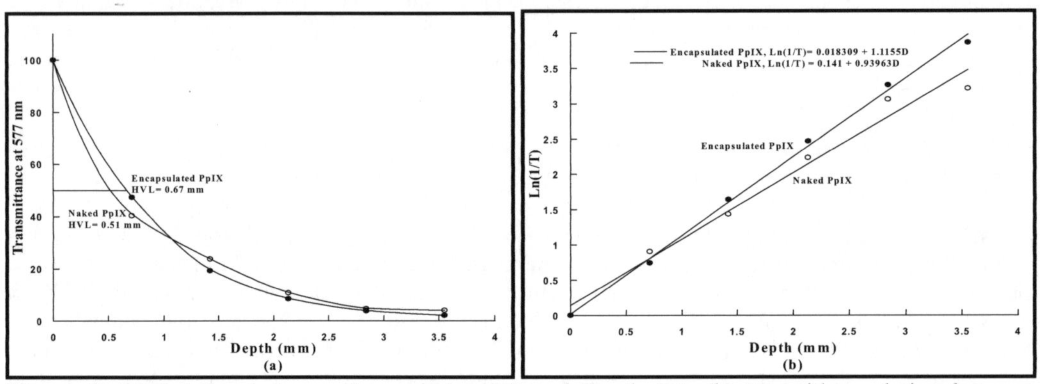

Figure 6. (a) Relation between Transmission (T) and Depth (D), (b) Logarithm relation between Transmission (T) and Depth (D).

An evaluation of the HVL values for both samples showed no significant disparity between the samples treated with encapsulated and naked PpIX. It can be deduced from the results that the relationship between Transmission (T) and Depth (D) can be applied to determine the HVL for other samples without having to replicate experiments.

4. Conclusion

It can be inferred from the results that encapsulation considerably improves the efficiency and applicability of PpIX as a photodynamic agent in the treatment of artificial tissues compared to the naked PpIX. As demonstrated in the *in-vitro* studies above, encapsulated PpIX destroys more RBCs than the naked PpIX. In general, SiNPs exhibited a high efficacy of 91.5 % (percentage between t_{50} for naked and encapsulated PpIX).

References

[1] He, X., Wu, X., Wang, K., Shi, B., & Hai, L., Methylene blue-encapsulated phosphonate-terminated silica nanoparticles for simultaneous in vivo imaging and photodynamic therapy. Biomaterials, 2009. 30(29): p. 5601-5609.

[2] Huang, Z., A Review of Progress in Clinical Photodynamic Therapy. Technol Cancer Res Treat, 2005. 4(3): p. 283–293.
[3] M-Ali H. Al-Akhras, Khaled Aljarrah, Borhan Albiss,Duaa Al-Khalili1.The Influence of Iron Oxide Nanoparticles (Fe3O4) on the Red Blood Cells Photohemolysis Sensitized with Photosensitizers; Photofrin and Rose Bengal.Photodiagnosis and Photodynamic Therapy. (2017), 18: 111–118
[4] M.-Ali H. Al-Akhras and Duaa J. Al-Khalili.The Influence of Iron Oxide Nanoparticles (Fe3O4) on the Red Blood Cells Photohemolysis Sensitized with Photofrin: Temperature effect. Jordan Journal of Biological Sciences:8 (1), 2015, pp. 55-59.
[5] Hermes, J. and C. Leuschner, Nanofabrication Towards Biomedical Applications, 2005, Wiley-VCH, Weinheim.
[6] Bharali, Dhruba J., Ilona Klejbor, Ewa K. Stachowiak, Purnendu Dutta, Indrajit Roy, Navjot Kaur, Earl J. Bergey, Paras N. Prasad, and Michal K. Stachowiak, Organically modified silica nanoparticles: a nonviral vector for in vivo gene delivery and expression in the brain. Proceedings of the National Academy of Sciences of the United States of America, 2005. 102(32): p. 11539-11544.
[7] Kneuer, C., Sameti, M., Haltner, E. G., Schiestel, T., Schirra, H., Schmidt, H., & Lehr, C. M., Silica nanoparticles modified with aminosilanes as carriers for plasmid DNA. International journal of pharmaceutics, 2000. 196(2): p. 257-261.
[8] Zhao, Y., Sadtler, B., Lin, M., Hockerman, G. H., & Wei, A., Nanoprobe implantation into mammalian cells by cationic transfection. Chemical communications, 2004(7): p. 784-785.
[9] Santra, Swadeshmukul, Heesun Yang, Debamitra Dutta, Jessie T. Stanley, Paul H. Holloway, Weihong Tan, Brij M. Moudgil, and Robert A. Mericle, TAT conjugated, FITC doped silica nanoparticles for bioimaging applications. Chemical communications, 2004(024): p. 2810-2811.
[10] Lee, J. B., Choi, J. Y., Chun, J. S., Yun, S. J., Lee, S. C., Oh, J., & Park, H. R, Relationship of protoporphyrin IX synthesis to photodynamic effects by 5-aminolaevulinic acid and its esters on various cell lines derived from the skin. British Journal of Dermatology, 2008. 159(1): p. 61-67.
[11] Berlin, N., A. Neuberger, and J. Scott, The metabolism of δ-aminolaevulic acid. 1. Normal pathways, studied with the aid of 15N. Biochemical Journal, 1956. 64(1): p. 80.
[12] Kloek, J., W. Akkermans, and G.M. Henegouwen, Derivatives of 5-Aminolevulinic Acid for Photodynamic Therapy: Enzymatic Conversion into Protoporphyrin. Photochemistry and photobiology, 1998. 67(1): p. 150-154.
[13] Georgakoudi, I. and T.H. Foster, Singlet Oxygen-Versus Nonsinglet Oxygen-Mediated Mechanisms of Sensitizer Photobleaching and Their Effects on Photodynamic Dosimetry. Photochemistry and photobiology, 1998. 67(6): p. 612-625.
[14] Gross, G., Therapy of human papillomavirus infection and associated epithelial tumors. Intervirology, 1997. 40(5-6): p. 368-377.
[15] Beutner, M., R. Karl, and M. Ferenczy, Therapeutic approaches to genital warts. The American journal of medicine, 1997. 102(5): p. 28-37.
[16] M-Ali H. Al-Akhras, Aljarrah, K, Al-omari, A., H. M. Al-Khateeb, B. A. Albiss, K. Azez,G. Makhadmeh.Influence of Extremely Low Energy Radiation on Artificial Tissue: Effects on Image quality and superficial dose. *Spectroscopy: An international journal* 22 (2008) pp 419-42826.
[17] Makhadmeh, Ghaseb Naser, Azlan Abdul Aziz, and Khairunisak Abdul Razak. "Artificial tissue sensitized with encapsulated methylene blue encapsulated by silica nanoparticles in photodynamic therapy." Artificial cells, nanomedicine, and biotechnology, 2016. 44.5: p. 1285-1289.

[18] Makhadmeh, Ghaseb N., Azlan Abdul Aziz, and Khairunisak Abdul Razak. "Semi-rigid artificial tissue sensitized by encapsulated Cichorium Pumilum in silica nanoparticles for photodynamic therapy application." AIP Conference Proceedings. Vol. 1657. No. 1. AIP Publishing, 2015.

[19] Dev Kumar Chatterjee , L.S.F., Yong Zhang, Nanoparticles in photodynamic therapy: An emerging paradigm. Advanced Drug Delivery Reviews, 2008. 60: p. 1627–1637.

[20] Jaradat, Adnan, M-Ali H. Al-Akhras, G. Makhadmeh, K. Aljarrah, A. Al-Omari, Z. Ababneh, Majed M. Masadeh, H. M. Al-Khateeb, B. A. Albiss, and M. Alshorman., Artificial semi-rigid tissue sensitized with natural pigments: Effect of photon radiations. Journal of Pharmacy And Bioallied Sciences, 2011. 3(2): p. 266.

[21] M-Ali H. Al-Akhras. A New Application of Gompertz Function in Photohemolysis: The Effect of Temperature on Red Blood Cell Hemolysis Photosensitized by Protoporphyrin IX. *Medical & Biological Eng& Computing*. 44 (2006) pp 703-710

IOP Conf. Series: Materials Science and Engineering **305** (2018) 012015 doi:10.1088/1757-899X/305/1/012015

Stress Relaxation in Porous GaN Prepared by UV Assisted Electrochemical Etching

K. Al-Heuseen, M. Kh. Alquran

Ajloun University College, Al-Balqa Applied University, Jordan

kalhussen@yahoo.com

Abstract: This study describes the stress relaxation in porous GaN grown on Al_2O_3 substrates. The results indicate that the stress relaxation has taken place in the samples which increases by increasing the etching current. As compared to the as-grown GaN films, porous GaN exhibits substantial photoluminescence (PL) intensity enhancement with red-shifted band-edge peaks associated with the relaxation of the compressive stress. The red shifted phonon energy peak (E_2) in the Raman spectra of the porous GaN films confirms further such a stress relaxation. Scanning electron microscopy (SEM) demonstrates that, the current density has significant effect on the size and shape of the pores.

Keywords: Stress relaxation; Porous GaN; Red shift; Raman spectroscopy.

1. Introduction

Among all the III-V nitrides, GaN is considerably the most intensely studied. GaN is a direct and wide band gap semiconductor (3.39 eV) at room temperature. With its superior radiation hardness and chemical stability, together with its large band gap characteristic, these properties have made GaN a suitable semiconductor material for device applications in the high-temperature and caustic environment as well as in space applications [1]. Producing and fabricating high quality electronic devices such as LEDs, LDs, photo detectors etcetera, requires growing high quality GaN doped thin films. Selecting appropriate substrate to deposit the film on it is one of the main factors that affect GaN growth techniques. The most common substrates used to grow GaN are; sapphire (Al_2O_3) and Si substrates. Sapphire is an interesting choice because it is a semi-insulating, high growth temperature and it is relatively cheap, but it is a large lattice mismatch (13.8%) [2], large thermal expansion coefficient mismatch (34%) and low thermal conductivity. The second choice is Si substrate which still has a large thermal expansion coefficient mismatch (56%) and large lattice mismatch (17%) [3]. The film/substrate mismatching produces high dislocation densities and high residual stress. Several techniques have been developed in order to reduce the dislocation density and residual stress including porous method.

One of the reasons for using a porous semiconductor as a substrate for depositing semiconductor layers is that, the nanopatterned porous structure can lead to a reduced extended defect density [4]. The interest in porous semiconductor arises from the fact that they can act as a sink for threading dislocations and accommodate the strain. Hence, they are widely used as a buffer or intermediate layer in epitaxial growth to obtain a subsequent layer with lowered strain and dislocation densities [5,6]. Mynbaeva et al. [7] proposed that the growth of GaN on porous GaN could lead to high-quality strain-released epilayers.
Porous GaN is a promising simple method to achieve low dislocation density and low residual stress on the whole area of a wafer. This nanostructure was first developed by Bardwell et al. by metal-assisted electroless chemical etching technique [8]. To date, several

IOP Conf. Series: Materials Science and Engineering **305** (2018) 012015 doi:10.1088/1757-899X/305/1/012015

additional techniques such as UV-assisted electrochemical etching and electrochemical etching have been developed [9-11]. In this method, the important parameters are applied voltage and current, electrolyte composition and illumination. These parameters control etch rate, morphology and optical properties.
In this work, we report the stress relaxation of GaN during etching process under different current densities of J= 5, 10 and 20 mA/cm^2 for 20 min. We investigated the surface morphology and the optical properties of porous GaN samples by scanning electron microscopy, and by high-spatial resolution photoluminescence and Raman spectroscopy.

2. Experiment

The samples used in this study were commercial (Si doped) n-GaN grown by metalorganic chemical vapor deposition (MOCVD) on Al_2O_3 substrates. The electron concentration obtained by Hall measurements was n=1x10^{17} cm^3. The samples were cleaned first with acetone and methanol, then cleaned in 1:20 NH_4OH:H_2O for 10 min, followed by cleaning in 1:50 HF:H_2O solution to remove the surface oxides. This was followed by cleaning in 3:1 HCl:HNO3 at 80 C° for 10 min. Between the cleaning steps, the samples were rinsed in deionized water. Aluminum was then partly evaporated on the surface of GaN using a thermal evaporation system, which provided an ohmic contact. By using a homemade Teflon-cell, we fixed the GaN sample as an anode and Pt wire as a cathode (fig.1). The electrolyte was a mixture of aqueous HF 49% and ethanol C2H5OH 95%, 1:4 by volume. We used constant current densities of J= 5, 10 and 20 mA/cm^2 for 20 min, (supplied by a Keithly 220 programmable current source) and a UV lamp in the electrochemical etching process. After the etching, the samples were rinsed in deionized water, and dried in ambient air. The optical properties of the porous GaN samples were investigated by high-spatial resolution Raman and PL spectroscopy.

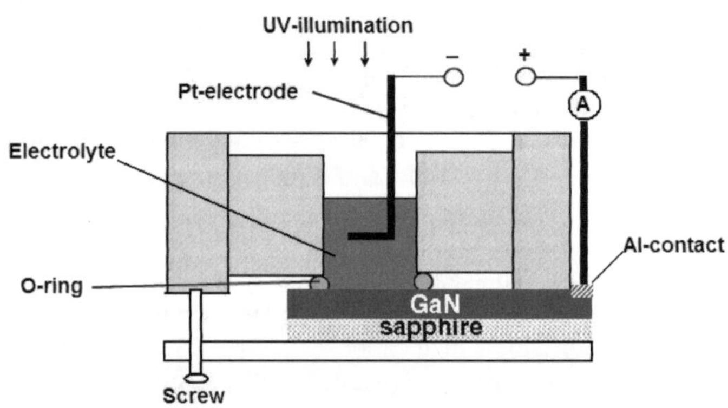

Fig. 1. The photoelectrochemical etching experimental setup used to generate porous GaN.

3. Results and Discussion

Figure 2 shows the scanning electron micrographs (SEM) of the morphology of the as-grown and porous GaN samples. The as-grown GaN possessed a smooth surface morphology (Figure. 2a). The evolution of the porous morphology with increase in etching current density is apparent from Figures 2(b-d). The figures show that the etching current density has significant effect on the size and shape of the pores. The pores stated to form in the sample etched with 5 mA/cm^2 as shown in Figure 2b. The sample etched with 10 mA/cm^2 (Figure 2c) shows that, the pore sizes appears to be around 30 to 40 nm and the distance between

IOP Conf. Series: Materials Science and Engineering **305** (2018) 012015 doi:10.1088/1757-899X/305/1/012015

them is around 50 to 60 nm. However, the sample etched with 20 mA/cm^2 (Figure 2d) shows that, the pore sizes were observed to increase which become around 50 to 60 nm and some areas were destroyed, indicating that the GaN layer was very thin [11,12].

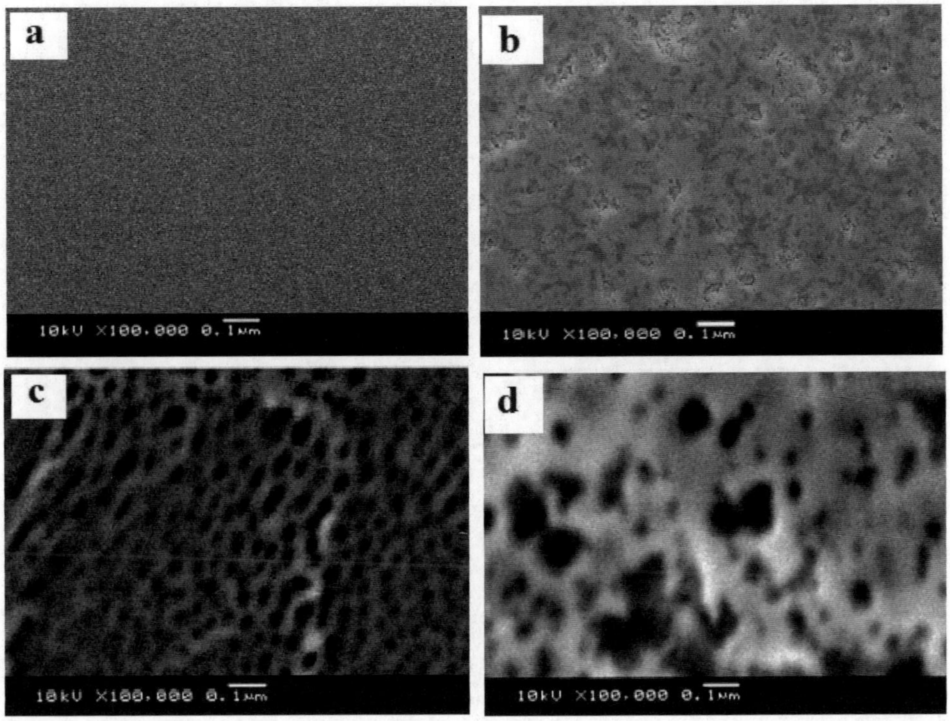

Fig. 2. SEM images of as-grown and porous GaN samples prepared with different etching current densities; a) as-grown b) 5 mA/cm^2, c) 10 mA/cm^2, and d) 20 mA/cm^2.

Figure 3 shows the room temperature PL spectra and the stress relaxation of porous GaN samples etched with various current densities. The spectra were observed to be slightly red-shifted for all samples (relative to the spectrum of the as-grown sample). Similar red-shifted PL from porous GaN has been reported before [13]. The red-shift can be attributed to the relaxation of the compressive stress in the porous samples. The energy shift between porous and the as-grown GaN sample can be used to quantify the amount of stress relaxation. The amount of stress relaxation can be calculated using the expression: $\Delta = K\sigma$, where K is the proportionality factor and σ is the stress. Using the K value reported in literature, i.e., K = 21.273.2 meV/GPa [14], using this value of the proportionality factor for the stress-induced PL peak shift, a compressive stress relaxation of about 0.12, 0.17 and 0.32 GPa is estimated in 5, 10 and 20 mA/cm^2 respectively.

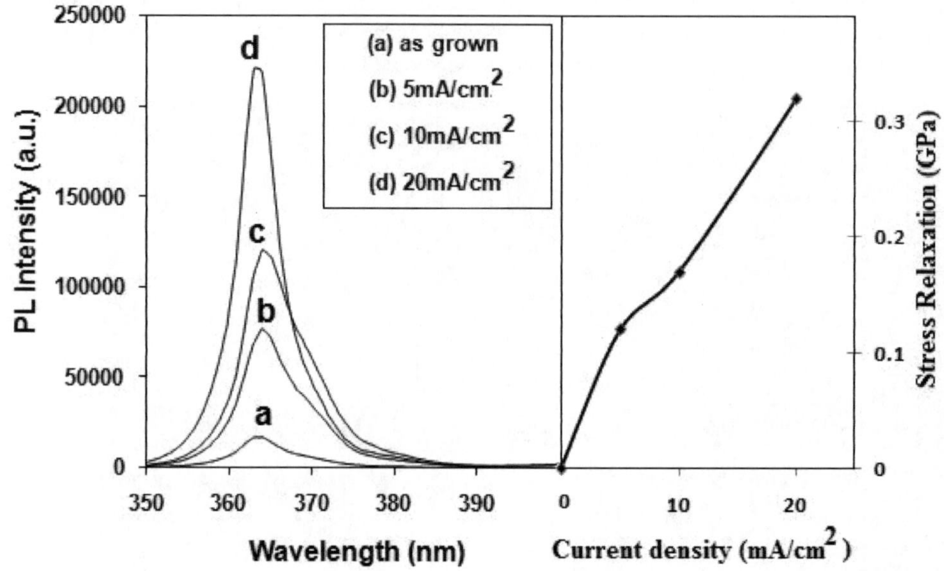

Fig. 3. Photoluminescence spectra and stress relaxation of porous GaN etched with different current densities: (a) as-grown, (b) 5 mA/cm^2, (c) 10 mA/cm^2 and (d) 20 mA/cm^2.

Small deviations in the peaks position indicating that the change of pore size has no considerable influence on the PL peak shift. However, the PL peak intensity of the porous samples has increased compared to that of the as-grown sample. The PL spectra for porous GaN samples show an increase in the intensity while increasing the current density. For example, at 10 mA/cm^2 and 20 mA/cm^2 current densities, the PL peak intensity was improved by factor of about 7 and 13, respectively, compared to that for the as-grown sample. The intensity of emitted light is proportional to the number of photons emitted. This means that the number of photons emitted is much higher in porous GaN than that in as-grown GaN. This could be attributed to the increase in the anti-reflectivity due to the roughening of the surface and the high surface area of porous structure.

Figure 4 shows Raman spectra of the as-grown and the porous GaN samples. The spectra exhibit silent phonon mode E_2 (high) and transverse optical mode A_1 (LO), and relatively small peak of A_1 (TO), also figure shows the increase in peaks intensity with etching current density. The frequencies of all the observed modes were in good agreement with the results of other researchers [15-17]. The phonon peak position, intensity and peak shift for GaN samples etched with different current densities are summarized in Table 1.

Circle inset in figure 4 shows the Raman spectra on E_2 (high) together with that of the as-grown one. Moreover, the E_2 Raman mode of as grown GaN appears at an identical wave number of 566.2 cm^{-1}. We take this number as a reference for unstressed films.

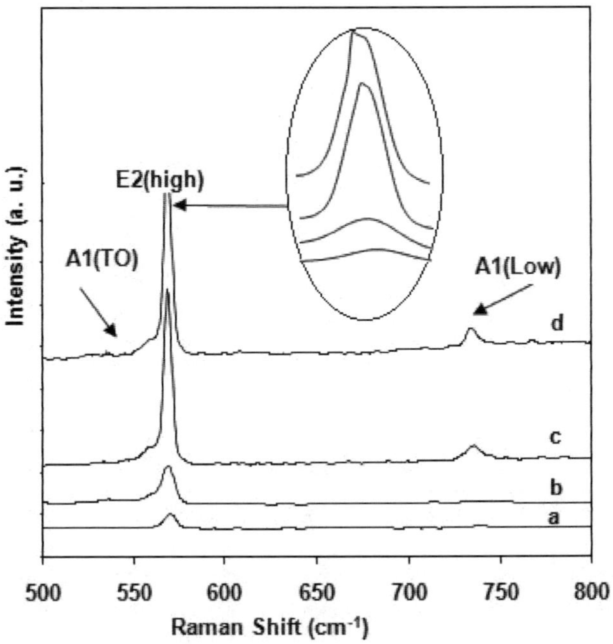

Fig. 4. Raman spectra of samples etched with different current densities; as-grown GaN, 5 mA/cm^2, 10 mA/cm^2 and 20 mA/cm^2, indicated by (a), (b), (c) and (d) respectively. (circle inset showing the small shift of E_2 (high) for porous samples compared to that of as-grown.

All the porous samples prepared with etching current densities of 5, 10 and 20 mA/cm^2 show an E_2 (high) shifted to slightly lower frequency than that of the as-grown one. Using the proportionality factor of 4.2 cm^{-1}/GPa for hexagonal GaN [16,18], the total stress in the as-grown GaN is 0.85GPa compared to the unstressed sample (free stress). While the relaxation of stress took place in the porous sample, and by using the same proportionality factor, the relaxation of stress was found to be 0.25 GPa for the 5 mA/cm^2 sample, 0.38 GPa for the 10 mA/cm^2 sample and 0.51 GPa for the 20 mA/cm^2 sample. These results indicate the stress relaxation has taken place in the samples and it increases by increasing the etching current density as seen in Figure 5.

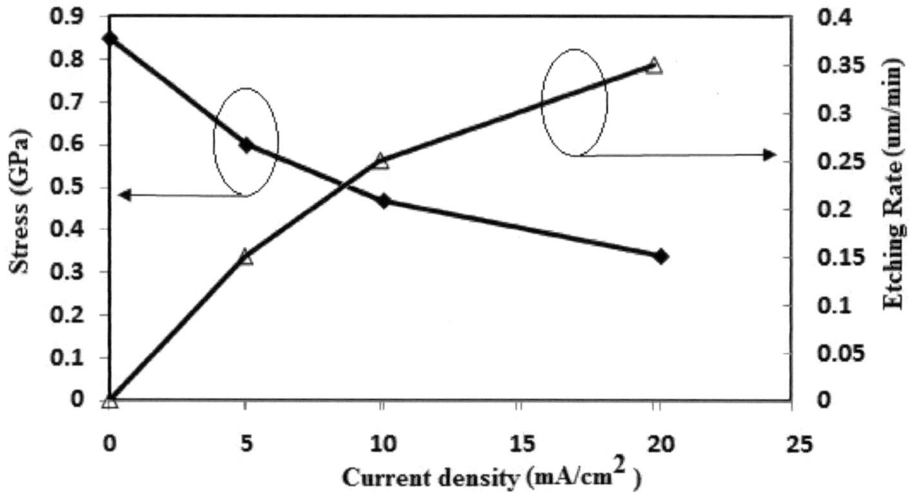

Fig. 5. The stress and the etching rate as a function of the current density.

Figure 5 also shows the dependence of the etching rate with respect to the current density. From this figure, we can conclude that the etch rate increases with the current density by an exponential proportionality. Another conclusion is that we can obtain from Figure 5 that, for high current densities, the higher etch rate hinders the fine control of the layer thickness in the case of thin layer.

Table 1. The phonon modes detected in the Raman spectra.

Phonon mode		**E_2(low)**	**A_1(TO)**	**A_1(LO)**	**E_2(high)**	**FWHM of E_2(high) (cm^{-1})**
as-grown (a)	Peak position (cm^{-1})	147.474	523.897	738.366	569.794	5.850
	Intensity (a.u)	41.377	37.222	45.742	102.757	
5 mA/cm^2 (b)	Peak position (cm^{-1})	143.517	536.200	738.880	568.734	5.570
	Intensity (a.u)	49.878	70.927	62.857	233.544	
	Shift (cm^{-1})	-3.960	12.300	0.514	-1.060	
10 mA/cm^2 (c)	Peak position (cm^{-1})	142.950	532.990	735.742	568.199	2.990
	Intensity(a.u)	40.484	51.185	126.711	871.553	
	Shift (cm^{-1})	-4.524	9.090	-2.624	-1.595	
20 mA/cm^2 (d)	Peak position (cm^{-1})	143.517	535.669	734.690	567.688	2.900
	Intensity (a.u)	65.000	163.712	283.981	1039.340	

Shift (cm^{-1})	-3.960	11.772	-3.676	-2.126

The full width at half maximum (FWHM) value of E_2 phonon for the as-grown and the porous samples at 5, 10 and 20 mA/cm^2 are also shown in Table 1, indicating decreasing FWHM value with increasing etching current density. The lower value of the FWHM represents good crystalline quality. This suggests an improvement in the crystalline structure of porous GaN with increasing etching current density [19].

Generally, the porous samples exhibited broader FWHM than the as grown sample for (0002) a diffraction planes, this could be due to the relatively wide statistical size distribution of the pores, however for porous samples, no relation between the FWHM and the average pore size was observed.

4. Conclusions

Porous n-GaN was prepared by photoelectrochemical etching under ultraviolet light using various current densities in electrolytes consisting of aqueous HF and ethanol electrolyte at room temperature. All porous samples showed increased PL and Raman intensities compared with that of the as-grown. PL and Raman results confirms that porous GaN layer possess less compressive stress than that for the as-grown GaN layer and the quality of the porous sample is as good as the as-grown sample. The peaks of E_2 (high) in Raman spectra for all the porous samples suggest that the crystallite quality and stress relaxation increases with increasing the current density. The relaxation of stress is 0.25 GPa for the 5 mA/cm^2 sample, 0.38 GPa for the 10 mA/cm^2 sample and 0.51 GPa for the 20 mA/cm^2 sample. These results indicate that the stress relaxation has taken place in the samples and increasing with higher etching current density. This indicates that, it is possible to fabricate high quality porous GaN layer with tunable stress, suitable for growing or depositing further high quality active layers with minimized lattice mismatch.

References:

[1] H.D. Xiao, J.Q. Liu, C.N. Luan, Z.W. Ji, J.S. Cui, *Appl Phys Lett* **100**, 213101 (2012).
[2] S. C. Binari, K. Doverspike, G. Kelner, H. B Dietrich, A. E. Wickenden, Solid-State Electron, **41**, 177(1997).
[3] T.L. Tansley, E.M. Goldys, M. Godlewski, B. Zhou and H. Y. Zuo, (1997) GaN and Related Materials, (S. J. Pearton ed., Gordon and Breach Science Publishers, Amsterdam.
[4] L. Beji, B. Ismail, L. Sfaxi, F. Hassen, H. Maaref, H. Ben Ouada, J. Cryst. Growth. **258**, 84 (2003).
[5] M. Mynbaeva, A. Titkov, A. Kryzhanovski, I. Kotousova, A.S. Zubrilov, V.V. Ratnikov, V.Y. Davydov, N.I. Kuznetsov, K. Mynbaev, D.V. Tsvetkov, S. Stepanov, A. Cherenkov, V.A. Dmitriev, MRS Internet J. Nitr. Semicond. Res. **4,** 14 (1999).
[6] R.S. Qhalid Fareed, V. Adivarahan, C.Q. Chen, S. Rai, E. Kuokstis, W. Yang, M. AsifKhan, J. Caissie, R.J. Molnar, Appl. Phys. Lett. **84**, 696 (2004).
[7] M. Mynbaeva, A. Titkov, A. Kryganovskii, V. Ranikov, K.H. Mynbaev, R. Huhtinen, Laiho, V. Dmitriev, Appl. Phys. Lett. **76**, 1113 (2000).
[8] J.A. Bardwell, I.G. Foulds, J.B. Webb, H. Tang, J. Fraser, S. Moisa, S.J. Rolfe, J. Electron, Mater. **28**, 24 (1999).

IOP Conf. Series: Materials Science and Engineering **305** (2018) 012015 doi:10.1088/1757-899X/305/1/012015

[9] M. Ikram, Md Taib, N. Zainal, Z. Hassan , M. Abu Bakar, ECS J. Solid State Sci. Technol. **5**, 10, (2016).
[10] D. Chen, H.D. Xiao, J. Han, J. Appl. Phys. **112**, 064303 (2012).
[11] Yusuke Kumazaki , Satoru Matsumoto , Taketomo Sato, J. Electrochem. Soc. **164**,7, H477-H483 (2017).
[12] H. Hartono, C. B.Soh , S. J. Chua, E. A. Fitzgerald, Phys. Stat. Sol. **4**, 2572 (2007) .
[13] K. Al-Heuseen, M.R. Hashim, and N.K. Ali, *J. Electrochem. Soc.* 158, D240 (2011).
[14] D.G. Zhao, S.J. Xu, M.H. Xie, S.Y. Tong, and H. Yang, Appl. Phys. Lett. **83**, 677 (2003).
[15] H. Harima, J. Phys. Cond. Matter. **14**, R967 (2002) .
[16] C. Kisielowski, J. Kruger, S. Ruvimov, T. Suski, J. W. Ager, E. Jones, Z. Liliental, M. Rubin, E. R. Weber, Phys. Rev. B **54**, 17745 (1996).
[17] P. Perlin, C.J. Carillon, J.P. Itie, A.S. Miguel, I. Grezgory, A. Polian, Phys. Rev. B **45**, 83(1992).
[18] M. Kuball, Surf. Interface Anal., 31, 987 (2001).
[19] A.P. Vajpeyi, S. Tripathy, S. J. Chua, E. A. Fitzgerald Physica E, **28**, 141 (2005).

The high-pressure stability of Ni_2In-type structure of ZrO_2 with respect to OII and Fe_2P-type phases: A *first-principles* study

H. Al-Taani[1*], K. Tarawneh[2], Y. Al-Khatatbeh[2] and B. Hamad[3,4]

[1] *School of Basic Sciences and Humanities, German Jordanian University, Amman 11180, Jordan*
[2] *Department of Basic Sciences, Princess Sumaya University for Technology, Amman 11941, Jordan*
[3] *Physics Department, The University of Jordan, Amman 11942, Jordan*
[4] *Physics Department, University of Arkansas, Fayetteville, AR 72701, USA*

hussein.taani@gju.edu.jo

Abstract: The density-functional theory is used to investigate the stability of the Ni_2In-type hexagonal structure with the space group: $P6_3/mmc$ at high pressures and compared to the orthorhombic OII and Fe_2P-type phases of zirconia (ZrO_2). The calculations showed that the high-pressure phase transition sequence in ZrO_2 is as follows: OII ➔ Fe_2P ➔ Ni_2In, which is consistent with the recent measurements that are observed in Fe_2P as a post-OII phase of ZrO_2. We obtained a very small volume change across OII ➔ Fe_2P transition, whereas an appreciable volume change is found across the Fe_2P ➔ Ni_2In transition. The compressibility of Ni_2In phase is found to be high as compared to the other two high-pressure phases.

1. Introduction

The transition-metal dioxide zirconia (ZrO_2) is involved in many industrial applications [1–7] due to its important properties. As a result of the bonding nature of this dioxide, ZrO_2 has received an obvious attention regarding its structural and mechanical properties, where many experimental (e.g., Refs. [8–13]) and theoretical (e.g., Refs. [8, 11, 14–17]) studies have focused on investigating the high-pressure behavior of ZrO_2 including the determination of the EOS of its various polymorphs, phase stability, and mechanical strength. At ambient temperature, the well-known high-pressure phase transition sequence [8, 12, 14, 15, 18, 19] is as follows: MI ➔ OI ➔ OII, where MI has a monoclinic structure (space group: $P2_1/c$) and both OI and OII have an orthorhombic structure (space groups: *Pbca* and *Pnma*, respectively). Recent measurements and calculations have confirmed the discovery of Fe_2P-type structure (Hexagonal, space group: $P\bar{6}2m$) as a post-OII phase for ZrO_2 at ultrahigh pressures [11], and thus, the most recent high-pressure sequence becomes as follows: MI ➔ OI ➔ OII ➔ Fe_2P (Fig. 1). We should mention that Fe_2P-type structure has been also discovered as a post-OII phase in the similar dioxides TiO_2 [20] and HfO_2 [21]. Recently, after this discovery, another theoretical study [16] has proposed a new hexagonal phase (Ni_2In-type structure) of ZrO_2 (Fig. 1) to be the most stable phase at pressures greater than 380 GPa. However, in this work [16], the transition to Ni_2In phase is claimed to likely occur from OII phase rather than Fe_2P phase, indicating that this study [16] suggests the following transition sequence: MI ➔ OI ➔ OII ➔ Ni_2In. In fact, ignoring Fe_2P phase from the high-pressure phase transition sequence is not consistent with the recent experimental and theoretical studies [11, 20, 21] that have confirmed the phase stability of Fe_2P as a post-OII phase. Thus, the reasonable expectation is that any new proposed phase transition in ZrO_2 should occur from the experimentally observed Fe_2P phase.

In our study, we use DFT calculations to test the stability of Ni_2In-type structure at high pressures with respect to OII and Fe_2P phases. The main goal in this work is to investigate the upper part of high-pressure phase transition sequence of ZrO_2 in an effort to better understand the high-pressure phase diagram of this dioxide.

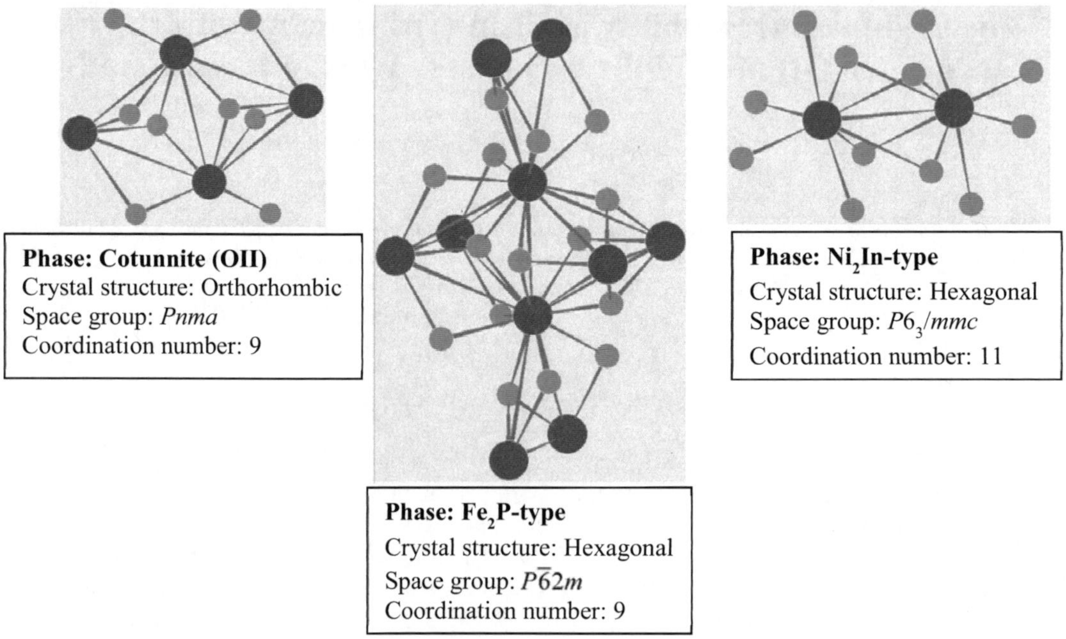

Figure 1: Crystal structures of ZrO_2 phases. The larger dark spheres represent the Zr atom, while the smaller light spheres represent the O atom.

2. Theoretical Methods

To investigate the high-pressure phase transitions and the EOSs within the three ZrO_2 phases, we used density-functional theory (DFT) [22] based static *first-principles* computations. The projector-augmented wave (PAW) formalism [23, 24] was used to treat the interactions between the atoms having a core radii of 2.500 Bohr for hafnium Hf (with the valence configuration of $4s^2 4p^6 5s^2 4d^2$) and 1.520 Bohr for oxygen O (with the valence configuration of $2s^2 2p^4$). The generalized gradient approximation (GGA) [25] was used to treat the electronic exchange and correlation effects. We performed our calculations using the VASP software package [26–29] with an energy cutoff of 600 eV and Γ-centered *k*-point meshes [30]. For all phases, total energies and pressures were converged to better than ~0.1 meV/atom and ~0.1 GPa, respectively. The Brillouin zone integration was performed using the following *k*-point meshes for the ZrO_2 phases: 4x8x4 for OII, 6x6x10 for Fe_2P, and 10x10x8 for Ni_2In. For a fixed volume, all internal degrees of freedom and unit-cell parameters of the structure were optimized simultaneously during the geometry optimizations. The ground-state energy for each phase was determined for 13-20 volumes. Up to highest pressure achieved in our calculations, all ZrO_2 phases remain insulators.

3. Results and Discussion

3.1. Bulk modulus determination

To study the compressibility of the three phases, we used Birch-Murnaghan (BM-) EOS to determine the bulk modulus for each phase. In the third-order BM-EOS [31], the pressure *P* is given by

$$P(V) = \frac{3}{2} K_0 \left[\left(\frac{V}{V_0} \right)^{-\frac{7}{3}} - \left(\frac{V}{V_0} \right)^{-\frac{7}{3}} \right] \left[1 + \frac{3}{4} \left((K_0' - 4) \left(\left(\frac{V}{V_0} \right)^{-\frac{2}{3}} - 1 \right) \right) \right] \quad \text{Eqn. (1)}$$

where V is the volume at pressure P, V_0 is the zero-pressure volume, K_0 is the zero-pressure bulk modulus, and K_0' is the first pressure derivative of the bulk modulus at zero pressure. From the thermodynamic relationship: $P(V) = -\frac{\partial E}{\partial V}$, the total energy E in the third-order form of BM-EOS is written as

$$E(V) = \frac{9K_0V_0}{2}\left[\frac{1}{2}\left(\left(\frac{V}{V_0}\right)^{-\frac{2}{3}} - 1\right)\right]^2\left[1 + (K_0'-4)\left(\frac{1}{2}\left(\left(\frac{V}{V_0}\right)^{-\frac{2}{3}} - 1\right)\right)\right] + E_0 \qquad \text{Eqn. (2)}$$

Where E_0 is the total energy at zero pressure. One can obtain the second-order BM-EOS by substituting $K_0' = 4$ in Eqns. 1 and 2. The EOS parameters for each ZrO_2 phase were obtained by fitting the total energy as a function of volume to the second- and third-order BM-EOS [31] (Table I). We determined the EOSs for the three phases and summarized them in Table I. We note that our bulk modulus (K_0) for both OII and Fe_2P phases is in excellent agreement with previous studies [8, 11, 16]. On the other hand, our computed K_0 for Ni_2In phase is ~16% less than previous results [16]. However, we should note that BM-EOS is sensitive to both V_0 and K_0' [31], where K_0 decreases with increasing V_0 and/or K_0'. Although our K_0 value (200 GPa) of Ni_2In is less than the previously reported one (239 GPa) [16], V_0 (31.81 Å^3) and K_0' (4) are greater than the reported values [16] of V_0 (29.21 Å^3) and K_0' (3.86).
The EOS calculations show that the change in K_0 across the phases is as follows: OII ➔ Fe_2P (K_0 increases) and Fe_2P ➔ Ni_2In (K_0 decreases), in agreement with previous studies [8, 11, 16]. In details, using the second-order BM-EOS, K_0 increases by ~5.8% across OII ➔ Fe_2P and decreases by ~26% across Fe_2P ➔ Ni_2In. This clearly indicates that Ni_2In-ZrO_2 phase is noticeably more compressible than OII and Fe_2P phases.

Table I: Theoretical EOSs of various ZrO_2 phases using the second- and third-order BM-EOS [31]. For comparison, we list other theoretical results [8, 11, 16]. 1σ uncertainties are given in parentheses.

Phase	Equation of state			Reference
	V_0 (Å^3)	K_0 (GPa)	K_0'	
OII	31.35 (0.04)	251 (3)	4 (fixed)	Ref. [8]
	30.66	258	4 (fixed)	Ref. [11]
	30.78	242	4.24	Ref. [11]
	30.88	211	4.01	Ref. [16]
	31.28 (0.04)	257 (2)	4 (fixed)	This work
	31.44 (0.03)	234 (3)	4.43 (0.06)	This work
Fe_2P	30.34	272	4 (fixed)	Ref. [11]
	30.43	260	4.18	Ref. [11]
	30.17	248	3.76	Ref. [16]
	30.94 (0.03)	272 (2)	4 (fixed)	This work
	31.03 (0.01)	255 (1)	4.31 (0.01)	This work
Ni_2In	29.21	239	3.86	Ref. [16]
	31.81 (0.13)	200 (5)	4 (fixed)	This work
	32.18 (0.03)	151 (2)	5.10 (0.04)	This work

To explain the noticeable compressibility in Ni_2In phase, we have investigated the change in the lattice parameters (a and c) of this phase with increasing pressure. Figure 2 clearly shows that up to a few tens

IOP Conf. Series: Materials Science and Engineering **305** (2018) 012016 doi:10.1088/1757-899X/305/1/012016

of GPa, the *c*/*a* ratio sharply decreases as pressure increases. This indicates that at relatively low pressure the lattice parameter *a* is much more incompressible than the parameter *c*, which likely explains the low K_0 value of Ni_2In phase.

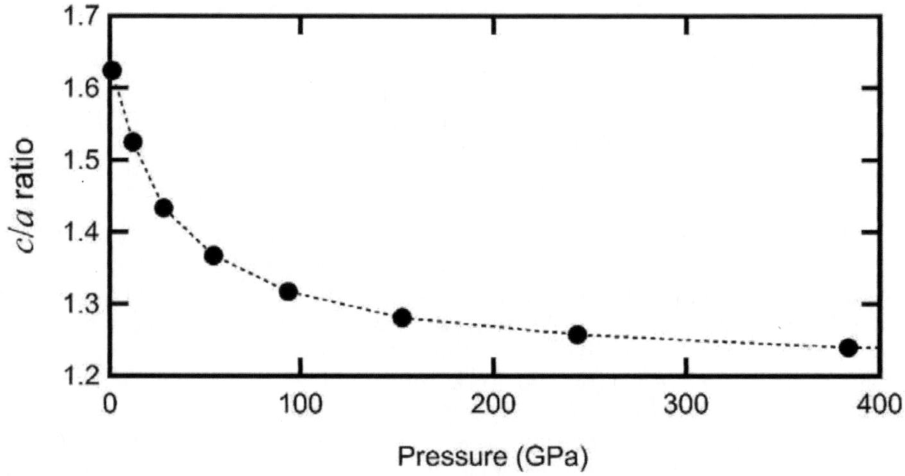

Figure 2: *c*/*a* ratio for Ni_2In-ZrO_2 phase as a function of pressure. The lattice parameter *a* is more incompressible than the parameter *c*, especially at low pressures.

3.2. Phase stability

Figure 3 shows the change in enthalpy of Fe_2P and Ni_2In phases with respect to OII phase. Our calculated transition pressures are as follows: across OII ➔ Fe_2P is 94 GPa (99 GPa) using the second- (third-) order BM-EOS [31], and across Fe_2P ➔ Ni_2In is 317 GPa (311 GPa) using the second- (third-) order BM-EOS [31]. Although our calculated transition pressure across OII ➔ Fe_2P of ~26-30% is less than previously reported results [11, 16], this transition is consistent with the recent measurements and calculations, where Fe_2P is observed to be the post-OII phase in ZrO_2 [11] and in the similar dioxide TiO_2 [20], and thus, any high-pressure phase transition to a new structure should occur from Fe_2P phase. However, a recent theoretical study [16] has concluded that the transition to Ni_2In phase is likely to occur from OII phase and not from Fe_2P phase which is not consistent with the recent experimental and theoretical results [20, 11]. These studies have obviously shown that Fe_2P phase is more stable at high pressures compared to OII phase. On the other hand, our high-pressure phase transition sequence for ZrO_2 (OII ➔ Fe_2P ➔ Ni_2In) is in reasonable agreement with previous experimental studies [20, 11] as we propose the transition to Ni_2In to occur from Fe_2P rather than from OII. Even though Ni_2In-ZrO_2 phase has not been experimentally observed yet, likely due to the extreme pressure-temperature conditions required to stabilize it, our calculations show that it is the most stable phase of ZrO_2 at ultra-high pressures in agreement with previous calculations [16].

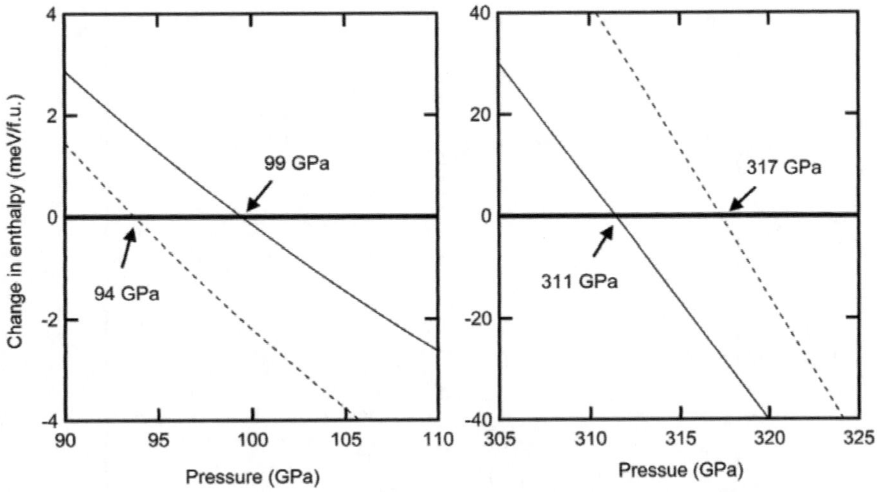

Figure 3: Change in enthalpy with respect to OII phase versus pressure of one ZrO_2 formula unit. The transition pressures from OII to Fe_2P (**left**) and from Fe_2P to Ni_2In (**right**) using a second- (dashed curve) and third- (solid curve) order BM-EOS [31] are given.

3.3. Volume change across transitions

In this section we discuss the volume decrease across each transition of the phase sequence OII ➔ Fe_2P ➔ Ni_2In. Figure 4 shows the volume of each ZrO_2 phase as a function of pressure as well as the volume decrease across each transition. Our calculations predict a small volume decrease of ~0.4% across OII ➔ Fe_2P, while we predict a large volume change of ~3.6% across Fe_2P ➔ Ni_2In. However, we should note that the coordination number (CN) across the transition OII ➔ Fe_2P remains unchanged, which explains the small volume decrease across this transition. On the other hand, the Fe_2P ➔ Ni_2In transition is associated with a CN increase from 9 to 11, and therefore, a large volume decrease is found as expected. This large volume decrease is consistent with previous studies on ZrO_2 [8] and similar dioxides TiO_2 [32] and HfO_2 [33] that have shown a large volume collapse when the transition across the phases is associated with a CN increase.

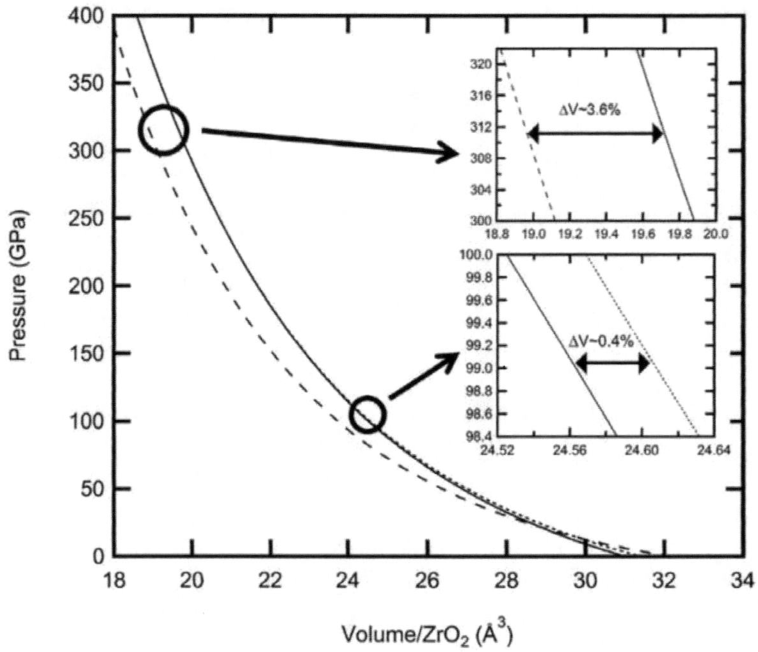

Figure 4: Pressure versus volume of one ZrO_2 formula unit using the third-order BM-EOS [31]. The dotted line represents the OII phase, and the solid line represents the Fe_2P phase, whereas the dashed line represents the Ni_2In phase. The volume change at the transition pressure is shown (**upper inset**: across Fe_2P ➔ Ni_2In, **lower inset**: across OII ➔ Fe_2P).

4. Conclusions

In summary, using DFT calculations, we investigated the stability of Ni_2In phase at high pressures with respect to OII and Fe_2P phases for ZrO_2. Our calculations showed that the predicted high-pressure phase sequence across these phases is as follows: OII ➔ Fe_2P ➔ Ni_2In, where Ni_2In phase is stable at pressures greater than ~317 GPa. Furthermore, we computed BM-EOS for the three phases, and found that the compressibility of Ni_2In phase is obviously high when compared to the other two high-pressure phases. Finally, we have predicted a very small volume change across OII ➔ Fe_2P and a noticeably large volume change across Fe_2P ➔ Ni_2In transition.

Acknowledgements

This work was completed utilizing the Holland Computing Center of the University of Nebraska.

References

[1] D. Panda and T.-Y. Tseng, "Growth, dielectric properties, and memory device applications of ZrO_2 thin films," *Thin Solid Films*, vol. 531, pp. 1–20, 2013.
[2] Y. H. Wu, L. L. Chen, W. C. Chen, C. C. Lin, M. L. Wu, and J. R. Wu, "MOS devices with tetragonal ZrO_2 as gate dielectric formed by annealing ZrO_2/Ge/ZrO_2 laminate," in *Microelectronic Engineering*, 2011, vol. 88, no. 7, pp. 1361–1364.
[3] I. J. Berlin and K. Joy, "Optical enhancement of Au doped ZrO_2 thin films by sol-gel dip coating method," *Phys. B Condens. Matter*, vol. 457, pp. 182–187, 2015.
[4] B. Lee *et al.*, "A novel thermally-stable zirconium amidinate ALD precursor for ZrO_2 thin films," *Microelectron. Eng.*, vol. 86, no. 3, pp. 272–276, 2009.
[5] A. G. Sato, D. P. Volanti, D. M. Meira, S. Damyanova, E. Longo, and J. M. C. Bueno, "Effect of the ZrO_2 phase on the structure and behavior of supported Cu catalysts for ethanol conversion," *J. Catal.*, vol. 307, pp. 1–17, 2013.
[6] T. Van Gestel, D. Sebold, H. Kruidhof, and H. J. M. Bouwmeester, "ZrO_2 and TiO_2

membranes for nanofiltration and pervaporation. Part 2. Development of ZrO_2 and TiO_2 toplayers for pervaporation," *J. Memb. Sci.*, vol. 318, no. 1–2, pp. 413–421, 2008.

[7] I. A. Siddiquey, T. Furusawa, M. Sato, N. M. Bahadur, M. N. Uddin, and N. Suzuki, "A rapid method for the preparation of silica-coated ZrO_2 nanoparticles by microwave irradiation," *Ceram. Int.*, vol. 37, no. 6, pp. 1755–1760, 2011.

[8] Y. Al-Khatatbeh, K. K. M. Lee, and B. Kiefer, "Phase relations and hardness trends of ZrO_2 phases at high pressure," *Phys. Rev. B*, vol. 81, p. 214102, 2010.

[9] J. Haines, J. M. Leger, and A. Atouf, "Crystal structure and equation of state of cotunnite-type zirconia," *J. Am. Ceram. Soc.*, vol. 78, no. 2, pp. 445–448, 1995.

[10] J. M. Leger, P. E. Tomaszewski, A. Atouf, and A. S. Pereira, "Pressure-induced structural phase transitions in zirconia under high pressure," *Phys. Rev. B*, vol. 47, no. 21, pp. 14075–14083, 1993.

[11] D. Nishio-Hamane, H. Dekura, Y. Seto, and T. Yagi, "Theoretical and experimental evidence for the post-cotunnite phase transition in zirconia at high pressure," *Phys. Chem. Miner.*, vol. 42, no. 5, pp. 385–392, 2015.

[12] O. Ohtaka, D. Andrault, P. Bouvier, E. Schultz, and M. Mezouar, "Phase relations and equation of state of ZrO_2 to 100 GPa," *J. Appl. Crystallogr.*, vol. 38, pp. 727–733, 2005.

[13] O. Ohtaka, H. Fukui, K. Funakoshi, W. Utsumi, T. Irifune, and T. Kikegawa, "Phase relations and EOS of ZrO_2 and HfO_2 under high-temperature and high-pressure," *High Press. Res.*, vol. 22, pp. 221–226, 2002.

[14] J. E. Jaffe, R. A. Bachorz, and M. Gutowski, "Low-temperature polymorphs of ZrO_2 and HfO_2: A density-functional theory study," *Phys. Rev. B*, vol. 72, p. 144107, 2005.

[15] G. Fadda, L. Colombo, and G. Zanzotto, "First-principles study of the structural and elastic properties of zirconia," *Phys. Rev. B*, vol. 79, p. 214102, 2009.

[16] M. Durandurdu, "Novel high-pressure phase of ZrO_2: An ab initio prediction," *J. Solid State Chem.*, vol. 230, no. July, pp. 233–236, 2015.

[17] H. Ozturk and M. Durandurdu, "High-pressure phases of ZrO_2: An ab initio constant-pressure study," *Phys. Rev. B*, vol. 79, p. 134111, 2009.

[18] Y. Al-Khatatbeh and K. K. M. Lee, "From superhard to hard: A review of transition metal dioxides TiO_2, ZrO_2, and HfO_2 hardness," *J. Superhard Mater.*, vol. 36, no. 4, pp. 231–245, 2014.

[19] O. Ohtaka, T. Yamanaka, S. Kume, E. Ito, and A. Navrotsky, "Stability of monoclinic and orthorhombic zirconia: Studies by high-pressure phase equilibria and calorimetry," *J. Am. Ceram. Soc.*, vol. 74, no. 3, pp. 505–509, 1991.

[20] H. Dekura, T. Tsuchiya, Y. Kuwayama, and J. Tsuchiya, "Theoretical and Experimental Evidence for a New Post-Cotunnite Phase of Titanium Dioxide with Significant Optical Absorption," *Phys. Rev. Lett.*, vol. 107, p. 45701, 2011.

[21] Y. Al-Khatatbeh, K. Tarawneh, H. Taani, and K. K. M. Lee, "Unpublished data."

[22] P. Hohenberg and W. Kohn, "Inhomogeneous electron gas," *Phys. Rev.*, vol. 136, no. 3B, pp. B864–B871, 1964.

[23] G. Kresse and D. Joubert, "From ultrasoft pseudopotentials to the projector augmented-wave method," *Phys. Rev. B*, vol. 59, no. 3, pp. 1758–1775, 1999.

[24] P. E. Blochl, "Projector augmented-wave method," *Phys. Rev. B*, vol. 50, no. 24, p. 17953, 1994.

[25] J. P. Perdew, K. Burke, and M. Ernzerhof, "Generalized gradient approximation made simple," *Phys. Rev. Lett.*, vol. 77, no. 18, pp. 3865–3868, 1996.

[26] G. Kresse and J. Furthmuller, "Efficiency of ab-initio total energy calculations for metals and semiconductors using a plane-wave basis set," *Comput. Mater. Sci.*, vol. 6, no. 1, pp. 15–50, 1996.

[27] G. Kresse and J. Furthmuller, "Efficient iterative schemes for ab initio total-energy calculations using a plane-wave basis set," *Phys. Rev. B*, vol. 54, no. 16, pp. 11169–11186, 1996.

[28] G. Kresse and J. Hafner, "Norm-conserving and ultrasoft pseudopotentials for first-row and transition elements," *J. Phys. Condens. Matter*, vol. 6, no. 40, p. 8245, 1994.

[29] G. Kresse and J. Hafner, "Ab initio molecular dynamics for open-shell transition metals," *Phys. Rev. B*, vol. 48, no. 17, pp. 13115–13118, 1993.
[30] H. J. Monkhorst and J. D. Pack, "Special points for Brillouin-zone integrations," *Phys. Rev. B*, vol. 13, no. 12, pp. 5188–5192, 1976.
[31] F. Birch, "Elasticity and constitution of the Earth's interior," *J. Geophys. Res.*, vol. 57, pp. 227–234, 1952.
[32] Y. Al-Khatatbeh, K. K. M. Lee, and B. Kiefer, "High-pressure behavior of TiO_2 as determined by experiment and theory," *Phys. Rev. B*, vol. 79, p. 134114, 2009.
[33] Y. Al-Khatatbeh, K. K. M. Lee, and B. Kiefer, "Phase diagram up to 105 GPa and mechanical strength of HfO_2," *Phys. Rev. B*, vol. 82, p. 144106, 2010.

Determination of Magneto-crystalline Anisotropy Energy (MAE) Of ordered $L1_0$ CoPt and FePt nanoparticles

A. Alsaad[1], A.A. Ahmad[1] A.A. Shukri[2] and O.A. Bani-Younes[1]

[1]*Department of Physics, Jordan University of science and technology, P. O. Box 3030, Irbid 22110, Jordan.*
[2]*Department of Physics, Tafila Technical University, At-Tafilah, Jordan.*

alsaad11@just.edu.jo

Abstract: The structural and magnetic properties of both L10 ordered FePt and CoPt nanoparticles make them potential candidates for optical-electronic and magneto-optical devices. First, we carried out *an ab initio* total energy minimization study to find the geometrical optimization of both L10 phases of FePt and CoPt nanoparticles. Then, we investigated the magnetocrystalline anisotropy energy (MAE) of both systems along special line joining the points of high symmetry (A,B and C points) using super-cell slap approach with alternating layers Fe/Co and Pt along the (001) direction. We found that the point (A) has the highest MAE value for both systems, where the value of MAE in FePt is 8.89×10^7 erg/cm^3 and in CoPt is 6.40×10^7 erg/cm^3. Our spin density based calculations indicate that large spin-orbit interaction and the hybridization between Pt 5d states and Fe/Co 3d states are the dominant factors in determining the MAE in both systems.

Keywords*:* Magneto-crystalline Anisotropy Energy (MAE); $L1_0$ CoPt and FePt nanoparticles; Spin density functional theory; Slap approach; Spin-orbit interaction

1. Introduction

A lot of efforts have been devoted to explore the physics of tiny structure for magnetic materials that can help to utilize its desirable properties [1-3]. In recent years, applications of magnetic nanoparticles have arisen, particularly those that have a high anisotropic energy [1, 4-7]. Because of their potential applications such as data storage, optical filters, and biomedical applications, some of magnetic nanoparticles gain crucial importance [8-10]. The value of magneto-crystalline anisotropy energy (MAE) of a magnetic material indicates the magnetic stability in response to the orientation of the magnetic spin moment.The storage devices based on magnetic materials are of great interest from fundamental and technological points of view due to their low cost [11-26]. When the thermal energy is large enough and able to overcome the anisotropy energy, the phenomenon of super paramagnetic occurs [29]. As a result, the magnetization is not stable which causes the damage and loss of the storage data [30]. To overcome this issue, we must look for magnetic materials that have high anisotropic energy [10, 31-32]. The FePt and CoPt nanoparticles adopting the face center tetragonal phase are considered among promising candidates for the field of data storage due to their high magnetic anisotropy energies [2, 16, 26]. Among the variant magnetic nanoparticles and their alloys, ordered ferromagnetic L10 FePt and CoPt alloys have desirable structural and magnetic properties. For instance, they exhibit good response to an external magnetic field; also, they show high magnetocrystalline anisotropy in bulk phase [2, 12]. In addition, The L10 FePt and L10 CoPt structures have huge values of MAE compared to the other magnetic nanoparticles. The MAE'S values lie in the range (7-11 $\times 10^7$ erg/cm^3) for L10 FePt structure [11, 13-14] and approximately 5 $\times 10^7$ erg/cm^3 for L10 CoPt [7, 14-15] structure. The CoPt binary exhibits a ferromagnetic behavior in its disordered A1 phase, while an ordered L12 CoPt phase possesses a cubic anisotropy and is said to be magnetically soft. On the other hand, the L10 FePt and CoPt phases are well known to have very high uniaxial magnetocrystalline anisotropy. The origin of this large anisotropy mainly comes from the significant spin-orbit coupling and the large hybridization between Co (Fe) 3d bands and Pt 5d states [36]. In this work, we attempt to present very accurate calculations of MAE in L10 CoPt and FePt nanoparticles using slap approach and spin density functional theory. The MAE of bulk FePt and CoPt is defined to be the energy difference, E100 - E001 between the two magnetic configurations in which the magnetic moments are parallel or perpendicular to the Fe (Co) or Pt (0 0 1) planes.

2. Methodolgy and Calculations

The methodolgy of the work includes calculations within the spin density functional theory (SDFT) using a plane wave basis set and the Projector augmented wave (PAW) pseudopotentials that are implemented in Vienna ab initio simulation package (VASP) [37-40]. The PAW pseudo potentials include the following valence electrons: Fe: $3p^6 3d^7 4s^1$ and Pt: $5d^9 5s^1$. The generalized gradient approximation (GGA) of Perdew–Burke–Ernzerhof (PBE) form was used for the exchange-correlation functional. The used parameters in calculating MAE of both FePt and CoPt SYSTEMS are as following: the plane wave cut-off energy is 400 eV, a 12 x 12 x 1 and 24×24×1 Monkhorst–Pack k-point meshes are used for both FePt and CoPt, respectively. The relaxed atomic positions were followed by minimizing the total energy (i.e. 0.003eV), and the Hellmann–Faynmann forces. These forces were as small as (0.003eV/Å) at convergence in the unit cell of all different electronic structure relaxations that have been performed. Finally, in order to evaluate the MAE results for both systems we performed self-consistence calculations for total energy for two different magnetic configurations. One configuration is demonstrated by allowing the magnetic moments to be oriented parallel, E(||), and the other by forcing the magnetic moments to be aligned perpendicular, E($\perp$), to the easy axis (c-axis). And then MAE can be considered as the energy difference between the two configurations,

$$MAE = E(||) - E(\perp)$$

3. Results

Crystallography Aspects Of The L10 Structure

The prototype structure for CoPt and FePt structures is the CuAu (I) type structure. The CoPt and FePt exhibit an A1 phase (The structure with A1 phase possess a disordered face centered cubic (fcc) structure) at high temperatures and they exhibit an ordered L10 or L12 phase at low temperatures. Figure 1 represents the unit cell of ordered L10 phase of both FePt and CoPt nanoparticles. In L10 FePt and CoPt structures, there are two types of atoms, one type of atoms occupy the corners and corresponding to [0 0 0] and [1/2 1/2 0] lattice sites, whereas the second type of atoms are located at the other two face centers with [1/2 0 1/2] and [0 1/2 1/2] site positions. In L10 structure, the arrangement of alternately aligned Co or Fe and Pt planes along the c-axis [001] results in the distortion of the cubic symmetry, this symmetry distortion leads to a slight tetragonal distortion in the unit cell. Hence this structure is referred to as an ordered face centered tetragonal (fct) L10 phase. The space group of the L10 structure is P4/mmm. We performed an *ab initio* total energy minimization study to find the geometrical optimization of both L10 phases of FePt and CoPt nanoparticles. The calculated values of the equilibrium lattice parameters are a_0 = 3.880 Å, c_0 =3.780 Å and a_0 = 3.814 Å, c_0 =3.710 Å for FePt and CoPt, respectively. These results are in agreement with data reported in the literature (2, 15-17, 31, 33).

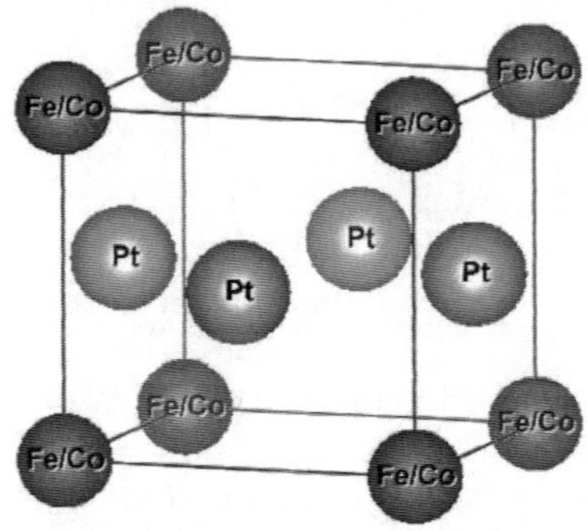

Fig.1.Unit cell of ordered L10 phase of both FePt and CoPt nanoparticles.

Determination of Magneto-crystalline Anisotropy Energy (MAE)

In magnetic anisotropic materials, the magnetic moment tends to be oriented in the energetically favorable direction, this direction is called "easy axis". By definition the magneto-crystalline Anisotropy Energy (MAE) is defined as the total energy E_{tot} associated with the change in the direction of the magnetization

IOP Conf. Series: Materials Science and Engineering **305** (2018) 012017 doi:10.1088/1757-899X/305/1/012017

with respect to the easy axis [34-35]. When the magnetic material has one easy axis, the material is called uniaxial. The L10 ordered FePt and CoPt structures are familiar examples of uniaxial magnetic materials. The MAE for L10 (face centered tetragonal) phase can be defined and calculated as the energy difference between two magnetic orientations as shown in equation [2, 16, 24],

$$\Delta E = E_{100} - E_{001,}$$

we used super-cell slap approach to compute the MAE of the L10 FePt and CoPt systems. The super cell approach is constructed from a relatively atomic layers (slap) separated by vacuum layers. The super cell is repeated throughout all space using periodic boundary conditions as shown in the Figure 2. Here, each slap in the super cell is constructed from ten atoms of Fe and Co and the eight Pt atoms with vacuum layers of 30 Å in z direction.

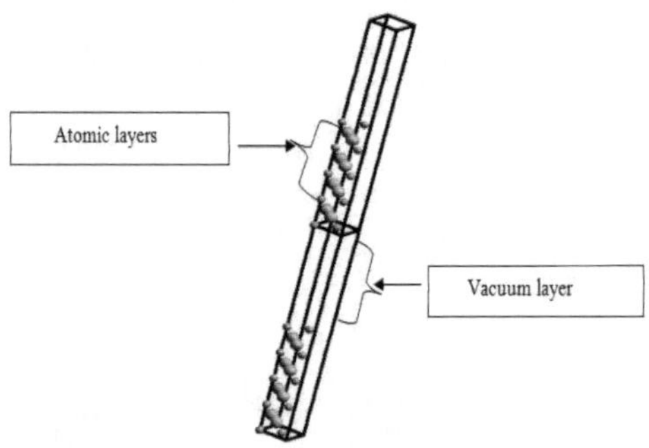

Figure 2: Super-cell slap approach which explain the atomic and vacuum layers. The small spheres denote the Fe/Co atoms while large spheres refer to Pt atoms.

We use this approach in order to compute the MAE of the FePt and CoPt along the high symmetry points. These points are symbolized by A, B, and C as shown in Figure3. The point A represents the regular L10 structure whereas point (B) is obtained by shifting the Fe or Co atoms by a displacement of (0.25, 0.25, 0) while the Pt atoms are shifted alternately between x and y coordinates by (0.75, 0.25,0) and (0.25, 0.75, 0). At point C with the shift of (0.5, 0, 0) the crystallites arranged atoms of Pt directly on top of each other across the interface in such a way that it corresponds to ideal FePt and CoPt lattices where one Fe or Co layer is removed. The structure of point C corresponds to a crack with one nearest neighbor across the interface.

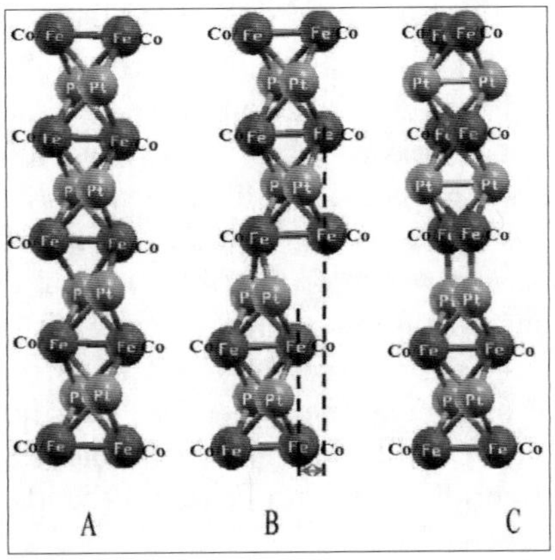

Figure 3: The super-cell slap approach of the ordered $L1_0$ CoPt and FePt systems, the high symmetry points A, B and C from left to right in regular super-cell structure, red spheres are used for Fe or Co atoms.

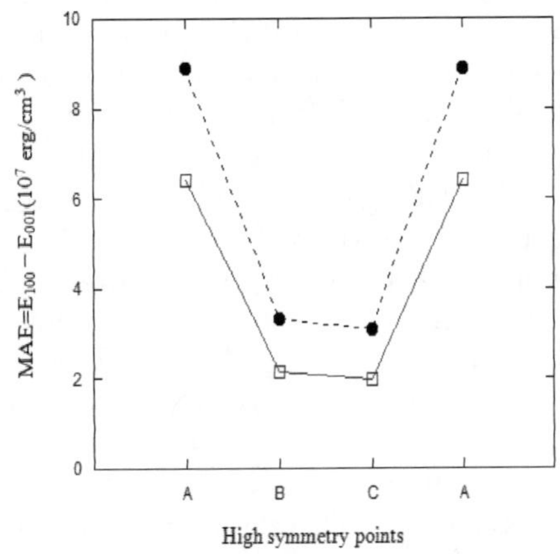

Fig. 4: Magnetocrystalline anisotropy energy (MAE) of L10 FePt (Black-dotted) and CoPt (Blue-solid) nanoparticles Plotted versus slip step along high symmetry points A-B-C-A of the regular lattice. Note that 1 J/m^3 is equivalent to 1 x 10^{-5}Merg/cm^3.

Figure 4, represents the MEA per unit volume of $L1_0$ FePt and $L1_0$CoPt at the high symmetry points. In FePt, point A has the highest MEA (8.89 $\times 10^7$ erg/cm^3), followed by point B (3.314$\times$ 10^7 erg/cm^3), and point C has the lowest MEA (3.087 $\times 10^7$ erg/cm^3). In CoPt system, the corresponding MAE values are at point A (6.40 $\times$ 10^7erg/cm^3), at point B (2.15 $\times$ 10^7erg/cm^3), and at point C (1.98 $\times$ 10^7erg/cm^3). It is obvious from Figure 3 the MAE decreases in along the line A$\rightarrow$

$B \rightarrow C \rightarrow A$ in both $L1_0$ FePt and $L1_0$ CoPt nano systems. The physical interpretation of our striking result comes from the spin-polarized band calculations including spin-orbit interaction for $L1_0$-FePt and CoPt ordered alloys performed within the framework of local spin density functional approximation. We have found that strong magnetocrystalline anisotropy of both ordered systems along the $A \rightarrow B \rightarrow C \rightarrow A$ line is due to a large spin-orbit coupling of Pt atom with Fe and Co atoms and a strong hybridization of Pt d bands with highly polarized Fe (Co) d bands. These two effects decrease along the $A \rightarrow B \rightarrow C \rightarrow A$ line for both ordered systems. This can be regarded as being associated with the behavior that the MAE of both systems decrease with decreasing of the strength of the spin-orbit coupling of Pt d-band with Fe (Co) bands, as well as, a significant decrease of Pt d-band hybridization with Fe (Co) d-bands.

4. Conclusion

We report our results on the structural and magnetic properties of $L1_0$-FePt and CoPt ordered systems using spin density functional theory. The spin-polarized band calculations including spin-orbit interaction for $L1_0$-FePt and CoPt ordered systems have been performed within local spin density functional approximation. Magneto-crystalline Anisotropy Energy (MAE) of L10 ordered FePt and CoPt nanoparticles were computed using super-cell slap approach within the framework of local spin density approximation. The (MAE) was computed at high symmetry points that are carefully selected in L10 ordered FePtand L10 CoPt systems. The points were chosen because of their importance from fundamental and experimental points of view. We found that MAE's of L10 FePt ordered structures decreases from 8.89×10^7 erg/cm^3 to 3.087×10^7 erg/cm^3 and that of L10 CoPt decreases from 6.40×10^7erg/cm^3 to 1.98×10^7erg/cm^3 along the $A \rightarrow B \rightarrow C \rightarrow A$ line joining points of high symmetry. This significant decrease in MAE's values of both ordered systems can be explained as being related with the observed decrease of the strength of the spin-orbit coupling of Pt d-band with Fe (Co) bands, as well as, a drastic decrease of Pt d-band hybridization with Fe (Co) d-bands. Our results are quite important for magnetic devices whose functionalities depend extremely on the magnetic reversal. This finding is very important and could be a guide for experimental research groups who are interested in growing both L10 FePt and L10 CoPt nano systems along the prescribed line joining the symmetry points. Our results could be used as guidelines for studying cracks that are formed during the growth of L10 FePt and L10 CoPt nanoparticles.

Acknowledgements

The authors would like to acknowledge university of Nebraska for providing us an access for the Holland computing center to perform our calculations. We would like to acknowledge the financial and technical support provided by Jordan University of Science and Technology. Our gratitude's go to Professor Renat Sabirianov of university of Nebraska at Omaha and Dr. Nabil Al-Aqtash of Hashemite university/Jordan for useful discussions.

IOP Conf. Series: Materials Science and Engineering **305** (2018) 012017 doi:10.1088/1757-899X/305/1/012017

References

[1] Kabir A, Hu J, Turkowski V, Wu R, Camley R, Rahman TS. Effect of structure on the magnetic anisotropy of L 1 0 FePt nanoparticles. Physical Review B. 2015 **92(5)** 054424.

[2] Alsaad A, Al-Aqtash N, Sabirianov RF. Generalized stacking fault in FePt nanoparticles and effects of extended defects on magnetocrystalline anisotropy energy. Journal of Magnetism and Magnetic Materials. 2015 **374** 525-9.

[3] Bell AT. The impact of nanoscience on heterogeneous catalysis. Science. 2003 **299(5613)** 1688-91.

[4] Stampfl C, Mannstadt W, Asahi R, Freeman A. Electronic structure and physical properties of early transition metal mononitrides: Density-functional theory LDA, GGA, and screened-exchange LDA FLAPW calculations. Physical Review B. 2001 **63(15)** 155106.

[5] McHenry M, Laughlin D. Nano-scale materials development for future magnetic applications. Acta materialia. 2000 **48(1)** 223-38.

[6] Gao J, Gu H, Xu B. Multifunctional magnetic nanoparticles: design, synthesis, and biomedical applications. Accounts of chemical research. 2009 **42(8)** 1097-107.

[7] Barmak K, Kim J, Lewis L, Coffey K, Toney M, Kellock A, et al. On the relationship of magnetocrystalline anisotropy and stoichiometry in epitaxial L10 CoPt (001) and FePt (001) thin films. Journal of Applied Physics. 2005 **98(3)** 033904.

[8] Lu AH, Salabas EeL, Schüth F. Magnetic nanoparticles: synthesis, protection, functionalization, and application. Angewandte Chemie International Edition. 2007 **46(8)** 1222-44.

[9] Issa B, Obaidat IM, Albiss BA, Haik Y. Magnetic nanoparticles: surface effects and properties related to biomedicine applications. International journal of molecular sciences. 2013 **14(11)** 21266-305.

[10] Gubin SP. Magnetic nanoparticles: John Wiley & Sons; 2009.

[11] Lyubina J, Opahle I, Müller K-H, Gutfleisch O, Richter M, Wolf M, et al. Magnetocrystalline anisotropy in L10 FePt and exchange coupling in FePt/Fe3Pt nanocomposites. Journal of Physics: Condensed Matter. 2005 **17(26)** 4157.

[12] Alam A, Kraczek B, Johnson DD. Structural, magnetic, and defect properties of Co-Pt-type magnetic-storage alloys: Density-functional theory study of thermal processing effects. Physical Review B. 2010 **82(2)** 024435.

13. Sun S, Murray CB, Weller D, Folks L, Moser A. Monodisperse FePt nanoparticles and ferromagnetic FePt nanocrystal superlattices. Science. 2000 **287(5460)** 1989-92.

[14] Barmak K, Kim J, Lewis L, Coffey K, Toney M, Kellock A, et al. On the relationship of magnetocrystalline anisotropy and stoichiometry in epitaxial L 1 0 CoPt (001) and FePt (001) thin films. Journal of Applied Physics. 2005 **98(3)** 033904.

[15] MacLaren JM, Duplessis RR, Stern RA, Willoughby S. First principles calculations of FePt, CoPt, Co 3 Pt, and Fe 3 Pt alloys. Magnetics, IEEE Transactions on. 2005 **41(12)** 4374-9.
[16] Han L. FePt and CoPt nanoparticles prepared by micellar method: Effects of A1→ L10 transition on oxidation resistance and magnetic properties: Ulm, Univ., Diss., 2011.
[17] Laughlin DE, Srinivasan K, Tanase M, Wang L. Crystallographic aspects of L1 0 magnetic materials. Scripta Materialia. 2005 **53(4)** 383-8.
[18] Sun S. Recent advances in chemical synthesis, self-assembly, and applications of FePt nanoparticles. Advanced Materials. 2006 **18(4)** 393-403.
[19] Klemmer T, Shukla N, Liu C, Wu X, Svedberg E, Mryasov O, et al. Structural studies of L10 FePt nanoparticles. Applied Physics Letters. 2002 **81(12)** 2220-2.
[20] Maxwell DO. Structural and Electronic properties of β-Al2X3 (X= O, Se, Te): University of Toledo; 2011.
[21] Chinnasamy C, Huang J, Lewis L, Latha B, Vittoria C, Harris V. Direct chemical synthesis of high coercivity air-stable SmCo nanoblades. Applied Physics Letters. 2008 **93(3)** 032505.
[22] Aroyo M, Perez-Mato J, Orobengoa D, Tasci E, De La Flor G, Kirov A. Crystallography online: Bilbao crystallographic server. Bulg Chem Commun. 2011 **43(2)** 183-97.
[23] Galanakis I, Alouani M, Dreyssé H. Calculated magnetic properties of low-dimensional systems: the AuCu-and AuCu 3-type ferromagnets. Journal of Magnetism and Magnetic Materials. 2002 **242** 27-32.
[24] Halilov S, Perlov AY, Oppeneer P, Yaresko A, Antonov V. Magnetocrystalline anisotropy energy in cubic Fe, Co, and Ni: Applicability of local-spin-density theory reexamined. Physical Review B. 1998 **57(16)** 9557.
[25] Takahashi Y, Koyama T, Ohnuma M, Ohkubo T, Hono K. Size dependence of ordering in FePt nanoparticles. Journal of Applied Physics. 2004 **95(5)** 2690-6.
[26] Reiss G, Hütten A. Magnetic nanoparticles: applications beyond data storage. Nature materials. 2005 **4(10)** 725-6.
[27] Kardelky S, Gebert A, Gutfleisch O, Hoffmann V, Schultz L. Prediction of the oxidation behaviour of Sm–Co-based magnets. Journal of Magnetism and Magnetic Materials. 2005 **290** 1226-9.
[28] Su Y-K, Qin D-H, Zhang H-L, Li H, Li H-L. Microstructure and magnetic properties of bamboo-like CoPt/Pt multilayered nanowire arrays. Chemical physics letters. 2004 **388(4)** 406-10.
[29] Sun S. Recent advances in chemical synthesis, self-assembly, and applications of FePt nanoparticles. Advanced Materials. 2006 **18(4)** 393-404.
[30] Entel P, Gruner M. Large-scale ab initio simulations of binary transition metal clusters for storage media materials. Journal of Physics: Condensed Matter. 2009 **21(6)** 064228.
[31] Yang Y. High anisotropy CoPt (L10 & L11 phase) magnetic recording media films: NATIONAL UNIVERSITY OF SINGAPORE; 2012.

IOP Conf. Series: Materials Science and Engineering **305** (2018) 012017 doi:10.1088/1757-899X/305/1/012017

[32] Frey NA, Sun S. Magnetic nanoparticle for information storage applications. Inorganic Nanoparticles: Synthesis, Applications, and Perspectives, CRC Press, Boca Raton, USA. 2010 33-67.

[33] Manchanda P, Kumar P, Kashyap A, Lucis M, Shield JE, Mubarok A, et al. Intrinsic Properties of Fe-Substituted Magnets. Magnetics, IEEE Transactions on. 2013 **49(10)** 5194-8.

[34] Steneteg P. Development of molecular dynamics methodology for simulations of hard materials. 2012.

[35] Rajalingam S. Theoretical investigations of β-Ga_2O_3, β-Ga_2Se_3 and β-Ga_2Te_3 using ab-initio calculations. 2012.

[36] Nakajima N, Koide T, Shidara T, Miyauchi H, Fukutani H, Fujimori A, Iio K, Katayama T, Nývlt M and Suzuki Y, Phys. Rev Lett. 1998 **81** 5229

[37] Kresse G and Hafner H. Ab initio molecular dynamics for liquid metals. Phys. Rev. B, 1993 **47** 558 [38] Kresse G and Hafner H. Ab initio molecular-dynamics simulation of the liquid-metal-amorphous-semiconductor transition in germanium. Phys. Rev. B, 1994 **49** 14251

[39] Kresse G and and Furthmüller J. Efficiency of ab-initio total energy calculations for metals and semiconductors using a plane-wave basis set. Comput. Mat. Sci., 1996 **6** 15

[40] Kresse G and and Furthmüller J. Efficient iterative schemes for ab initio total-energy calculations using a plane-wave basis set. Phys. Rev. B, 1996 **54** 11169

Fenton-like Degradation of Phenol Catalyzed by a Series of Fe-Containing Mixed Oxides Systems

T T Alhmoud, S S Mahmoud, A Y Hammoudeh*

Chemistry Department, Yarmouk University, Irbid-Jordan

ayman@yu.edu.jo

Abstract: In our attempts to develop a solid catalyst to degrade organic pollutants in wastewater via the Fenton-like reaction, six Fe-containing mixed oxide systems were prepared by means of the sol-gel auto-combustion method to have the following stoichiometries: $CuFe_{1.2}O_{2.8}$, $BaFe_{7.2}O_{11.8}$, $BaFe_{7.2}Cu_2O_{13.8}$, $BaFe_{5.4}V_3O_{16.6}$, $BaFe_{4.8}Cu_2V_3O_{17.7}$ and $Ag_2Fe_{5.4}V_3O_{16.6}$. The prepared systems were thermally treated at 550°C, 650°C, 800°C and 1100°C, and then characterized by XRD to identify the present phases. The systems were tested with respect to their catalytic efficiency in the degradation of phenol (200 ppm) in water where $CuFe_{1.2}O_{2.8}$ was found to be the most reactive one (80% removal in 60 min). It showed thereby first-order kinetics and an enhanced behavior under irradiation with a 30-W LED light source. The positive role of irradiation was most obvious in the case of $Ag_2Fe_{5.4}V_3O_{16.6}$ in which almost complete conversion was achieved in 120 min compared to only 45% in the same period but without irradiation. However, increasing the temperature at which thermal treatment is performed was found to suppress the catalytic activity of the system. Due to their high efficiency and rather low leaching rates of constituents, $CuFe_{1.2}O_{2.8}$ or $Ag_2Fe_{5.4}V_3O_{16.6}$ seem to be very promising in the Fenton-like degradation of organic pollutants.

Keywords: Degradation of Organic Pollutants, Fenton Reaction, Phenol, Iron oxide, Advanced Oxidation Processes, Peroxide Oxidation.

1. Introduction

Advanced Oxidation Processes (AOP) represent a class of attractive and promising wastewater treatment methods based on the oxidation of organic pollutants by *in-situ* generated hydroxyl radicals. They are appealing because of the possibility to gain complete oxidation or mineralization of organic contaminants through a process that operates near ambient temperature and pressure [1]. Hydrogen peroxide can thereby be used as a powerful oxidant and the classical Fenton reaction catalyzed by Fe^{2+} ions can be used to decompose it generating the desired hydroxyl radicals. The process is highly efficient, rather simple and H_2O_2 is non-toxic [2-7]. It suffers however from several drawbacks including the high operational cost, difficulties in recycling the catalyst (Fe^{2+}), the limited pH range of operation (optimum at pH 3) and the large volume of iron sludge produced [8, 9]. One way to overcome these disadvantages is by the application of heterogeneous catalysts that can mediate the Fenton reaction over a wide pH range with limited leaching into solution and can be easily separated from the reaction mixture and then recycled [10, 11]. Moreover, irradiation can also elevate the reaction efficiency and in the past two decades, interest in the replacement of UV radiation with solar energy in wastewater treatment increased [12]. Nevertheless, heterogeneous catalysts show lower rates of oxidation when compared to the homogeneous reaction due to the small number of active sites at the catalyst surface, the reason why intensive research is being conducted to develop highly active solid catalysts for this purpose [13].

A wide variety of solid materials have been tested as Fenton catalysts among which iron based systems gained great attention. The latter vary in shape and form from zero-valent iron [14,15], iron oxides [16,17], to clay supported catalysts [11,18]. Modification of such systems has also been attempted, for example, by doping with sulfur [19] or by *in-situ* growing of a Fe_3O_4/SiO_2 ceramic coating on a metal substrate to maintain its strong mechanical adhesion for repeated use [20]. Exploitation of some industrial waste products, such as coal fly ash, to catalyze the Fenton reaction has been also reported where the catalytic activity of coal fly ash has been attributed to its content of Fe [21,22]. Ferrites were also widely investigated [12 and references therein]. The redox properties of the ferrites are strongly affected by the presence of different metals in the lattice structure, their nature, sites and amounts [23-25]. Ferrites have also the merit of a band gap capable of absorbing visible light and enhanced efficiency due to the available extra catalytic sites [26]. In virtue of their magnetic properties, they can also be easily separated [27]. Baldrain et al. reported the decolorization of a variety of dyes by hydrogen peroxide over $MO\text{-}Fe_2O_3$ mixed oxide catalysts (spinel ferrites) with M= Fe, Cu, Co, Mn [28]. Systems such as $Fe_{2.25}Co_{0.75}O_4$ and $Fe_{2.47}Mn_{0.53}O_4$ have been reported to be very efficient (99% removal) in the degradation of methylene blue [23]. Another type of an iron containing mixed oxide, namely $BiFeO_3$ with perovskite structure, was shown to achieve 75% degradation of Rhodamine B in 90 min in dark [29].

In this work, six Fe-containing mixed oxide systems, $CuFe_{1.2}O_{2.8}$, $BaFe_{7.2}O_{11.8}$, $BaFe_{7.2}Cu_2O_{13.8}$, $BaFe_{5.4}V_3O_{16.6}$, $BaFe_{4.8}Cu_2V_3O_{17.7}$ and $Ag_2Fe_{5.4}V_3O_{16.6}$, were prepared, characterized and investigated with respect to their efficiency in the Fenton-like degradation of phenol, a typical bio-refractory compound with high toxicity level and can therefore be used as a model of such contaminations in industrial wastewater.

2. Experimental:

The mixed oxide systems have been prepared via the sol-gel auto-combustion method to have the following stoichiometries: $CuFe_{1.2}O_{2.8}$, $BaFe_{7.2}O_{11.8}$, $BaFe_{7.2}Cu_2O_{13.8}$, $BaFe_{5.4}V_3O_{16.6}$, $BaFe_{4.8}Cu_2V_3O_{17.7}$ and $Ag_2Fe_{5.4}V_3O_{16.6}$. $Fe(NO_3)_3.9H_2O$ (Alfa Aesar), $Cu(NO_3)_2.4H_2O$ (BDH), NH_4VO_3 (Riedel-deHaěn), CH_3COOAg (PHARMACOS LTD), $Ba(NO_3)_2.2H_2O$ (BDH) and citric acid (AVONCHEM) were used as starting materials. The sol was prepared by dissolving the proper ratio of metal nitrates and ammonium metavanadate in 100 ml of water. Citric Acid (CA) with a 3:1 CA/Fe^{3+} molar ratio was separately dissolved in minimum amount of water. The two solutions were then combined and heated to reach a temperature of 80°C. The pH was adjusted to 8 using aqueous ammonia solution (Xllong chemical Industry) before heating on a hot plate for few hours with constant stirring during which the water content was reduced continuously by evaporation, thus, increasing the viscosity of the sol and indicating gel formation. The gel was left overnight. Next day, it was heated on the hot plate until auto-ignition started forming loose powder. The prepared systems were thermally treated for 4 hours at 550°C under flowing oxygen and at 650°C, 800°C and 1100°C under open atmosphere. The prepared systems are designated by their composition followed by the temperature at which thermal treatment has been carried out, for example, $CuFe_{1.2}O_{2.8}$-550 and $Ag_2Fe_{5.4}V_3O_{16.6}$-800. All systems were characterized with respect to their structure by X-ray diffraction (XRD) using a Shimadzu LabX device with a Cu-Kα-tube (λ=1.5418 Å) in the range of 2θ=20-80° with a scanning rate of 0.5°/min.

The reaction system for the phenol degradation consists of a 500-ml conical flask, magnetic stirrer (STUAART Scientific Co), peristaltic pump (Gilson-Model 312), a temperature controlled water bath (MA 6 LAUDA and B LAUDA) and an LED (30W, NWELIHT, IP66) as a light source. In a typical experiment, 0.25 g catalyst is added to 250 mL of 200 ppm aqueous phenol solution (CDH laboratory Reagent). The reaction solution is stirred at constant temperature (313 K) for 15 minutes. A zero-time sample is withdrawn before the continuous addition of 3% hydrogen peroxide solution (GFS CHEMICALS) is started by means of the peristaltic pump at a flow rate of 20 mL/h. If applied, the reaction solution is exposed perpendicularly to the light produced by the LED source positioned 30 cm above the reaction vessel. The progress of reaction was monitored by high-performance liquid chromatography, HPLC, (Dionex, Ultimate 3000 LC System) equipped with a variable wavelength detector. Samples were withdrawn every 30 min and immediately centrifuged to remove suspended particles. The analysis method was as follows: 250×4.6 C-18 column; mobile phase 60% methanol - 40% water at a flow rate of 1.20 mL/min with constant wavelength detection at 270 nm.

3. Results and Discussion

3.1. X-ray Diffraction

Figure 1 represents the XRD patterns of $CuFe_{1.2}O_{2.8}$-U, $CuFe_{1.2}O_{2.8}$-550, $CuFe_{1.2}O_{2.8}$-650, $CuFe_{1.2}O_{2.8}$-800 and $CuFe_{1.2}O_{2.8}$-1100. The major diffraction lines coincide with those of tetragonal $CuFe_2O_4$ as indicated by the red vertical lines of the standard pattern (JCPDS 00-034-0425). The tetragonal phase is the more stable Cu ferrite phase [30]. A deficiency of Cu^{2+} cations in the B sublattice leads however to the absence of cooperative Jahn-Teller distortion and thus to the transformation of the crystal phase into a cubic one [31], but no indication for formation the cubic phase is obvious in figure 1 (black vertical lines corresponding to the standard pattern JCPDS 00-025-0238). The presence of monoclinic copper (II) oxide is confirmed by the diffraction lines at 38.9, 48.6° and the shoulder at ~35.6° coinciding with those of CuO standard pattern (JCPDS 00-045-0937) as indicated by the blue vertical lines in Figure 1. Figure 1 shows also that as the temperature at which the thermal treatment was performed is increased, the diffraction lines become sharper corresponding to increased degree of crystallinity.

Figure 2 represents the XRD patterns of $Ag_2Fe_{5.4}V_3O_{16.6}$-U, $Ag_2Fe_{5.4}V_3O_{16.62}$-550, $Ag_2Fe_{5.4}V_3O_{16.6}$-650, $Ag_2Fe_{5.4}V_3O_{16.6}$-800 and $Ag_2Fe_{5.4}V_3O_{16.6}$-1100. The major diffraction lines coincide with those of α-Fe_2O_3 as indicated by the black vertical lines of the standard pattern (JCPDS card 00-033-0664) in the Figure. A series of diffraction features most obvious in the 2θ-range 25-32° seem to belong to a divanadate phase. Due to the non-availability of XRD reference patterns for barium or iron vanadate, the reference pattern (red vertical lines) for $Mg_2V_2O_7$ (JCPDS card 00-031-0816) was inserted in the figure with the purpose of identifying the divanadate phase. The good agreement in the positions of the diffraction lines of $Mg_2V_2O_7$ with those experimentally observed in the 2θ-range 25-32° makes us believe that iron is incorporated in this divanadate phase because the ionic radius of Fe^{2+} (0.780 Å) is closer to that of Mg^{2+} (0.720 Å) than Ba^{2+} (1.35 Å). The divanadate diffraction features are however not pronounced in the thermally untreated samples. As the temperature at which the thermal treatment has been performed is raised (especially ≥ 650°C), the diffraction lines of both phases (α-Fe_2O_3 and divanadate) become sharper and more pronounced indicating increased crystallinity and crystal growth. In addition, crystalline orthorhombic V_2O_5 can be recognized to be present, especially after thermal treatment at 650°C or higher temperatures (diffraction patterns c-e in Figure 2), as indicated by the vertical blue lines of the standard (JCPDS card 00-045-1074). Below 650°C (i.e. in $Ag_2Fe_{5.4}V_3O_{16.6}$-550), tetragonal V_2O_5 (JCPDS card 00-041-1426) is the dominant vanadium(V) oxide phase. For the purpose of clarity, its main diffraction lines are assigned by asterisks in Figure 2. The results of the XRD characterization of the other systems are summarized in Table 1.

IOP Conf. Series: Materials Science and Engineering **305** (2018) 012018 doi:10.1088/1757-899X/305/1/012018

Table 1: Summary of XRD characterization of the investigated systems.

System	Dominant Phase(s)	Minor Phase(s)
$BaFe_{7.2}O_{11.8}$-U	γ-Fe_2O_3 (maghemite)	BaO
$BaFe_{7.2}O_{11.8}$-550	γ-Fe_2O_3 (maghemite)	BaO
$BaFe_{7.2}O_{11.8}$-650	γ-Fe_2O_3 (maghemite)	α-Fe_2O_3 (hematite); orthorhombic $BaFe_2O_4$
$BaFe_{7.2}O_{11.8}$-800	γ-Fe_2O_3 (maghemite)	α-Fe_2O_3 (hematite); orthorhombic $BaFe_2O_4$
$BaFe_{7.2}O_{11.8}$-1000	M-type Ba-hexaferrite	α-Fe_2O_3 (hematite); orthorhombic $BaFe_2O_4$
$BaFe_{7.2}Cu_2O_{13.8}$-U; -550; -650; -800	M-type Ba-hexaferrite orthorhombic $BaFe_2O_4$ cubic $CuFe_2O_4$ monoclinic CuO	α-Fe_2O_3 (hematite)
$BaFe_{5.4}V_3O_{16.6}$-U; -550; -650; -800	α-Fe_2O_3 (hematite) divanadate	
$BaFe_{4.8}Cu_2V_3O_{17.7}$ -U; -550; -800	α-Fe_2O_3 (hematite) divanadate	Orthorhombic V_2O_5 orthovanadate

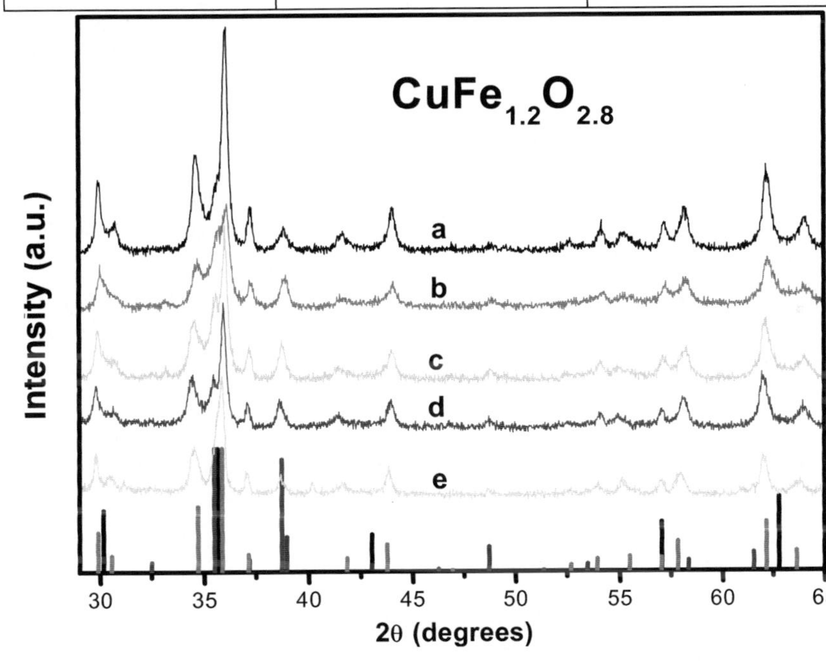

Fig. 1: XRD patterns of a) $CuFe_{1.2}O_{2.8}$-U, b) $CuFe_{1.2}O_{2.8}$-550, c) $CuFe_{1.2}O_{2.8}$-650, d) $CuFe_{1.2}O_{2.8}$-800 and e) $CuFe_{1.2}O_{2.8}$-1100. The vertical lines correspond to the standard patterns of tetragonal $CuFe_2O_4$ (red), cubic $CuFe_2O_4$ (black) and monoclinic CuO (blue).

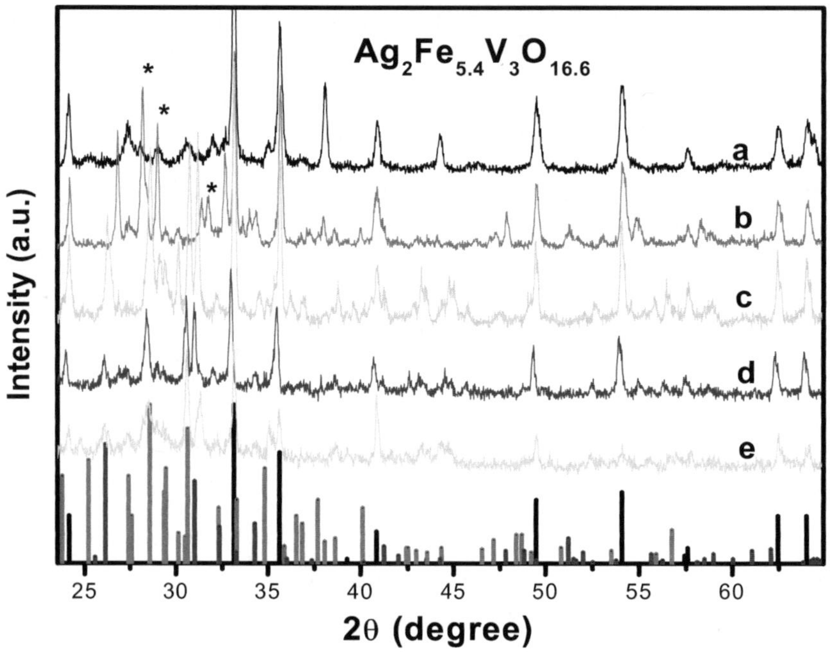

Fig. 2: XRD patterns of a) $Ag_2Fe_{5.4}V_3O_{16.6}$-U, b) $Ag_2Fe_{5.4}V_3O_{16.6}$-550, c) $Ag_2Fe_{5.4}V_3O_{16.6}$-650, d) $Ag_2Fe_{5.4}V_3O_{16.6}$-800 and $Ag_2Fe_{5.4}V_3O_{16.6}$-1100.

3.2. Phenol degradation

Figure 3 represents the progress of phenol degradation over the thermally untreated forms of all six systems with (right) and without irradiation (left). Without irradiation, $CuFe_{1.2}O_{2.8}$-U (thermally Untreated) is by far the most active system accomplishing 95% degradation in two hours. This high activity of $CuFe_{1.2}O_{2.8}$-U can be correlated with the dominance of the spinel copper ferrite phase reported to show a rather high activity in the degradation of phenol [32]. On the other hand, $BaFe_{7.2}O_{11.8}$-U seems to be completely inactive. The partial substitution of Fe(III) by Cu(II) does not help much as only a slight decrease in the phenol concentration is observed after 180 min ($BaFe_{7.2}Cu_2O_{13.8}$-U). It is however obvious from Figure 3 that incorporating vanadium(V) in the system has a positive effect and the best behavior for the thermally untreated systems was observed for $BaFe_{4.8}Cu_2V_3O_{17.7}$ achieving almost 90% conversion in 150 min. The higher efficiency of $BaFe_{4.8}Cu_2V_3O_{17.7}$ compared to $BaFe_{5.4}V_3O_{16.6}$ reflects the importance of having Cu(II) ions incorporated in the system.

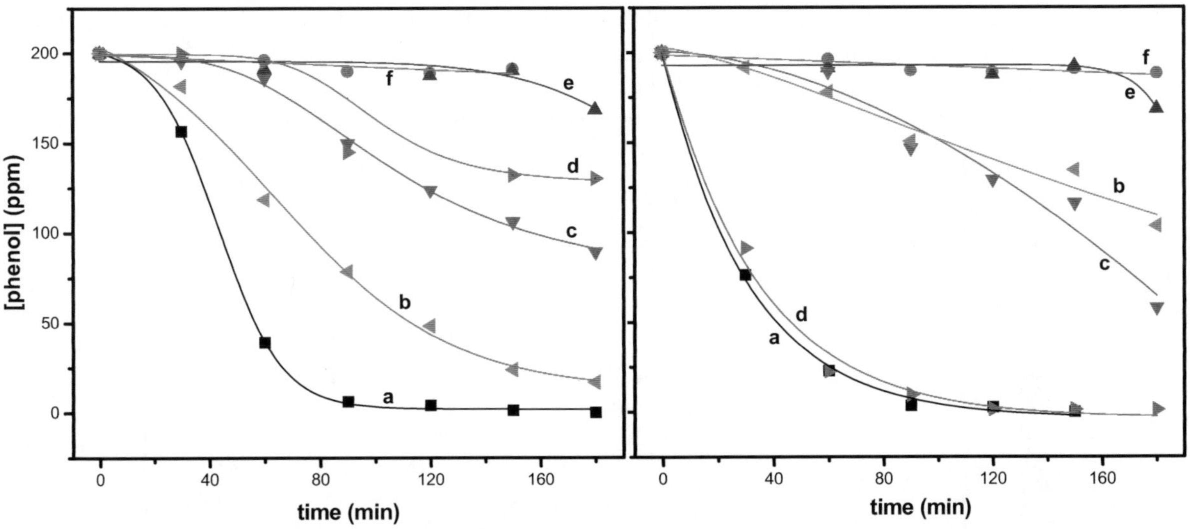

Fig. 3: Degradation of Phenol over 0.25 g of the thermally untreated forms of a) $CuFe_{1.2}O_{2.8}$, b) $BaFe_{4.8}Cu_2V_3O_{17.7}$, c) $BaFe_{5.4}V_3O_{16.6}$, d) $Ag_2Fe_{5.4}V_3O_{16.6}$, e) $BaFe_{7.2}Cu_2O_{13.8}$ and f) $BaFe_{7.2}O_{11.8}$ with (right) and without irradiation (left); $[phenol]_o$=200 ppm, T=40°C, 20 mL 3% H_2O_2/h.

An important aspect of the degradation profile seen in Figure 3 is its sigmoidal shape most obvious in the case of $CuFe_{1.2}O_{2.8}$-U and $BaFe_{4.8}Cu_2V_3O_{17.7}$ where the degradation of phenol does not start immediately and the concentration of phenol drops only slowly at the beginning of reaction characterizing a so-called induction period. This behavior is typical for a consecutive reaction and suggests that in the first step an active intermediate -in this case the hydroxyl radical as proposed by several studies [33-36]- is produced before it attacks in the second step the phenol molecules initiating thus its degradation. The extension of the induction period reflects thus how fast the hydroxyl radical is produced; the shorter the induction period, the more active the catalyst is.

Upon irradiation with a 30-W LED source (emitted radiation 400-750 nm) situated 30 cm from the reaction vessel, the catalytic performance of $CuFe_{1.2}O_{2.8}$-U has been improved as evidenced by the disappearance of the induction period as well as by the higher conversions achieved in a shorter time (> 96% in 90 min). The positive role of irradiation was most obvious in the case of $Ag_2Fe_{5.4}V_3O_{16.6}$-U in which almost complete conversion was achieved in 120 min compared to only 45% in the same period but without irradiation. Also in the case of $Ag_2Fe_{5.4}V_3O_{16.6}$-U the induction period disappears indicating the high production rate of the hydroxyl radicals. Whereas the catalytic activity of $BaFe_{7.2}O_{11.8}$-U and $BaFe_{7.2}Cu_2O_{13.8}$-U was not affected by irradiation, that of $BaFe_{4.8}Cu_2V_3O_{17.7}$-U was suppressed. One possible explanation for this unexpected behavior is that irradiation enhances the decomposition of hydrogen peroxide but in an undesired way. In a previous study [37], it was concluded that parallel to the hydroxyl radical formation, H_2O_2 may decompose in an undesired route yielding oxygen and that catalysts active in the latter route do not achieve high phenol degradation rates.

Figure 4 shows that beyond the induction period, the phenol degradation obeys first-order kinetics regardless whether the reaction system was irradiated or not. This is in agreement with literature reports [38] as well as kinetic models for the heterogeneously catalyzed degradation of phenol [39]. A closer look in Figure 4, for example to the effect of LED radiation on the activity of $CuFe_{1.2}O_{2.8}$-550, indicates that irradiation not only decreases the extent of the induction period but increases also the first-order rate constant of degradation by almost 50% from 0.031 min^{-1} to

0.048 min^{-1} suggesting an increased concentration of the hydroxyl radicals. The same trend has been also observed in the other systems $CuFe_{1.2}O_{2.8}$-U and $Ag_2Fe_{5.4}V_3O_{16.6}$-U, $BaFe_{4.8}Cu_2V_3O_{17.7}$-U being thereby an exception as discussed above.

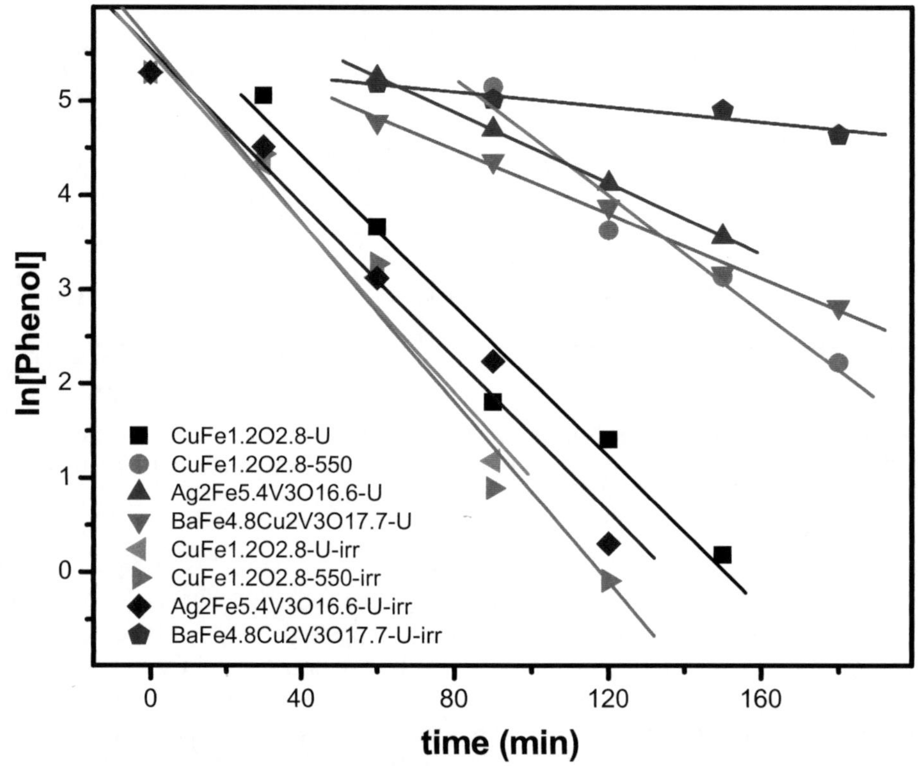

Fig. 4: First-order kinetics of the degradtion of phenol over selected catalytic systems with and without irradiation: $[phenol]_o$=200 ppm, T=40°C, 20 mL 3%H_2O_2/h.

The effect of temperature at which the thermal treatment was performed on the catalyst efficiency in the degradation of phenol is shown in Figure 5 for the $CuFe_{1.2}O_{2.8}$ as an example. In general, the activity decreases as the thermal treatment temperature is increased but the $CuFe_{1.2}O_{2.8}$-1100 represents an exception to the observed trend as its activity is larger than that of $CuFe_{1.2}O_{2.8}$-800. Under irradiation, the same general trend has been observed ($CuFe_{1.2}O_{2.8}$-U ≈ $CuFe_{1.2}O_{2.8}$-550 > $CuFe_{1.2}O_{2.8}$-650 > $CuFe_{1.2}O_{2.8}$-1100 > $CuFe_{1.2}O_{2.8}$-800). Again, the activity of $CuFe_{1.2}O_{2.8}$-1100 is larger than that of $CuFe_{1.2}O_{2.8}$-800. The other systems follow also in general that trend with exceptions, for example $BaFe_{4.8}Cu_2V_3O_{17.7}$-U > $BaFe_{4.8}Cu_2V_3O_{17.7}$-550 ≈ $BaFe_{4.8}Cu_2V_3O_{17.7}$-800 > $BaFe_{4.8}Cu_2V_3O_{17.7}$-650. Although systems treated thermally at higher temperatures are usually expected to be less active than the corresponding ones treated at lower temperatures due to surface area loss [40] and consequently active sites loss [41], previous studies in our laboratories show that the $Ni_xCu_{1-x}Fe_2O_4$ systems thermally treated at 800°C were more active in the phenol degradation than when treated at 400°C despite the fact that the latter were much more active in the H_2O_2 decomposition into water and oxygen. The behavior can be attributed to the higher ability of the 400-series catalysts to catalyze the decomposition of HOO· radicals into O_2 and H^+, as has been reported to proceed by Fe^{3+} [12,42], wasting thus the valuable oxidant.

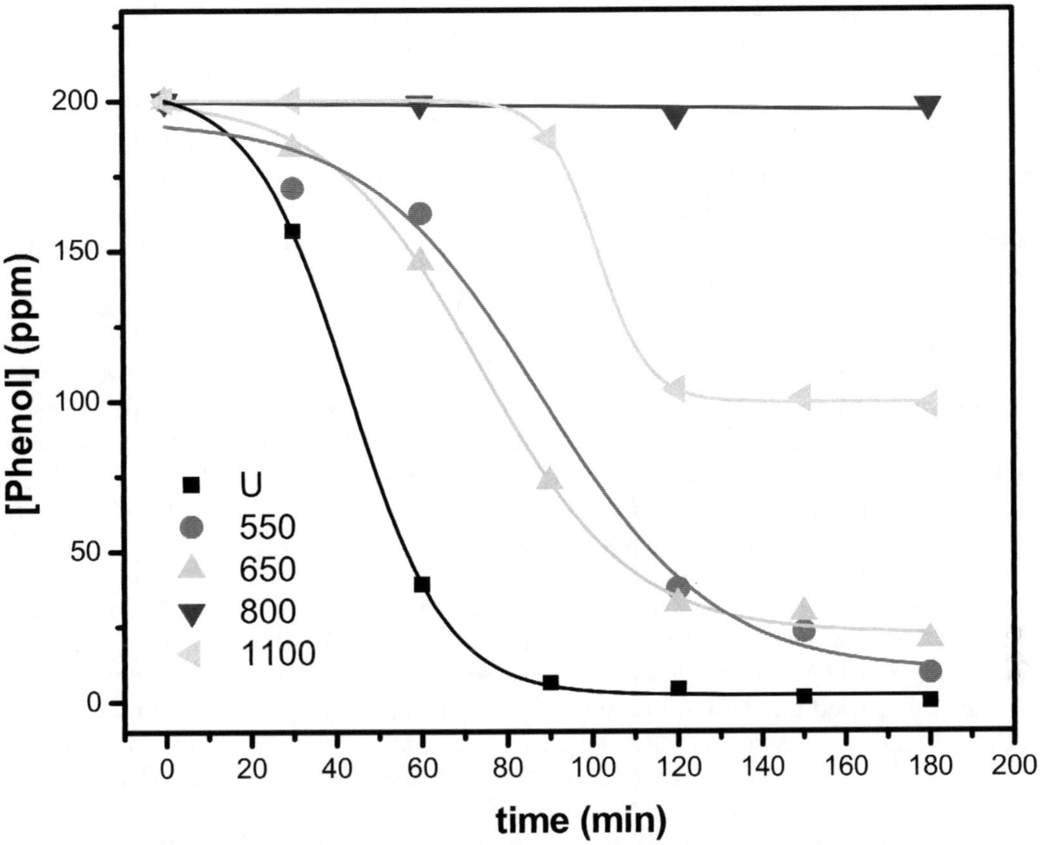

Fig. 5: The effect of thermal treatment temperature on the degradtion of phenol over $CuFe_{1.2}O_{2.8}$: $[phenol]_o$=200 ppm, T=40°C, 20 mL 3%H_2O_2/h.

3.3. Leaching of catalysts components into solution

In addition to the efficiency of the catalyst in degrading phenol and formed intermediates, it is also important to check whether the catalyst components do leach under reaction conditions into solution in form of ions because these ions can in turn be themselves toxic. The most active catalyst found in this work ($CuFe_{1.2}O_{2.8}$-550 under irradiation) was tested with this respect and the concentration of Cu^{2+} and Fe^{3+} ions that possibly leach in water during the reaction was determined by means of Atomic Absorption Spectroscopy. The results are summarized in Table 2.

Table 2: The concentration of Cu^{2+} and Fe^{3+} ions leaching into solution during the degradation of phenol over $CuFe_{1.2}O_{2.8}$ -550 under irradiation: $[phenol]_o$=200 ppm, T=40°C, 20 mL 3%H_2O_2/h.

time (min)	[Fe^{3+}] (ppm)	[Cu^{2+}] (ppm)
150	0.49	0.83
180	0.80	2.51
overnight	1.60	4.53

Taking into consideration that only 90 min are needed to achieve more than 98% degradation of phenol, the results of the Cu^{2+} concentrations leaching into solution presented in Table 2 suggests the safe application of the above catalyst as it satisfies the WHO safety guidelines for copper in drinking water (2 mg/L) [43]. Moreover, the concentration of leached iron from $CuFe_{1.2}O_{2.8}$ after three hours of reaction (0.80 ppm) is below the limit established by The Illinois Environmental Protection Agency (IEPA) for iron in drinking water at 1.0 mg/l [44]. The low concentrations of leached Fe^{3+} and Cu^{2+} ions into solution, responsible for the homogeneously mediated decomposition, confirm the heterogeneity of the catalytic process where the decomposition of H_2O_2 takes place at the solid surface.

4. Conclusions:

In our attempts to develop a solid catalyst to degrade organic pollutants in wastewater via the Fenton-like reaction, six Fe-containing mixed oxide systems were prepared by the sol-gel auto-combustion method to have the following stoichiometries: $CuFe_{1.2}O_{2.8}$, $BaFe_{7.2}O_{11.8}$, $BaFe_{7.2}Cu_2O_{13.8}$, $BaFe_{5.4}V_3O_{16.6}$, $BaFe_{4.8}Cu_2V_3O_{17.7}$ and $Ag_2Fe_{5.4}V_3O_{16.6}$. The prepared systems were thermally treated at 550°C, 650°C, 800°C and 1100°C, and then characterized by XRD to identify the present phases. The systems were tested with respect to their catalytic efficiency in the degradation of phenol (200 ppm) in water where $CuFe_{1.2}O_{2.8}$ was found to be the most reactive one (80% removal in 60 min). The rate of phenol degradation was enhanced under irradiation with a 30-W LED light source. The positive role of irradiation was most obvious in the case of $Ag_2Fe_{5.4}V_3O_{16.6}$ in which almost complete conversion was achieved in 120 min compared to only 45% in the same period but without irradiation. The phenol degradation was found to obey first-order kinetics and it is believed that the first step of degradation process is that in which hydroxyl radicals are produced from H_2O_2. Analysis of the kinetic data suggests that the role of irradiation lies in increasing the rate of production of hydroxyl radicals.

The Fenton-like degradation of phenol using $CuFe_{1.2}O_{2.8}$ or $Ag_2Fe_{5.4}V_3O_{16.6}$ as a catalyst seems to represent an effective method of degradation due to the high activity of these catalysts and due also to the fact that leaching of constituting ions of these catalysts into water is within the acceptable health safety range.

Acknowledgement

The authors acknowledge the generous fund of the Deanship of Scientific Research at Yarmouk University, Irbid-Jordan.

References

[1] Murray C A and Parsons S A 2004 *Chemosphere* **54** 1017
[2] Zelmanov G and Semiat R 2008 *Water Res.* **42** 3848
[3] De Laat J, Le T G and Legube B 2004 *Chemosphere* **55** 715
[4] De Laat J and T. G. Le T G 2005 *Environ. Sci. Technol.* **39** (2005) 1811
[5] Ensing B, Buda F and Baerends E J 2003 *J. Phys. Chem.* A 5731
[6] Neyens E and Baeyens J 2003 *J. Hazard. Mater.* **98** 33
[7] Pignatello J J 1992 *Environ. Sci. Technol.* **26** 944
[8] Yuan S H, Gou N, Alshawabkeh A N and Gu A Z 2013 *Chemosphere* **93** 2796
[9] Wang N, Zheng T, Zhang G and Wang P 2016 *J. Environ. Chem. Eng.* **4** 762
[10] Pliego G, Zazo J A, Garcia-Munoz P, Munoz M, Casas J A and Rodriguez J J 2015 *Crit. Rev. Env. Sci. Technol.* **45** 2611
[11] Garrido-Ramírez E G, Theng B K G and Mora M L 2010 *Applied Clay Science* **47** 182
[12] Casbeer E, Sharma V K and Li X-Zh 2012 *Sep. Purif. Technol.* **87** 1
[13] Matavos-Aramyan S and Moussavi M 2017 *Int. J. Environ. Sci. Nat. Res.* **2** ID 555594

[14] Xu L and Wang J 2011 *J. Hazard. Mater.* **186** 256
[15] Hansson H, Kaczala F, Marques M and Hogland W 2012, *Int. J. Photoenergy* Article ID 531076
[16] Ruales-Lonfat C, Barona J F, Sienkiewicz A, Bensimon M, Vélez-Colmenares J, Benítez N and Pulgarín C 2015 *Appl. Catal. B Environ.* **166-167** 497
[17] Pastrana-Martínez L M, Pereira N, Lima R, Faria J L, Gomes H T and Silva A M T 2015 *Chem. Eng. J.* **261** 45
[18] Kurian M and Babu R 2013 *J. Environ. Chem. Eng.* **1** 86
[19] Du J, Bao J, Fu X, Lu C, Kim C H 2016 *Appl. Catal.* B: *Environ.* **184** 132
[20] Wang J, Yao Zh, Wang Y, Xia Q, Chu H and Jiang Zh 2017 *Environ. Pollut.* in press.
[21] Wang N, Zhaob Q and Zhang A 2017 *RSC Adv.* **7** 27619
[22] Khatri C and Rani A 2008 *Fuel* **87** 2886
[23] Costa R C C, Lelis M dF F, Oliveira L C A, Fabris J D, Ardisson J D, Rios R R V A, Silva C N and Lago R M 2003 *Catal. Commun.* **4** 525
[24] Soliman Selim M, Turky G, Shouman M A and El-Shobaky G A 1999 *Solid State Ionics* **120** 173
[25] Dom R, Subasri R, Radha K and Borse P H 2011 *Solid State Commun.* **151** 470
[26] Wulfsberg G 2002 *Inorganic Chemistry* (University Science Books)
[27] Sun S, Yang X, Zhang Y, Zhang F, Ding J, Bao J and Gao Ch 2012 *Prog. Nat. Sci.-Mater.* **22** 639
[28] Baldrian P, Merhautová V, Gabriel J, Nerud F, Stopka P, Hrubý M and Benes M J 2006 *Appl. Catal.* B*: Environ.* **66** 258
[29] Qi Y, Liu Zh, Lu Ch and Zhang T 2017 *Sci. Adv. Mater.* **9** 501
[30] Estrella M, Barrio L, Zhou G, Wang X, Wang Q, Wen W, Flanson J C, Irenkel A I and Rodriguez J A 2009 *J. Phys. Chem.* **113** 1441
[31] Monova E, Tsonchera T, Ponerce D, Popora M, Velinor N, Kuner B, Tencher K and Mihor I 2011 *Solid State Chem.* **184** 1153
[32] Hamdan N, Abu Haija M, Banat F and Eskhan A 2017 *Desalin. Water Treat.* **69** 268
[33] Tan X, Li G, Zhao Y and Hu C 2009 *Mater. Res. Bull.* **44** 2160
[34] Zenchenko T A and Morozov V N 1995 *Protein Sci.* **4** 251
[35] Kremer M L 1962 *J. Catal.* **1** 351
[36] A. de Luis, J. I. Lombrana, F. Varona, and A. Menendez, *Korean J. Chem. Eng.* **26** (2009) 48
[37] Magaireh N, Master thesis, Yarmouk University 2013
[38] Babuponnusami A and Muthukumar K 2012 *Chem. Eng. J.* **183** 1
[39] de Luis A, Lombrana J I, Varona F and Menendez A 2009 *Korean J. Chem. Eng.* **26** 48
[40] Dumitrescu A M, Samoila P M, Nica V, Iordan A R and Palamaru M N 2012 *Rev. Roum. Chim.* **57** 843
[41] Gogate M R, Spivey J J and Zoeller J R 1997 *Catal. Today* **36** 243
[42] Babuponnusami A and Muthukumar K 2014 *J. Environ. Chem. Eng.* **2** 557
[43] Fitzgerald D J 1998 *Am. J. Clin. Nutr.* **67** 1098S
[44] http://www.idph.state.il.us

IOP Conf. Series: Materials Science and Engineering **305** (2018) 012019 doi:10.1088/1757-899X/305/1/012019

Graphene-Based Transparent Electrodes for Dye Sensitized Solar Cells

Nathir A.F. Al-Rawashdeh[1, 2], Borhan A Albiss[3] and Mo'ath H.I. Yousef[1]

[1]*Jordan University of Science and Technology, Faculty of Science and Arts, Chemistry Department, P.O. Box 3030, Irbid 22110, Jordan.*
[2] *ETS, Math & Natural Science (Chemistry); RAS Al-Khaimah Women's Campus; Higher Colleges of Technology, P.O. Box 4792, Ras Al Khaimah, UAE*
[3]*Jordan University of Science and Technology, Faculty of Science and Arts, Physics Department, P.O. Box 3030, Irbid 22110, Jordan.*

nathir@just.edu.jo

Abstract: Several Zinc Oxide (ZnO) photo-anodes were prepared with different morphologies. For each morphology, two composites containing graphene oxide (GO) were prepared. ZnO sheet-flowers attained the highest efficiency among control samples, owing to the light diffraction that may be caused by such morphology. On the other hand, ZnO rods achieved lower performance than ZnO sheet-flowers, but higher than ZnO flowers, due to their porosity and structure, which may scatter light effectively. The effect of including GO in the photoanode matrix was studied and the results demonstrate a significant increase in short circuit current density (J_{SC}). The addition of GO suggested an overall positive effect on cell performance, where samples of ZnO rods and Flowers had the most significant increase in their performance, due to the inhibition of charge recombination by GO.

Keywords: Zinc Oxide; graphene oxide; photoanode; dye sensitized solar cell.

1. Introduction

Several power sources have been explored combat the dependence on fossil fuels. Solar power is a prime candidate, due to the low pollution associated with solar cells, and their easy fabrication. Many types and generations of photovoltaic devices were studied. However, this study focuses on Dye Sensitized Solar Cells (DSSCs). DSSCs are a third generation solar cells that have been invented by Michael Grätzel in 1991 [1]. Dye Sensitized Solar Cells attract interest due to their cost effectiveness, their easy fabrication, and the moderate performance they can provide. The highest performance of such cells was recorded at 14.3% [2]. Such cells are comprised of various parts, but this study focuses on photoanodes, where the size, morphology and the type of material plays a crucial role in performance. The type of the semiconductor is an important factor in cell efficiency, Titanium Oxide (TiO_2) is commonly used due to its high performance. However, alternatives such as Zinc Oxide (ZnO) are utilized in this study due to their easier preparation techniques.

Several approaches are taken to study and enhance photoanode performance. Such as altering morphology [3], doping [4], and using added materials to form composite photoanodes [5]. Generally, carbonaceous materials are common among the added materials to photoanodes, such as graphene-based materials. Graphene is a two dimensional, single layer of SP^2 carbons, which has been gaining attention, owing to its physical and electrical properties [6-8]. Since then, graphene based materials were employed in numerous applications such as catalyst supports [9], gas sensors [10], ultra-thin lenses [11] and as a photovoltaic material in this work.

This study aims to investigate the effect of changing the morphology of ZnO as photoanode material, and the effect of the addition of a graphene-based conductor material, mainly graphene oxide (GO), and hence investigate the changes in performance for each morphology, and study the changes brought upon the incorporation of graphene oxide.

2. Material and Methods

2.1. Materials and Instruments

Zinc nitrate hexahydrate ($Zn(NO_3)_2{\cdot}6H_2O$) (purity ≥ 98%) and chloroplatinic acid H_2PtCl_6 (≥37.50% Pt basis) was obtained from Aldrich, zinc acetate dihydrate ($Zn(CH_3COO)_2.2H_2O$) (purity ≥ 99%) was bought from Xilong Chemical industry, tri-sodium citrate ($C_6H_5Na_3O_7{\cdot}2\ H_2O$) (purity ≥ 99%) and potassium hydroxide (KOH) (purity ≥ 90%) were obtained from Sigma, while graphite powder was provided by Acros, potassium permanganate (purity ≥ 99.5%) ($KMnO_4$) was bought from Park Scientific, sodium hydroxide (NaOH) (purity ≥ 99.5%) from BBC chemicals, while triton-x surfactant (purity ≥ 99.5%) and iodine (I_2) (purity ≥ 99.5%) was provided by BDH chemicals, Polyethylene Glycol 35,000 was bought from Merk, Guanidine Thiocyanate (GuCN) (purity ≥ 95%) was bought from Combi-blocks, Hydrogen Peroxide (H_2O_2) (purity 30%) from Applichem, Sulfuric Acid (H_2SO_4)(purity 95-98%) from Schalru, and finally, potassium iodide (KI) (purity ≥ 99%) was provided by Janssen Chemica.

X-Ray diffraction (XRD) was performed on (XRD-6000, Shimadzu) with CuKα radiation λ=1.5405 Å, while Fourier transform infrared (FTIR) data were measured using Bruker Alpha. UV-Vis data was obtained using Shimadzu Double Beam UV-2500PC Series. The scanning electron microscope used was a Quanta FEI 450 SEM. The J-V characteristics were measured using a PECCEL PEC-l01 portable solar simulator, and Keithley 2425 sourcemeter, along with Labtracer software.

2.2. Synthesis of ZnO Rods

ZnO rods were prepared according the method reported in literature with minor modify-cations [12] , a 50 ml of a 1 M of $Zn(NO_3)_2{\cdot}6H_2O$ solution was prepared and stirred at room temperature for 10 minutes, until the solution was completely dissolved, next, a 1.5 M KOH solution was added dropwise to the zinc nitrate solution, until the pH reached 12, the resulting solution was transferred into a teflon lined stainless steel autoclave for 20 hours at 120 °C in an oven, the resulting precipitate was obtained by centrifugation and heated at 105 °C in an oven overnight for drying.

2.3. Synthesis of ZnO Sheet-Flowers

ZnO Sheet-flowers were prepared according to the literature [13], with a few modifications to produce higher quantities, a 400 ml solution of $Zn(CH_3COO)_2.2H_2O$ 0.125M and 0.3 M $C_6H_5Na_3O_7{\cdot}2\ H_2O$ was prepared and stirred for 15 minutes, resulting in a clear solution, then, a 100 ml of 2.5 M NaOH solution was prepared and stirred until it completely dissolved, finally, the alkaline solution was added to the first solution dropwise using a burette, the resulting solution was transferred to a water bath at 80 °C for 2 hours, the product was centrifuged and dried at 105 °C overnight.

2.4. Synthesis of ZnO Flowers

ZnO Flowers were prepared by making two solutions, the first solution was prepared by making a 200 ml of 0.3 M $Zn(CH_3COO)_2.2H_2O$ solution and stirring until the solution became transparent, next a 200 ml of an alkaline NaOH solution of 3 M was prepared and stirred until it completely dissolved. Next, the Zinc Acetate solution was added to the alkaline solution dropwise, the resulting mixture was transferred to a water bath at 90 °C for 3 hours; the resulting ZnO Flowers were separated by centrifugation and heated at 105 °C overnight for drying.

2.5. Preparation of Graphene Oxide

Graphene oxide was prepared via modified Hummers method with minor modifications [14], 2 g of graphite were added to 47 ml of concentrated H_2SO_4 and stirred vigorously in a conical flask, the flask was transferred to an ice bath and cooled to 0 °C, then 7 g of ($KMnO_4$) were added slowly and carefully,

IOP Conf. Series: Materials Science and Engineering **305** (2018) 012019 doi:10.1088/1757-899X/305/1/012019

while ensuring the temperature does not exceed 20 °C, after the addition of $KMnO_4$, the mixture was transferred to a 35 °C water bath and stirred for 2 hours. The resulting solution was diluted with 100 ml of Deionized water, and the temperature rose dramatically, then 10 ml of 30% Hydrogen peroxide (H_2O_2) was added and the solution became yellow. The mixture was centrifuged and washed with 10% HCl and twice with Deionized (DI) water to remove the metal ions. The resulting GO product was dried for 48 hours at 105 °C in an oven.

2.6. Preparation of Composite Pastes

Nine composites were prepared, each morphology three composites, which were summarized in Table 3.1. For the preparation of GO-free paste (S0, R0 and F0), 1 g of the prepared ZnO (Rods, Sheets, Flowers) was grounded with 0.25 g of Polyethylene Glycol 35,000 (PEG 35,000) to give consistent paste, followed by the addition of an ethanol/water solution (4:1 v:v) and finally 30 μL of Triton-x surfactant were added to enhance the adhesion to the substrate surface.
For the preparation of GO containing pastes (composites) the same previous procedure was used, however an appropriate amount of GO was added to the ethanol/water solution and sonicated for 10 minutes.

Table 1. Summary for morphology, ratio and abbreviation for the corresponding synthesized samples.

Morphology	%GO (wt%)	Abbreviation
Rods	0	R0
Rods	1	R1
Rods	3	R3
Sheet-Flowers	0	S0
Sheet-Flowers	1	S1
Sheet-Flowers	3	S3
Flowers	0	F0
Flowers	1	F1
Flowers	3	F3

2.7. Fabrication of the DSSCs

To prepare the photoanodes, the Fluorine Doped Tin Oxide (FTO) substrates were sonicated for five minutes with different solvents, first with soap, followed by sonication with isopropanol, and finally sonicated with acetone. The substrates were then dried in oven at 40 °C. Then, scotch tape was used to control the thickness of the photoanodes, the paste was deployed on the substrate by doctor-blading (screen printing), to create the photoanode films, the photoanodes were then annealed at 400 °C for 30 minutes, and allowed to cool in air, the area of the anodes was 1 cm^2. The cooled photoanodes were sensitized by soaking in a 0.2 mM N749 dye ethanolic solution for 24 hours. When the sensitization was complete, the Photoanodes were rinsed with isopropanol to remove the unattached dye and allowed to dry. On the other hand, the counter electrodes were prepared by dropping and spreading 20 μL of 5 mM of H_2PtCl_6 and allowing the electrodes to dry in air, next, the counter electrodes were annealed at 400 °C in a furnace for 30 minutes. To finally prepare the DSSCs, a counter electrode and a photoanode were clipped against each other using paper clips. To prepare the electrolyte, a solution containing 0.5 M Potassium Iodide and 0.1 M Iodine and 0.5 M Guanidine Thiocyanate (GuCN) in EG was stirred at room temperature for 30 minutes. The electrolyte was introduced into the cell via capillary action. Samples of the prepared samples are shown in Figure 1.

IOP Conf. Series: Materials Science and Engineering **305** (2018) 012019 doi:10.1088/1757-899X/305/1/012019

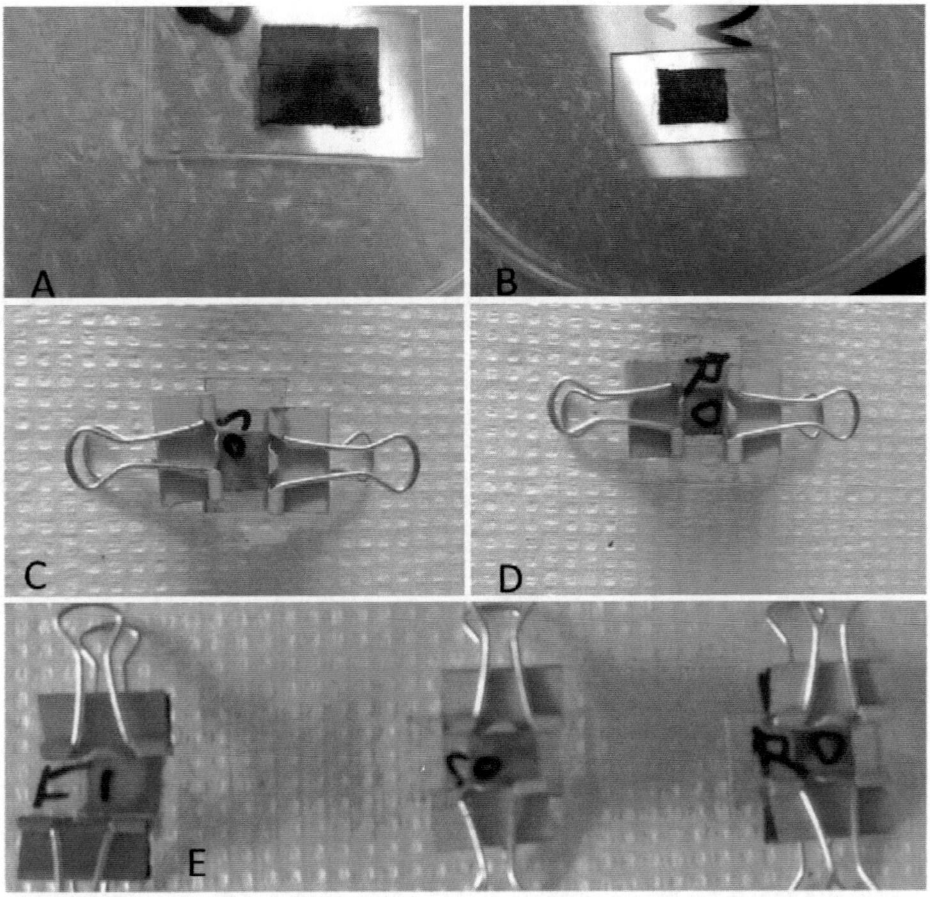

Figure 1. (a and b), Prepared photoanodes, (c-e), Pictures of prepared DSSCs.

3. Results and Discussion

3.1 X-Ray Diffraction Results

To investigate the crystallinity of the prepared ZnO morphologies, X-ray diffraction (XRD) was used, and judging by the resulting patterns, it is quite evident that the prepared ZnO composites are all of the same wurzite hexagonal structure, where the peaks at 31.75°, 34.42°, 36.24°, 47.52°, 56.59°, 62.84°, 66.32°, 67.93°, 69.03°, 72.59° and 76.90° corresponding to planes (100), (002), (101), (102) (110), (103), (200), (112), (201), (004) and (202), all correspond to wurzite structure (JCPDS card no. 01-080-0075), in addition to the high crystallinity as shown by the patterns in the Figure 2.

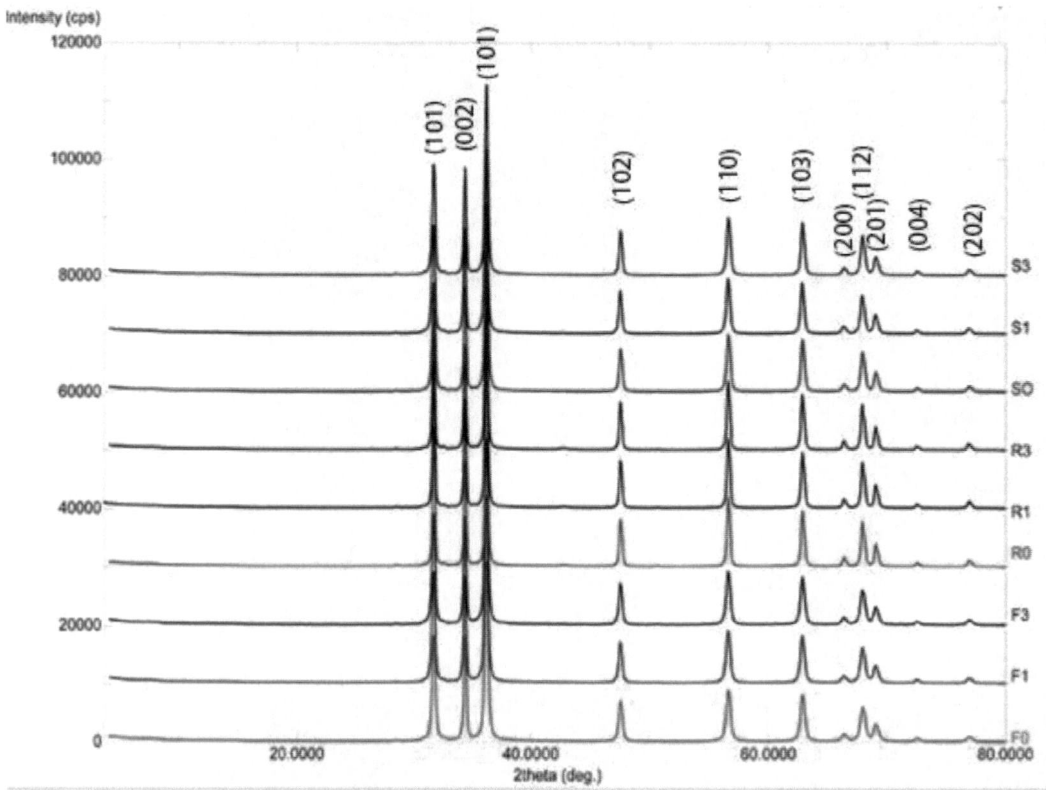

Figure 2. XRD of the prepared composites.

3.2 Morphology Analysis

To study the prepared GO and ZnO composites, several SEM images were taken, images in Figure 3 show the prepared GO by modified Hummers method, which shows minimal stacking because of the loss of many π bonds, reducing π- π stacking. A look at Figure 3(c) shows the improper stacking of the sheets, indicating successful GO production.

The role of morphology is imperative regarding photovoltaic performance, the SEM images shown in Figure 4 display ZnO crystals with a "Sheet-Flower" shape, which do indeed show high surface porosity and irregularities, which may scatter incoming light among the anode, indicating interesting properties. The large sheets in the image are graphene oxide sheets. The SEM images show flowers in the range of 1-2μm.

In addition to the previously prepared composites, ZnO "Nanorods" were also prepared, and are shown in Figure 5, the rods appear to be in sizes ranging between 0.1-1 μm, such shapes may contain large spaces between the crystals, which may be counterproductive to electron transport; however, effective paste for this material could overcome such issues. The large sheet which the ZnO rods lay upon is graphene oxide sheets.

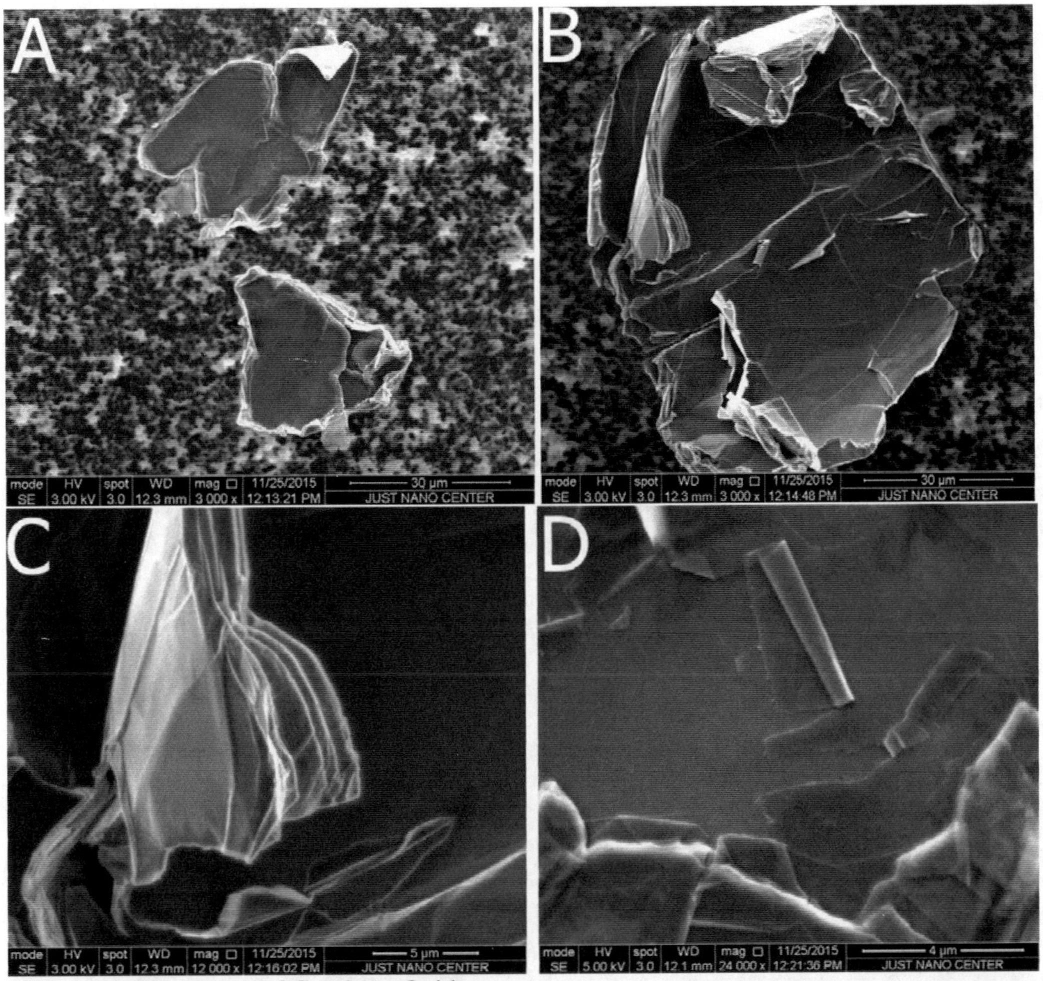

Figure 3. SEM images of Graphene Oxide.

Figure 4. SEM images of the prepared ZnO Sheet-Flowers and GO composite.

Figure 5. SEM images of the prepared ZnO Rods and GO composite.

Figure 6 shows several SEM images of the prepared ZnO Flowers that lay on GO sheets. The morphology of those sheets shows layers of thin sheets; however, their sizes stretch to more than 10 μm approximately. These sheets appear to have low surface area when compared to rods and flowers. In addition, this shape may have large spaces between the crystals, suggesting poor performance.

Figure 6. SEM images of the prepared ZnO Flowers and GO composite.

3.3. FT-IR Spectra

Figure 7 shows the FT-IR spectrum for GO, the spectrum shows peaks of oxygen containing groups, the peak at 3446 cm^{-1} belongs to hydroxyl groups (O-H), peak at 1385 cm^{-1} belongs to vibrations of C-OH bonds, moreover, the resulting peak at 1710 cm^{-1} and 1566 cm^{-1} corresponds to both the carbonyl (C=O) and The alkene (C=C) stretching vibrations, and the small peak at 1054 cm^{-1} corresponds to epoxide (C-O-C) vibrations, and finally the peak at 1476 cm^{-1} may be assigned to CH_2 bending, thus the formation of GO is confirmed.

IOP Conf. Series: Materials Science and Engineering **305** (2018) 012019 doi:10.1088/1757-899X/305/1/012019

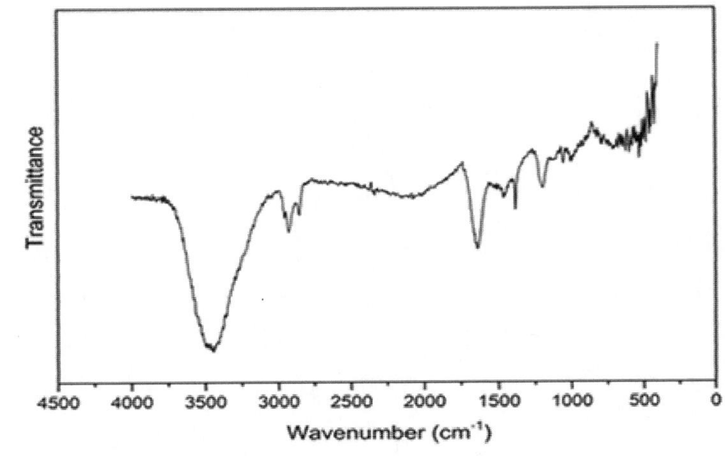

Figure 7. FTIR spectra for GO.

3.4. UV-Vis Analysis

To further study the composites, the UV-Vis absorption spectra were obtained and shown in Figure 8 (a-c). Figure 8 shows a characteristic absorption peak at 375nm for ZnO, furthermore the overall absorption of composites improves upon the addition of GO due to increasing of the surface electric charge, which was caused by the presence of GO. The presence of GO may increase porosity and may give rise to light trapping effect [15], allowing more light to be absorbed, caused by more scattering and irregularities in the matrix, indicating a decrease in reflectance and an increase in absorption.

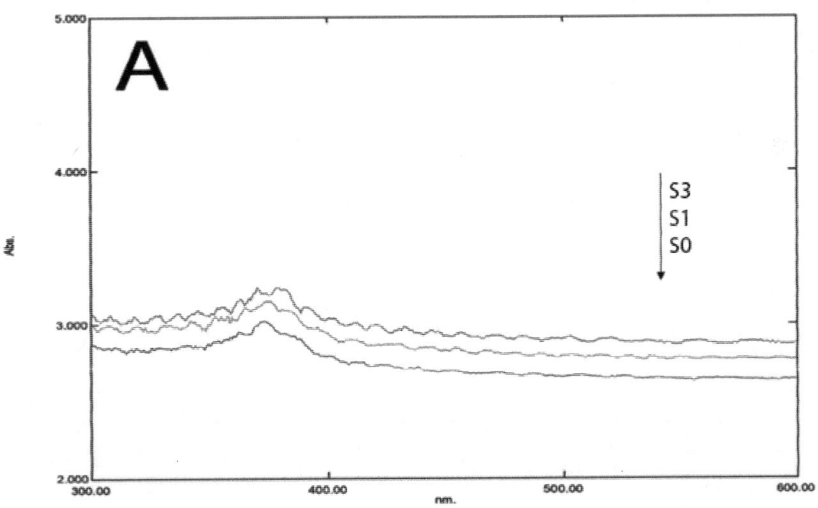

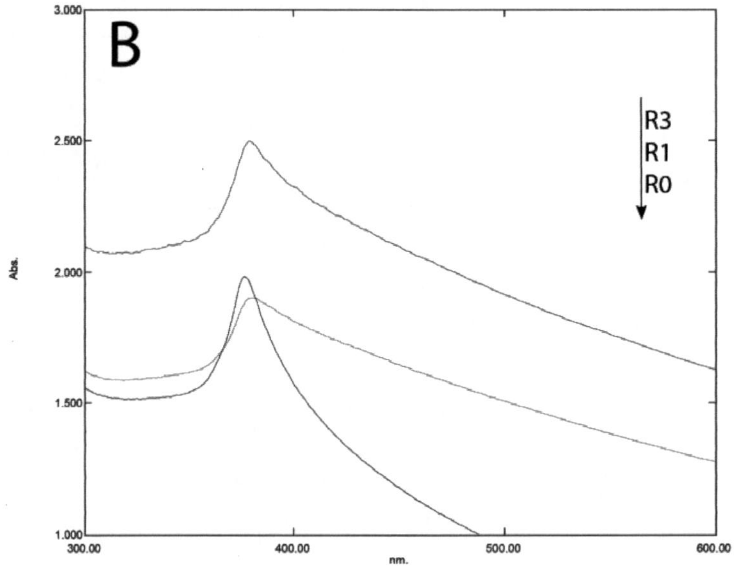

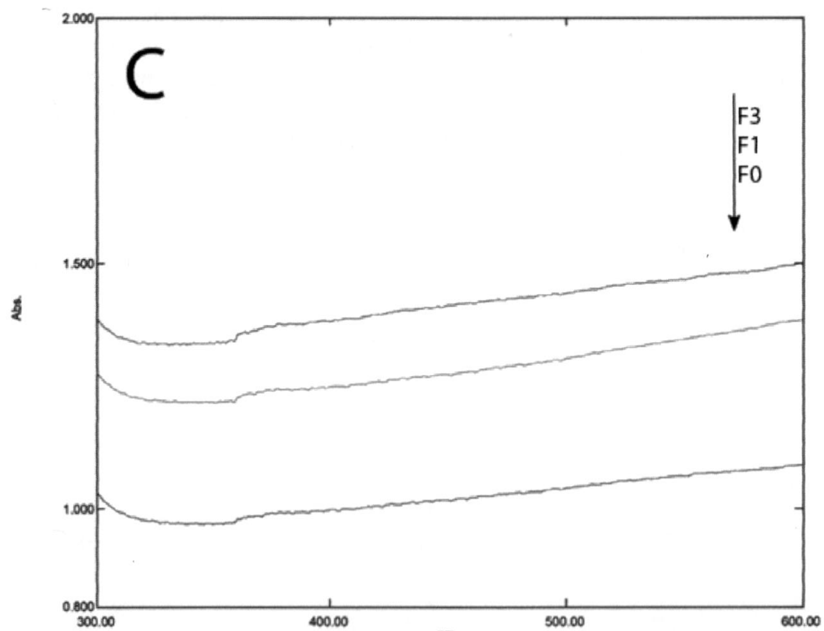

Figure 8. UV-Vis spectra of (a) S0-S3, (b) R0-R3, (c) F0-F3.

3.5. J-V Characteristics

To analyze the efficiency and the performance of the prepared samples, the Fill Factor (FF) and Photoconversion Efficiency (η %) were estimated using equation 1 and 2, respectively:

$$FF = \frac{Jmp \times Vmp}{Jsc \times Voc} \qquad 1$$

$$\eta\,\% = \frac{Jsc \times Voc \times FF}{Pin} \qquad 2$$

Where J_{mp}, V_{mp}, J_{sc}, Voc, and P_{in} are maximum power current, maximum power voltage, short circuit current, open current voltage and incident power respectively. The photovoltaic parameters measured were summarized in Table 2.

The corresponding η% for each sample is summarized in Table 2. Among samples with no GO, ZnO sheet-flowers have the highest performance, followed by ZnO rods and finally ZnO flowers. This behavior could be attributed to the structural morphology which plays a major role in light scattering and trapping. The arrangement and porosity of ZnO sheet-flowers may allow more photons to be scattered across the photoanode. ZnO rods have a higher efficiency than ZnO flowers which can be attributed to their irregular shapes and small sizes, which allows more incoming photons to be scattered and trapped across the photoanode matrix.

For ZnO flowers, the large crystal sizes, may play a negative role in the process of electron transport. It can also be observed from Table 2 that the efficiency of the composites is indeed higher than the samples that lack GO. Figures 9 and 10 show the J-V curves of the corresponding samples, sample R3 shows a significant increase in performance, much higher than the other composites, the added graphene based material to the photoanode matrix may indeed fill any gaps in the film, which in turn shuttles electrons across it, due to the high conductivity of graphene based materials, and finally reducing charge recombination. R1 shows a six-fold enhancement in performance compared to R0 and R3 shows twelve-fold increase in performance in comparison to R0.

Table 2. The photovoltaic parameters of DSSCs.

Samples	J_{SC} (mA/cm^2)	V_{OC} (V)	J_{mp} (mA/cm^2)	V_{mp}	Fill Factor	η %
F0	0.25	0.102	0.54	0.013	0.275	0.007
F1	0.20	0.255	0.13	0.149	0.389	0.020
F3	0.43	0.227	0.23	0.127	0.295	0.029
R0	0.56	0.382	0.22	0.173	0.174	0.037
R1	1.63	0.390	1.06	0.237	0.396	0.251
R3	2.59	0.389	1.80	0.250	0.447	0.450
S0	3.04	0.199	1.73	0.104	0.297	0.18
S1	2.81	0.329	1.40	0.188	0.285	0.26
S3	2.11	0.317	1.33	0.183	0.364	0.24

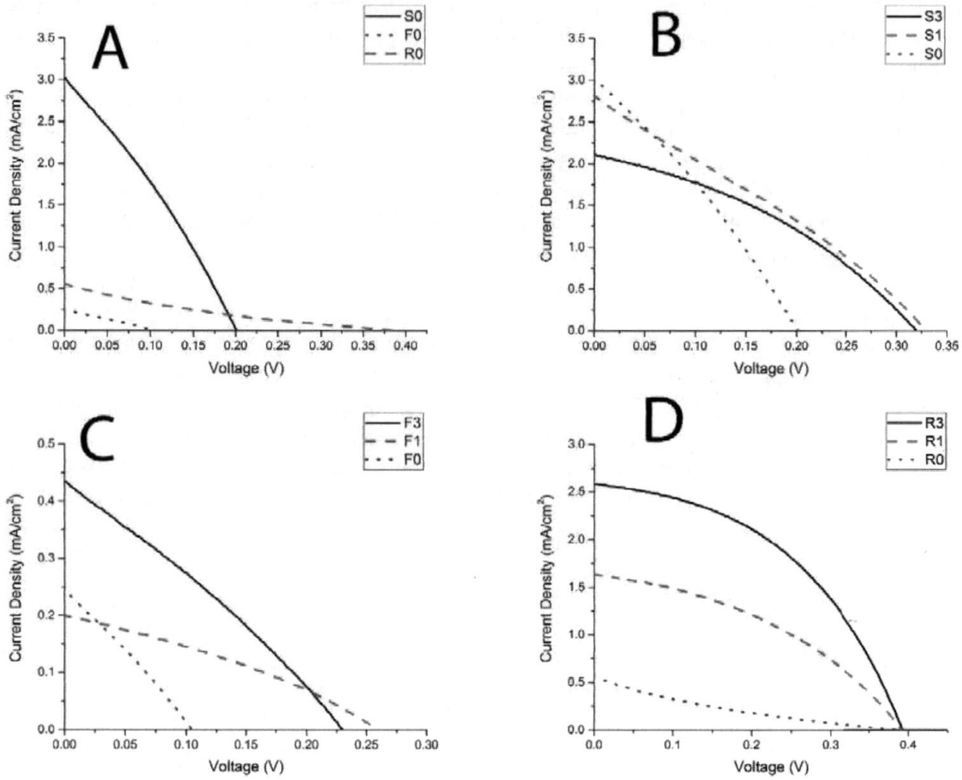

Figure 9. J-V characteristics for samples (a)S0, R0 and F0 (b)S0, S1 and S3 (c) F0, F1 and F3 (d) R0, R1 and R3.

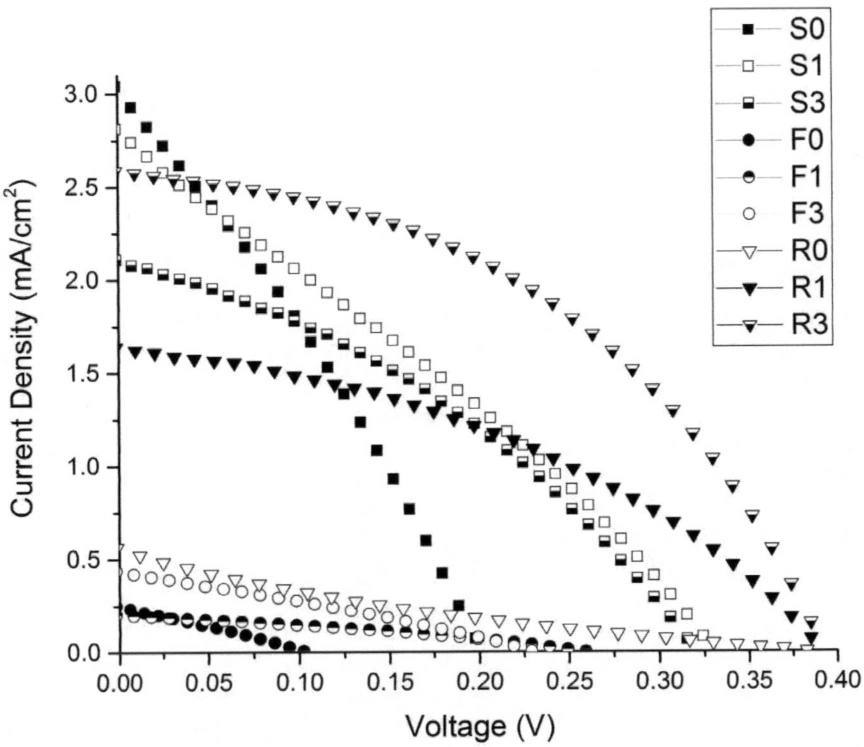

Figure 10. J-V characteristics for all samples.

On the other hand, samples S1 and S3 show a slight increase in performance when compared to S0, the added GO did not have a significant impact on the performance, the shape of ZnO sheet-flowers may not be compatible with GO, and the arrangement of GO sheets could be improper, and no remarkable change can be observed. Finally, the ZnO flowers based photoanodes had poor efficiency compared to the ZnO rods and sheet-flowers, nonetheless, ZnO flowers samples containing GO had a noteworthy rise in performance, however, the distinction is not significant, since ZnO flowers had poor performance originally.

4. Conclusion

Several ZnO samples were prepared, and the effect of adding GO to ZnO anodes with differing morphologies was studied, the added GO is reduced thermally during the photoanode preparation, restoring some of the double bonds in the sheets. Consequently, conductivity increases and thus charge recombination is reduced. According to J-V characteristics curve, the addition of GO to photoanodes enhanced the performance, however, for ZnO flowers, the addition of GO isn't quite notable, the performance of ZnO flowers is minimal due to the large sizes of the crystals, leaving very large gaps in the electron transfer paths, and the addition of GO did elevate performance, however, the enhancement is limited. In addition, ZnO rods showed low performance (R0), due to their large sizes, allowing spaces to be formed within the anode, interrupting electron transport, however, upon the addition of GO, the performance is increased dramatically, the ZnO rods may have bad electron transport, however, the added carbon-based materials minimized this drawback, and allowed for an interesting rise in efficiency. And finally, ZnO sheet-flowers demonstrated a good performance compared to the other morphologies, however, the addition of GO has indeed showed a direct effect on overall cell performance, nonetheless, the resulting increase in efficiency isn't as drastic as ZnO rods.

Acknowledgments
The authors acknowledge the financial support of the Jordan University of Science and Technology Research Fund. This research was funded by a grant from the Deanship of research-JUST, grant number 20160328.

References

[1] O'regan B, Grfitzeli M. 1991 Nature 353(6346) 737-740.
[2] Kakiage K, Aoyama Y, Yano T, Oya K, Fujisawa J-i, Hanaya M. 2015 C*hemical Communications* 51(88) 15894-15897.
[3] Zhang Y, Xie Z, Wang J. S 2009 *ACS applied materials & interfaces* 1(12) 2789-2795.
[4] Raj CJ, Prabakar K, Karthick S, Hemalatha K, Son M-K, Kim H-J. 2013 *The Journal of Physical Chemistry C.* 117(6) 2600-2607.
[5] Albiss BA, Al-Widyan MI, Obaidat I. 2016 *Ceramics International* 42(2) 3563-3568.
[6] Novoselov KS, Geim AK, Morozov S, Jiang D, Katsnelson M, Grigorieva I, et al. 2005 Nature 438(7065) 197-200.
[7] Park J, Ahn Y, Ruiz-Vargas C. 2009 *Nano letters* 9(5) 1742-1746.
[8] Zhang Y, Tan Y-W, Stormer HL, Kim P. 2005 *Nature* 438(7065) 201-204.
[9] Sun Y, Du C, An M, Du L, Tan Q, Liu C, et al. 2015 *Journal of Power Sources* 300 245-253.
[10] Srivastava V, Jain K. 2016 *Materials Letters* 169 28-32.
[11] Zheng X, Jia B, Lin H, Qiu L, Li D, Gu M. 2015 *Nature communications* 2015 6.
[12] Kiomarsipour N, Razavi RS. 2012 *Superlattices and Microstructures* 52(4) 704-710.
[13] Ma Q, Wang Y, Kong J, Jia H, Wang Z. 2015 *Superlattices and Microstructures* 84 1-12.
[14] Ma Y, Sun L, Huang W, Zhang L, Zhao J, Fan Q, et al. 2011 *The Journal of Physical Chemistry C* 115(50) 24592-24597.
[15] Moussa H, Girot E, Mozet K, Alem H, Medjahdi G, Schneider R. 2016 *Applied Catalysis B: Environmental* 185 11-21.

IOP Conf. Series: Materials Science and Engineering **305** (2018) 012020 doi:10.1088/1757-899X/305/1/012020

Effects of Substitutions on the Biodegradation Potential of Benzotriazole Derivatives

M A Abu-Dalo[1], I O'Brien[2] and M T Hernandez[3]

[1]*Jordan University of Science and Technology, Faculty of Science and Arts, Chemistry Department, P.O. Box 3030, Irbid 22110, Jordan*
[2]*Air Force Civil Engineer Center (AFCEC), San Antonio, TX, USA*
[3]*Department of Civil and Environmental Engineering, University of Colorado at Boulder, Campus Box 428, Boulder, CO 80309-0428, USA.*

maabudalo@just.edu.jo

Abstract: Fourteen benzotriazole derivatives were subjected to microcosm tests to study the influence of substitutions on their biodegradation potential. Methylated, nitrated, carboxylated, and propionated bezotriazoles, a heterocyclic triazole, as well as methylated benzimidazoles, were introduced to activated sludge and soil enrichment cultures as the only carbon source. Some of the enrichment cultures were derived from airport soils that had been previously contaminated with aircraft deicing fluids and subsequently enriched with the commercially significant corrosion inhibitor methylbenzotriazole. The 5-methylbenzotriazole and only the carboxylated derivatives were degraded by soil or activated sludge biomass regardless of acclimation conditions. Radiotracer studies of [U-^{14}C] 5-methylbenzotriazole, and [U-^{14}C] 5-carboxybenzotriazole confirmed that relatively high concentrations (25mg L^{-1}) of these derivatives can be completely mineralized in relatively short time frames by microbial consortia regardless of prior exposure. Observations suggested that the growth yield on these compounds is likely low. Biodegradation patterns suggested that carboxylated benzotriazole derivatives are more readily biodegradable than their more popular methylated counterparts.

1. Introduction

Benzotriazole (BT) and its derivatives are produced in large quantities for various well-known purposes such as the production of fuels, detergents and varnish stabilizers [1, 2]. With respect to environmental dissemination routes, methylbenzotriazole (MeBT) has been found as the most commonly occurring triazole-containing pollutant on waterways [3]. This corrosion inhibitor is commonly found in aircraft deicing fluid (ADF) formulations and other freeze point depressing additives. More than 7 billion gallons of stormwater contaminated with deicing chemicals are generated annually in the USA alone [4]. Accordingly, comprehending the mechanisms governing the fate and transport of the deicing components of benzotriazoles is essential for the evaluation of the potential terrestrial environmental impacts [5]. The environmental loadings of benzotriazole derivatives are relatively high and do last for long periods [3, 5, 6]. The storm runoff containing benzotriazoles can impose toxicity to model ecosystem receptors [7, 8]. Surveys in northern Europe [5] and central Europe [9] have already identified the benzotriazole derivatives with significant concentrations in the major drinking water sources.

The commercially significant benzotriazole derivatives are likely will not biodegrade under common water and soil conditions. However, benzotriazole is known to be light sensitive and photodegradable in certain environments [10]. Some direct evidence of biodegradation under specific environmental conditions is available, but albeit limited. It was claimed that of the two commercially significant methylbenzotriazole isomers, 5-MeBT is actually biodegradable, while 4-MeBT is recalcitrant in oxygen saturated water or soil with active microbial biomass [11, 12]. Benzotriazole (BT), 4-methyl- and 5-methyl-benzotriazle (4- and 5- MeBT) were suggested to not biodegrade under anaerobic conditions in the presence of different electron acceptors. Therefore, presumably, aerobic biodegradation is the major removal mechanism in water systems such as activated sludge systems [13]. 5-MeBT biodegradation was found to be enhanced through acclimation and nutrient supply [14]. Recent study found that the integration of the sunlight-induced processes (with TiO_2) with subsurface-flow treatment wetlands caused 42% further elimination of benzotriazole [15]. Methylbenzotriazole was found to degrade by the lignin peroxidase induced from the fungus *Phanerochaete chrysosporium* [16]. Therefore, it appears that 4- and 5- MeBT are recalcitrant under conditions void of oxygen and do not serve as substrates for denitrification, sulfate-reduction, or fermentation during conventional (activated sludge) wastewater treatment, or anaerobic digestion [17-19]. Other studies found that BT and 5-MeBT can biodegrade under aerobic and anaerobic conditions and their biodegradation efficiency and biodegradation products were dependent on the predominant terminal electron-accepting condition [20, 21].

The substitutions of certain functional groups on compounds containing single or multiple aromatic rings can facilitate the degradation of an otherwise recalcitrant substrate—such substitutions have been observed with other azo-containing compounds [22-24]. An explanation offered for these observations is that some substituted compounds become more accessible to the enzymatic attack of terrestrial microbes or those otherwise enriched in wastewater treatment plants [25]. The contribution of common functional constituents toward higher biodegradation potentials depends on the pathways available through microbial enzyme inventories expressed, as well as local environmental conditions. There have been few reports documenting enhancements to the biodegradation potential of aromatic conjugates with heterocyclic azides, but none have been extended to include those benzotriazoles that are commonly used as industrial corrosion inhibitors. Zimmermann and coworkers showed that some enzyme reaction rates increased with the substitution of electron-withdrawing substituents on phenyl rings; these substitutions, in turn, enhanced the degradation of some common azo dyes [26, 27]. Patil and coworkers [28] observed that the introduction of functional groups of known electronegative behaviors, particularly electron-withdrawing and lipophilic groups, could enhance the ability of benzyltriazole conjugates to bacterial biodegradation. These observations are in agreement with those reported on the biodegradation of selected azo dyes where electron-donating functional groups, particularly hydroxyl substituents in positions ortho to azo-nitrogen, contributed to increasing biodegradation potential of these compounds [24]. As a gross physical-chemical parameter, water solubility appears to be important to biodegradability, which is clearly impacted by substitution [29, 30]; however, this generality does not appear to apply to heterocyclic azo-containing compounds. In this context, prediction of xenobiotic biodegradation potential, based solely on gross chemical properties, has not been generally successful.

Prevention of environmental hazards and satisfying the impending regulatory pressures entail providing different benzotriazole requiring industries with cost effective and readily biodegradable corrosion inhibitors. In relevant to this, fourteen different benzotriazole derivatives (functional substitutions) were tested for their biodegradation potential using microcosms-containing acclimated and non-acclimated enrichment cultures of soil and activated sludge. In addition, using radio labeled substrates [U-^{14}C], the more readily biodegradable derivatives were examined for their potential to be completely mineralized. Radiography observations were leveraged to estimate effective biomass yields.

2. Material and Methods

2.1. Culture Media

Support media for microcosms slurries consisted of the following (mg per liter of MQ): NH_4SO_4, 600; K_2HPO_4, 440; KH_2PO_4, 340; $MgSO_4 \cdot 7H_2O$, 53; $CaCl_2 \cdot 2H_2O$, 28; $FeCl_3 \cdot 6H_2O$, 8.1; $ZnSO_4 \cdot 7H_2O$, 0.68; $CuSO_4 \cdot 5H_2O$, 0.17; $CoCl_2 \cdot 6H_2O$, 0.17; $NaMoO_4 \cdot 2H_2O$, 0.17; $MnSO_4 \cdot H_2O$, 0.15; H_3BO_3, 0.04. The pH of this solution was 7.0. All solutions were sterilized (autoclaved) prior to their use.

2.2. Benzotriazoles Derivatives

[U-^{14}C]-5-Methylbenzotriazole ([U-^{14}C] 5-MeBT) was synthesized from [U-^{14}C]-p-toluidine according to a zinc-catalyzed substitution pathway published in U.S. Patent 4061491. The radiochemical purity of the [U-^{14}C]-5-MeBT using high pressure liquid chromatography (HPLC) was greater than 98%, with a specific activity of c. 500 µCi g^{-1}. [U-^{14}C] 5-Carboxybenzotriazole ([U-^{14}C] 5-CBT) was synthesized from [U-^{14}C] 5-methylbenzotriazole by selective oxidation of the methyl moiety. One-half gram of [U-^{14}C] 5-MeBT was dissolved in 10 ml, 50% NaOH. The mixture was heated to between 50-60 °C, and 1.6 g $KMnO_4$ was gradually added (stirred) in small portions. The reaction mixture was incubated for 4 hours at 55°C. After this time, the mixture was filtered to remove any reduced manganese, which was present as MnO_2 (s). For purification, 0.5 g charcoal was then added to the filtrate and heated between 50-60 °C. The resulting black precipitates were filtered out (Whatman qualitative circles; Fisher Scientific Pittsburgh, PA, USA) and the filtrate discarded. The filter-retained solids were washed in concentrated HCl, and the pH of the wash solution subsequently adjusted to 7. This process yielded [U-^{14}C] 5-CBT with purity determined by HPLC to be > 98 %, and specific activity of c. 300 µCi g^{-1}. The scintillation cocktail used for all radioactivity measurements was Ultima Gold (Packard BioScience Company, Meriden, CT). All other benzotriazole derivatives used in this investigation were synthesized and obtained in cooperation with PMC Specialty Group Inc. (Cincinnati, OH, USA), or Sigma Chemical Company (St. Louis, MO, USA). The benzotriazoles derivatives chemical structures are shown in Figure 1.

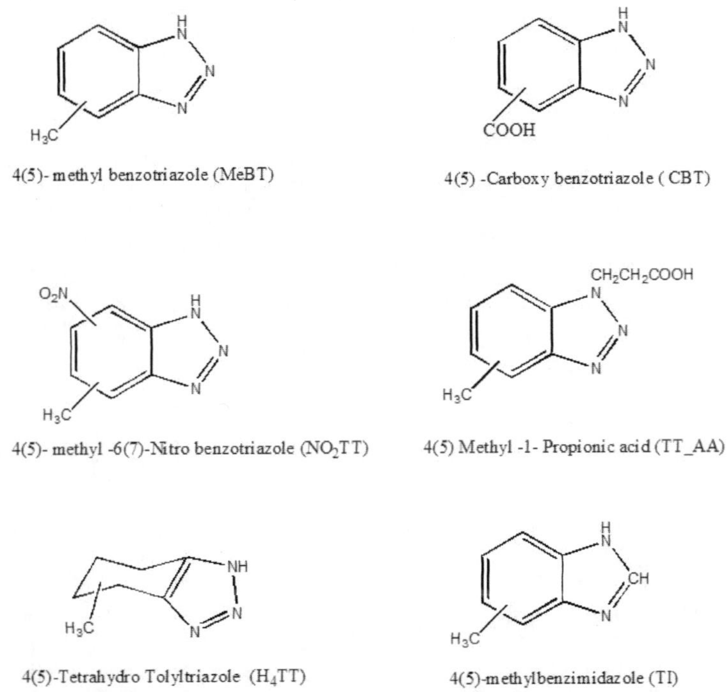

Figure 1. Chemical structures of benzotriazole derivatives observed in biodegradation studies

2.3. Microbial Enrichments

Enrichments consisted of both non-acclimated and acclimated consortia. The acclimated cultures for these experiments were enriched from shallow subsurface soils within two meters of the central deicing pad at Denver International Airport, Colorado, USA. Soils removed from this area are geotechnical grade fills, and have been comprehensively analyzed using accepted geotechnical methods [31]; soil fill characteristics, sampling, and microbial enrichment have been previously described [11]. This culture has been maintained

IOP Conf. Series: Materials Science and Engineering **305** (2018) 012020 doi:10.1088/1757-899X/305/1/012020

in batch reactors and exposed to 4-and 5-MeBT for extended periods (months at room temperature under aerobic conditions).

Non-acclimated cultures were obtained from a bench-scale sequencing batch activated sludge reactor. This system was initially inoculated from a local municipal wastewater treatment plant (Boulder, CO). The reactor was operating under aerobic conditions, at a residence time of 5 days with a glucose substrate, when biomass samples were removed for these experiments.

2.4. Biodegradation Potential Assays

Inocula for these experiments consisted of either acclimated or non-acclimated biomass, which was normalized by widely accepted gravimetric analysis [32] immediately prior to metabolic and radiotracer assays. Cells were harvested by centrifugation at 3500 rpm for 30 min, and washed three times in sterile 0.85% NaCl, before final suspension in culture media. All microcosm assays were initiated with a biomass concentration of 500 mg L^{-1} total suspended solids (TSS) including 25 mg L^{-1} of each benzotriazole derivative in sterile one-liter Erlenmeyer flasks. All flasks were covered with sterile gauze and incubated on a rotary shaker table at 180 rpm at 25 °C ± 2°C. Control flasks consisted of autoclaved inocula to quantify any potential abiotic transformations, or sorptive losses, but were maintained under otherwise identical conditions.

Samples for benzotriazole derivative analysis were periodically withdrawn with a sterile polypropylene pipette and centrifuged at 10,000 x g in polypropylene micro centrifuge tubes for 5 minutes to separate biomass. After centrifugation, the supernatant was transferred to 2.0 ml amber crimp top high pressure liquid chromatography (HPLC) vials (Agilent Technologies, # 5181-3376), stored at 4°C prior to analysis, and analyzed by HPLC.

2.5. Chemical and Physical Parameter Analysis

Benzotriazole Derivative Assay. All benzotriazole derivatives were measured using a HPLC fitted with an UV detector (λ = 254 nm) except for H_4TT (λ = 220 nm). Separation of the isomers was performed isocratically using two Zorbax Rx-C8 4.6 x 250 mm columns in series (MacMod Analytical, Inc., Chadds Ford, PA, USA). The eluent for this method consisted of a phosphate buffer mixed in a 70:30 ratio with HPLC grade acetylnitrile at a flow rate of 1.5 ml $minute^{-1}$ with a sample injection volume of 200 µL. All samples were centrifuged at 10,000 x g for 5 minutes prior to injection. The method detection limit was 0.2 mg L^{-1}, using accepted USEPA QA/QC methods [32].

2.6. Gravimetric Biomass Measurements

Total suspended solids (TSS) were measured according to standard method for the analysis of water and wastewater 2540D [32], with the exception that a 45-µm Durapore filter (Fisher Scientific, Pittsburgh, PA) was used instead of glass fiber filters.

2.7. Mineralization of 5-MeBT and 5-CBT

During radiotracer biotransformation assays, an overall mass balance of the sole organic carbon source supplied (as MeBT or CBT) was compiled by measuring the gaseous, soluble, and biomass fractions in a time series according to widely-accepted methods [33]. As a function of time, the degraded carbon source [U-^{14}C] 5-Methylbenzotriazole or [U-^{14}C] 5-Carboxybenzotriazole, C_T:

$$C_T = C_B + C_{CO2} + C_S$$

Where,

C_T = total amount of ^{14}carbon added to the culture solution

C_B = biomass ^{14}carbon

C_{CO2} = carbon recovered as $^{14}CO_2$

C_S = soluble ^{14}carbon (soluble biomass + HCO_3^- in solution + other soluble products)

The initial benzotriazole mass was normalized by its β-particle emissions, which were traced through metabolism (if it occurred), sorptive processes, gas phase partitioning, and biomass incorporation. In 250 mL borosilicate biometer flasks (Bellco Glass, Inc, Vineland, NJ), both ^{14}C labeled 5-MeBT and ^{14}C labeled 5-CBT (25 mg L^{-1}) were incubated under sustained aerobic conditions. Each microcosm was prepared in triplicate using acclimated and non-acclimated biomass as previously described. The reaction was held on a shaker table at room temperature 25°C ± 2°C. The potential for non-biological CO_2 evolution was evaluated by aerating a sterile control, which consisted of biomass inactivated with 0.2% sodium azide. Side arms of $^{14}CO_2$ traps contained 5 ml of 1M NaOH. At each sampling interval, all the NaOH solution was removed from the trap and replaced with fresh NaOH. The radioactivity retained by the NaOH in the trap was determined by diluting 1 ml aliquot in 15 mL of scintillation cocktail. The quantity of $^{14}CO_2$ evolved and subsequently trapped was measured by accepted liquid scintillation counting methods in accordance with the protocol outlined [34] with a Packard Instrument Model Tri-Carb 2300 TR Liquid scintillation counter (LSC) (Meriden, CT). Counts per minute (cpm) were converted to disintegrations per minute (dpm) by using an external standard (t-SIE) quench curve available from the scintillation manufacturer. When the rates of radiolabeled carbon dioxide evolution indicated that mineralization process reached a plateau, ^{14}C, which was respectively incorporated into biomass and present in solution, was determined by the following procedure: a 1 mL sample of suspended biomass was obtained from each biometer and centrifuged for 15 minutes at 12,000 x g. After centrifugation, the supernatant was carefully separated from the biomass pellet. The activity of the supernatant (C_S) was measured by LSC as described above. To the biomass pellet, 250 μL of a proprietary alkaline-surfactant tissue solubilizer (ScintiGest, Fisher Scientific, Pittsburgh, PA) was added to aid in sample digestion and solubilization of biomass components. The resulting mixture was then sonicated on ice for 2 hours, after which the β-emissions of the pellet (C_B) were also determined by LSC as described above. β-emissions from all experimental dilutions were measured within an hour of preparation.

3. Results and Discussion

3.1. Biotransformation of Benzotriazole Derivatives

Among the benzotriazoles tested, only those derivatives conjugated to a methyl group in the 5-position, or a carboxyl group in either the 4- or 5- positions were labile for the biodegradation under the conditions described. All three of these isomers were susceptible to biodegradation by either acclimated or non-acclimated microbial biomass, although a slower degradation rate (lag phase) was observed for the degradation of 4-CBT (Figure 2 and Figure 3). Replacement of a 5-methyl group by a carboxylic acid moiety increased biodegradation potential (rate and onset) of respective CBT and MeBT disappearance under identical conditions.

The substituted groups of nitro on benzene ring, the propionic acid on the triazole ring, and replacement of benzene ring with non-aromatic ring (cyclohexane) has been found to be inhibitory to biodegradation. The biodegradation patterns observed here suggest that the biodegradation potential depends mainly on the biodegradation pathways [24]. More derivatives should be tested in the future to explore the quantitative relationships between biodegradability of benzotriazole and their chemical structures.

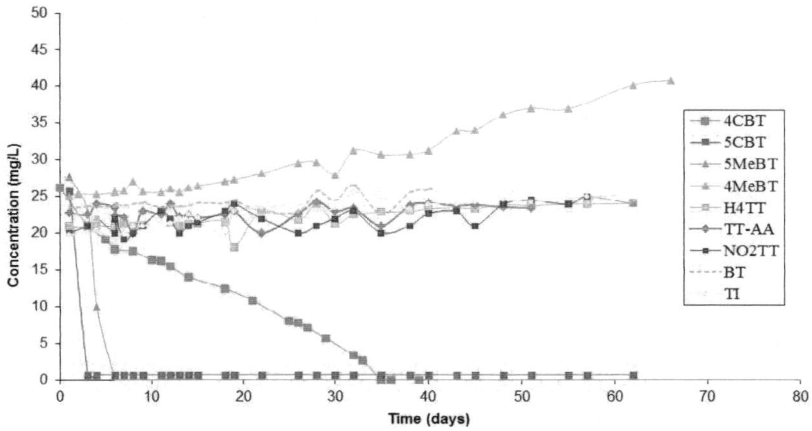

Figure 2. Concentrations of commercially significant and noble benzotriazole isomers in soil enrichment cultures acclimated to aircraft deicing fluids. CBT = carboxybenzotriazole, MeBT = methylbenzotriazole, H_4TT = tetrahydro tolyltriazole, TT-AA = 1-propionyl-5-methylbenzotriazole, NO_2TT=methyl-6(7)-nitrobenzotriazole, BT = benzotriazole, TI = methylbenzimidazole. Slight increases in concentration are due to evaporative losses.

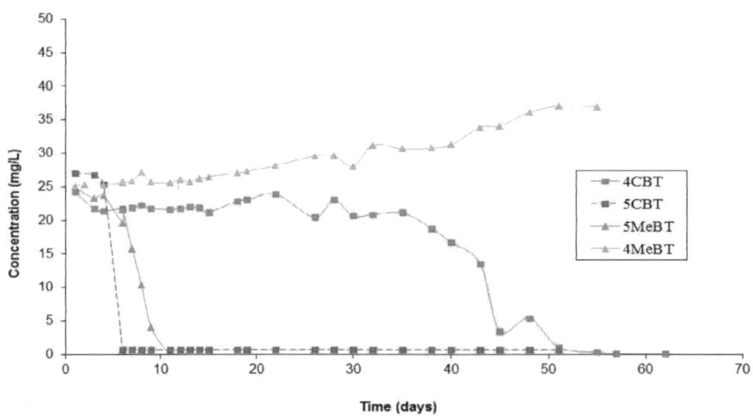

Figure 3. Concentration (mg/L) of methylbenzotriazole and carboxylated benzotriazole isomers introduced to activated sludge biomass (non-acclimated). CBT=carboxybenzotriazole, MeBT=methylbenzotriazole. Slight increases in concentration are due to microcosm evaporative losses.

3.2. Mineralization Potential of Methylbenzotriazole

Soil enrichment cultures, previously exposed to aircraft deicing components, including MeBT, demonstrated a significant aerobic mineralization response by degrading up to 25 mg L^{-1} [U-^{14}C]-5-MeBT within 5 days, most of which was recovered as ^{14}C CO_2. On average, final recoveries of 65% as $^{14}CO_2$ and 18% in biomass were observed for soil enrichment cultures (Table 1). Degradation of [U-^{14}C]-5-MeBT to $^{14}CO_2$ was entirely attributable to biological activity because no mineralization was observed in otherwise identical systems that had been inactivated with sodium azide. When using [U-^{14}C]-5-MeBT, the ^{14}C radioactivity left in solution accounted for 11% of the total ^{14}C beta emission originally added to the biometers (Table 1). As judged by HPLC, this ^{14}C could not be associated with any measurable amounts of 5-MeBT (< 0.2 mg L^{-1}), and is likely some other unknown byproduct of bacterial metabolism and/or colloidal, unfilterable biomass fragments and/or $^{14}CO_2$ (as bicarbonate alkalinity) in solution.

3.3. Mineralization of Carboxybenzotriazole

Replacement of a methyl group with a carboxylic acid group in either the 4- or 5- positions resulted in the enhancement of biodegradation potential with respect to the methylated derivatives due to the present carboxyl group (COOH) which enhances the primary degradation of aromatic rings under aerobic conditions. Regardless of acclimation condition or consortia of soil or sewage origin, significant aerobic biodegradation of carboxybenzotriazole was observed. Fifty-eight and fifty-seven percent of the radiolabeled carbon added as [U-^{14}C] 5-CBT were recovered as $^{14}CO_2$, whereas 15% and 14% were recovered in biomass. No mineralization was observed for otherwise identical systems containing sodium azide confirming biological activity. The amount of ^{14}C left in solution was 17% and 16% of the total carbon introduced to acclimated and non-acclimated biomass, respectively (Table 1).

The differences in biodegradation responses and the intermolecular architecture of 5- and 4-substituted benzotriazoles carry some resemblance to biodegradation and substitution patterns reported before when investigating a class or poorly degradable sulfonated azo-dyes [35]. Among the sulfonated azo dyes, the hydroxyl group in para position relative to the azo linkage were the most susceptible to degradation, such that there was no steric hindrance or potential for intermolecular azo(N)—hydrogen bonding. As with some azo dyes, differences in biodegradation potential of 4 and 5 substituted benzotriazoles are due to the steric hindrance of methyl and carboxy group, 4-MeBT was not biodegradable and 4 CBT was less susceptible to degradation than 5-CBTand lag times for degradation of 4 CBT was longer than 5 CBT. The group substituted closer to azo bond tends to hinder the steric hindrance that could prevent or slow the interactions between microbial enzymes and the methyl or carboxyl group in the substrate.

The results agree with Kulla and coworkers [36] who reported analogous patterns for Orange I and Orange II sulfonted and carboxylated azo dyes. Also, Pasti-Grigsby et al. [37] reported an enhancement of overall degradability by replacing selected moieties with a carboxyl group. In general oxygen – containing molecular fragments (such as OH, carboxy, and carbonyl groups) have been previously reported to enhance the biodegradability regardless of the molecular skeleton [38]. The biodegradation patterns observed here suggest that at least one enzyme system associated with the biodegradation of 4-carboxybenzotriazole is inducible. There is no direct evidence here, or in the literature, documenting the biodegradation potential of 4-methyl substituted benzotriazole [13].

3.4. Observed Yields

When challenged with [U-^{14}C]-5-MeBT, the amount of ^{14}C recovered from acclimated biomass was 18% of the total ^{14}C added (Table 1). This agrees with yield observations estimated using standard suspended solids measurements and biomass phospholipid increases (effective yield c. 0.20). When challenged with [U-^{14}C]-5-CBT, 15% and 14% percentages (Table 1) were recovered from the acclimated and the non-acclimated enrichment cultures respectively. This agrees with the yield observations estimated using the standard suspended solids measurements and the biomass phospholipid increases (effective yield c. 0.1). These observed yields are markedly lower than those commonly observed for the aerobic biodegradation of phenyl compounds and/or conjugates containing methyl or carboxyl moieties [39]. This may be due to the great deal of energy needed to break the azo ring before the substrate can be metabolized to produce energy.

Table 1. Triplicate means of the recovery of ^{14}C as percentage of total carbon added as either ^{14}C-MeBT or ^{14}C-CBT to soil and activated sludge enrichment cultures.

	Acclimated to aircraft deicing fluids (Soil enrichment cultures)	Non-acclimated to aircraft deicing fluids (Activated sludge cultures)

	^{14}C-MeBT	^{14}C-CBT	^{14}C-CBT
CO_2 Recovery	65±3.3	58±1.2	57±1.5
Soluble Recovery	11±0.4	17±0.1	16±0.1
Biomass Recovery	18±2.8	15±0.6	14±1.9
Total Recovery	94±6.5	90±0.6	87±0.4

4. Conclusions

Using acclimated soil enrichment cultures and non-acclimated activated sludge, only 5-MeBT, 4-CBT and 5-CBT appear labile to biodegradation at different degradation rates, and under aerobic conditions. Replacement of a methyl group with a carboxylic acid group in either the 4- or 5- positions resulted in the enhancement of biodegradation potential with respect to the methylated derivatives. The carbon in these benzotriazole derivatives was found to be mineralized to nontoxic products (CO_2) in both acclimated and non-acclimated cultures; Therefore, carboxylate derivatives may be a good industrial alternative to methylated counterparts and can lower environmental exposures and ecosystem risks. No clear correlation was found between aerobic biodegradation of benzotriazoles and their chemical structure.

Acknowledgments

We express our gratitude to Ronald L. Pearson for the constructive discussions regarding benzotriazoles derivative synthesis. This research was supported by a grant from the Cincinnati Specialties Inc. and a US National Science Foundation CAREER award BES-9702165.

References

[1] Huntscha S, Hofstetter T B, Schymanski E L, Spahr S and Hollender J 2014 *Environ. Sci. Technol.* **48** 4435-4443.

[2] Dummer N M 2014 *Rev. Environ. Sci. Biotechnol.* **13** 53-61.

[3] Kolpin D W, Furlong E T, Meyer M T, Thurman E M, Zaugg S D, Barber L B and Buxton H T 2002 *Environ. Sci. Technol.* **36** 1202-1211.

[4] USEPA 2000 Preliminary Data Summary - Airport Deicing Operations, US Environmental Protection Agency, Washington DC, p. 448.

[5] Weiss S, Jakobs J and Reemtsma T 2006 *Environ. Sci. Technol.* **40** 7193-7199.

[6] Kiss A and Fries E 2009 *Environ. Sci. Pollut. Res.* **16** 702-710.

[7] Cancilla D A, Holtkamp A, Matassa L and Fang X 1997 *Environ. Toxicol. Chem.* **16** 430-434.

[8] Cornell J S, Pillard D A and Hernandez M T 2000 *Environ. Toxicol. Chem.* **19** 1465-1472.

[9] Voutsa D, Hartmann P, Schaffner C and Giger W 2006 *Environ. Sci. Pollut. Res.* **13** 333-341.

[10] Hem L J, Weideborg M and Schram E 2000 Degradation and Toxicity of Additives to Aircraft De-Icing Fluids; The Effect of Discharge of Such Fluids to Municipal Wastewater Treatment Plants., WEF and Purdue University Industrial Wastes Technical Conference., Oslo, Noreay.

[11] Cornell J S 2002 The Environmental Chemistry of Aircraft Deicing Fluid (ADF) Component Chemicals, Civil, Environmental and Architectural Engineering, University of Colorado, Boulder.

[12] Rao N, M., Lu F F and Nhgiem N P 1996 Method of Preventing Yellow Metal Corrosion in Aqueous Systems with Superior Corrosion Performance in Reduced Environmental Impact, Nalco Chemical Company, USA.

[13] Herzog B, Lemmer H, Huber B, Horn H and Müller E 2014 *Environ. Sci. Pollut. Res.* **21** 2795-2804.

[14] Herzog B, Yuan H, Lemmer H, Horn H and Müller E 2014 *Bioresour. Technol.* **163** 381-385.

[15] Felis E, Sochacki A and Magiera S 2016 *Water Res.* **104** 441-448.

[16] Wu X, Chou N, Lupher D and Davis D 1998 Benzotriazoles: Toxicity and Degradation, in: Erikson L E,Rankin M M (Eds.), 13th Annual Conference on Hazardous Waste Research, Kansas State University, Manhattan, KS.
[17] Gruden C L 2000 Fate and Toxicity of aircraft deicing fluid additives through anaerobic digester, Department of civil, environmental, and archtectural engineering, University of colorado, Boulder.
[18] Gruden C L, Dow S M and Hernandez M T 2001 *Water Environ. Res.* **73** 72-79.
[19] O'Brien I 2002 The Environmental Chemistry of Aircraft Deicing Fluid (ADF) Component Chemicals, Civil, Environmental and Architectural Engineering, University of Colorado, Boulder.
[20] Liu Y-S, Ying G-G, Shareef A and Kookana R S 2013 *J. Contam. Hydrol.* **151** 131-139.
[21] Liu Y-S, Ying G-G, Shareef A and Kookana R S 2011 *Water Res.* **45** 5005-5014.
[22] Pasti-Grigsby M B, Burke N S, Goszczynski S and Crawford D L 1996 *Appl. Environ. Microbiol.* **62** 1814-1817.
[23] Pasti-Grigsby M B, Paszczynski A, Goszczynski S, Crawford D L and Crawford R L 1992 *Appl. Environ. Microbiol.* **58** 3605-3613.
[24] Suzuki T, Timofei S, Kurunczi L, Dietze U and Schuurmann G 2001 *Chemosphere* 45 1-9.
[25] Paszczynski A, Pasti M B, Goszczynski S, Crawford D L and Crawford R L 1991 *Enzyme microb. technol.* **13** 378-383.
[26] Zimmermann T, Kulla H G and Leisinger T 1982 *Eur. J. of Biochem.* **129** 197-203.
[27] Zimmermann T, Gasser F, Kulla H G and Leisinger T 1984 *Arch. Microbiol.* **138** 37-43.
[28] Patil S G, Nicholls P H and al e 1998 *Pestic. Sci.* **22** 333-342.
[29] Howard P H, Saxena J and Sikka H 1973 *Environ. Sci. Technol.* **12** 398-407.
[30] Alexander M 1973 *Biotechnol. Bioeng.* **15** 611-647.
[31] AASHTO 2003 Handbook Standard Recommended Practice for Conducting Geotechnical Subsurface Investigations, Thomson, Ann Arbor.
[32] APHA 1995 Standard Methods for the Examination of Water and Wastewater, American Public Health Association, Washington, DC.
[33] Raghavan D, Wagner G C and Wool R P 1993 *J. Environ. Polym. Degrad.* **1** 203-211.
[34] Schmidt S K and Scow K M 1997 Use of soil bioreactors and microcosms in bioremdiation research, in: Hurst C J (Ed.) Manual of Environmental Microbiology, ASM Press, Washington, DC, pp. pp. 822-829.
[35] Pasti-Grigsby M B, Lewis T A, Crawford D L and Crawford R L 1996 *Appl. Environ. Microbiol*. **62** 1120-1123.
[36] Kulla H G, Klausener F, Meyer U, Lüdeke B and Leisinger T 1983 *Arch. Microbiol.* **135** 1-7.
[37] Pasti-Grigsby M B, Burke N S, Goszczynski S and Crawford D L 1996 *Appl. Environ. Microbiol.* **62** 1814-1817.
[38] Alexander M 1999 Biodegradation and bioremediation, Gulf Professional Publishing.
[39] Grady L C P, Daigger G L and Lim H C 1999 Biological Wastewater Treatment, Marcel Dekker, New York.

Veratric acid removal from water by electrochemical oxidation on BDD anode

Inshad Jum'h[1,*], Arwa Abdelhay[1], Ahmad Telfah[2], M-Ali Al-Akhras[3], Akeel Al-Kazwini[1], Stefan Rosiwal[4]

[1]*German-Jordanian University, Amman 11180 Jordan. Tel. +962 (0)6 429 4242.*
[2]*Leibniz-Institut für Analytische, Wissenschaften - ISAS - e.V. 44139 Dortmund, Germany. Tel: +49 231 1392-269*
[3]*Jordan University of Science & Technology (JUST), Irbid 22110, JORDAN. Tel: +962779488104*
[4]*Institute of Metals Science and Technology (WTM), Friedrich-Alexander University, 5, 91058 Erlangen. Germany. Tel: +49 (0)9131 85- 27517.*

Inshad.yousef@gju.edu.jo

Abstract: The efficiency of boron doped diamond (BDD) in the electrochemical treatment of synthetically contaminated water with veratric acid (VA), one kind of polyphenolic type compounds, is investigated in this work. A BDD electrode was practically fabricated using hot filament chemical vapor deposition (HFCVD). Later on, the BDD electrode was implemented as an anode in a batch electrolytic reactor. The effect of operating factors such as the initial concentration of VA, NaCl addition, and supporting electrolyte type (H_2SO_4, H_3PO_4 and Na_2SO_4) was studied. The chemical oxygen demand (COD) measurements were conducted to study the VA electrolysis kinetics. The experimental data suggested that sodium sulfate was the best supporting electrolyte as the COD removal reached a percentage of 100% using 1 mmol/dm^3 as VA concentration. The kinetics of the COD decay of the VA electrolysis were found to obey the pseudo-first order model. Remarkably, the electrolysis process is significantly speeded up once chloride is added to the reaction. The complete COD removal was achieved in 60 minutes of treatment.

Keywords: Boron-doped diamond electrode, Electrochemical oxidation, Veratric acid, wastewater, Advanced oxidation processes

1. Introduction

Industrial effluents containing organic and refractory pollutants raise the call for efficient wastewater treatment methods. The increasing interest in such methods is due to the environmental damage and water quality deterioration caused by these pollutants [1]. Therefore, wastewater has to be treated before it is discharged into the drainage system [2]. A variety of conventional wastewater treatment methods have been reported in literature such as physical, chemical, and biological methods [3]. However, these methods have limitations in their operation. Biological processes need long hydraulic detention time and large reactor volumes [4]. Physical and chemical methods necessitate the handling of the large volume of sludge produced [5]. In this framework, electrochemical processes offer good opportunity for the remediation of recalcitrant and persistent organic pollutants [6-8]. Electrochemical technologies have emerged as novel treatment technologies for the elimination of broad-range organic contaminants from water. The most important advantages of electrochemical methods for the treatment of wastewater are their high efficiency, mild operating conditions, ease of automation, versatility, and low cost [9]. Anodic oxidation of organic pollutants is guaranteed by different types of electrodes on which pollutants are directly destroyed by the free hydroxyl radicals produced on their surface [10]. This anodic oxidation might lead to complete mineralization of the organic contaminants [11]. Among the different electrodes used, boron-doped diamond (BDD) electrodes possess novel and intrinsically outstanding properties with comparing to conventional electrodes such as glassy carbon, graphite or platinum. The BDD electrodes have a large potential window in aqueous solution with low background currents [12-14]. In addition, the BDD electrodes possess distinct physical properties including: high hardness, low environmental impact, high hole mobility, high thermal conductivity and excellent resistance to radiation damage [11]. Souza et al. tested the electrooxidation of phenol on the surface of BDD anode and reported significant COD removal from water samples [15]. Similarly, Carvalho de Almeida et al. treated Brazilian petrochemical effluents by anodic oxidation on BDD [16]. Results clearly showed that BDD promotes complete COD removal (98%) due to the high amounts of effective hydroxyl radicals and peroxodisulfates generated from water oxidation. Errami et al. treated wastewater Industrial Cartons electrochemically using a BDD electrode [17]. Medeiros de Araújo et al. studied the influence of the mediated oxidation on the removal of rhodamine B (xanthene dye) solutions with conductive-diamond electrochemical oxidation and reported total removal of organic pollutants [18]. It is true that the performance of the electrolysis process depends strongly on the type of electrode used as previously mentioned. However, further enhancement can be achieved by adding a supporting electrolyte or salt to the electrochemical cell not only to increase the solution conductivity, but also to reduce the cell potential in order to make the electrochemical process feasible in terms of energy consumption [15].

This work focuses on the electrochemical treatment of synthetically contaminated water with veratric acid VA as a model organic compound found in many wastewater effluents namely: olive mill, pulp and paper, detergent, pesticides, agro- industries and pharmaceutical industries [19]. The VA oxidation was carried out in a laboratory batch reactor provided with a BDD electrode. The experimental assays were conducted to study the kinetics of VA oxidation as well as to figure out the effect of different chemical reaction operating factors such as the electrolyte type, NaCl addition and the initial concentration of VA. Up to our best knowledge, one similar study [20] has targeted and approached this topic. However, it does not explore the effect of VA initial concentration as well as the effect of supporting electrolytes other than sulfate-containing supporting electrolytes.

2. Materials, methods and devices

2.1. Materials

Synthetically polluted water with VA samples were prepared by dissolving VA $C_9H_{10}O_4$ (Fig. 1) in distilled water in three different concentrations (1 mmol/dm^3, 2 mmol/dm^3, and 3 mmol/dm^3). Three supporting electrolytes were used: sodium sulfate Na_2SO_4 (50 mmol/dm^3- Sigma Aldrich), sulfuric acid H_2SO_4 (50 mmol/dm^3- Sigma Aldrich) and phosphoric acid H_3PO_4 (50 mmol/dm^3- Sigma Aldrich). Sodium chloride NaCl (Sigma Aldrich) was added to water (10 mmol/dm^3).

IOP Conf. Series: Materials Science and Engineering **305** (2018) 012021 doi:10.1088/1757-899X/305/1/012021

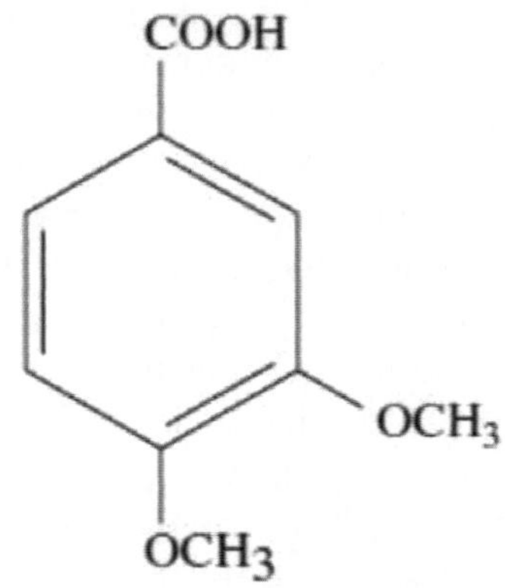

Fig.1. Molecular structure of the veratric acid.

2.2.Laboratory-scale experimental set up

Electrode preparation

The electrode substrate used in this work is made of an expanded grid of Niobium having a circular shape. The grid has a thickness of 2.8 mm and diameter of 90 mm. The expanded grid niobium, as opposed to sheets, is well flown through because of its open geometry of both gas bubbles and from the medium to be cleaned. Moreover, due to the high ratio of surface area to mass, the price per electrode substrate surface is relatively low.

The diamond coated electrodes on niobium substrate were produced by a hot filament chemical vapor deposition (HFCVD) process. Niobium substrates were pretreated by sand blasting in order to roughen the surface and therefore to enhance the adhesion strength between diamond and substrate, and meanwhile, the specific surface area was increased to expectably improve the electrochemical performance. The sand blasting was followed by twice ultrasonic cleaning in ethanol. Afterwards the substrates were seeded by ultrasonic seeding in a diamond powder suspension. The CVD-diamond coating took place in a CVD hot filament coating machine CC800/8 DIA with 1% methane CH_4 and a small amount of trimethyl borate $B(OCH_3)_3$) to enhance the electrical conductivity of diamond by boron-doping. The niobium grid was coated on both sides using two parallel filament rows for 50 hours. The film thickness measured in cross sections is 8 μm according to an identical test process. The surface topography, roughness, the grain size perpendicular to the growth direction and the morphologies of the diamond films were characterized using Scanning Electron Microscopy.

Electrochemical reactor

The experiments were performed in a batch electrochemical plexiglas reactor of 1000 cm^3 supplemented with two BDD electrodes separated by a distance of 2 mm and they have each a total area of 65 cm^2. The reactor was connected to an external DC digital power supply with maximum output current and voltage of 6 A and 30 V. The working current density was 50 mA/cm^2 applied for a maximum process time of 350 min.

Analytical procedures

The electrooxidation assays were run to test three operating factors: supporting electrolyte type, NaCl addition, and initial VA concentration. Samples were withdrawn at regular time intervals and collected in 20 cm^3 glass vials. All withdrawn samples were characterized by the following procedures:
The chemical oxygen demand (COD) experimental values of treated samples were determined using dichromate standard method **[21]** Each sample was introduced into commercially available digestion solution containing potassium dichromate (Hach-Germany), afterward; the solution was incubated at 148 $^{\circ}C$ for 120 minutes in a COD reactor (Hach Lange GMBH-Germany). COD measurements were performed by a spectrophotometer (DR Lange Lasa 100 Germany). Complete mineralization of VA was detected by UV spectrophotometry (UV-3600/UV-VIR spectrophotometer Shimadzu). The pH was measured using Jenway 3540 pH/conductivity meter. All samples prepared in this work have an initial pH in the range of 5.5 to 6.8.

Determination of the rate constant
In order to compare the kinetics of COD removal using the three supporting electrolytes, the pseudo-first order rate constant (k) was determined with exponential regression and least square method. The rate constant was determined based on the following equation [22]:

$$COD = COD_0\, e^{-kt}$$

Where COD_0 is the initial chemical oxygen demand at the beginning of the electrolysis process. COD is the chemical oxygen demand at time t. k is the electrolysis rate constant. t is the time of electrolysis.

Determination of the energy consumption during electrolysis
During the experiments carried out using the three different supporting electrolytes namely, H_2SO_4, H_3PO_4, and Na_2SO_4, the current was kept constant at 50 mA/cm^2. The specific energy consumption using each supporting electrolyte was calculated in kWh consumed per kg COD removed according to the following equation:

$$\text{Energy consumption (kWh/kg COD removed)} = \frac{A \times V \times t}{v \times \Delta COD}$$

Where t is the time of electrolysis (h); V and A are the average cell voltage and the electrolysis current; v is the reactor volume (dm3), and ΔCOD is the difference between the initial COD value at process time zero and the COD value at the end of the process time (mg O_2/dm^3).

3. Results and discussion

3.1. Testing BDD prepared electrode

Scanning electron microscopy (SEM) images and Raman spectrum of the BDD layers deposited on niobium substrate are shown in Fig.2. The visible roughness (Fig. 1 (a, b)) of several micrometers is a result of surface roughening pretreatment of the metallic substrate surfaces. The diamond layer surface shows well (111)-faceted crystals with an average grain size of 0.6 ± 0.3 μm and the diamond thickness is around 6-8 μm. Importantly, no cracks or holes were observed and the peeling occurrence did not take place which indicates the high quality of coating process, accordingly, the electrodes area completely contributes in the electrochemical oxidation (high efficiency factor).

(a) *(b)* *(c)*

Fig.2. *Scanning electron microscopy (SEM) images of the BDD layers deposited on niobium substrate with two different magnifications (a) 100 μm (b) 25μm (c)Raman spectrum of the BDD film.*

The quality of BDD layers deposited on niobium substrate was tested employing Raman analyses. The Raman spectrum of the diamond films (Fig. 1(c)) reveals a sharp peak at 1340 cm^{-1} that is usually associated with diamond sp3 bonding. The boron could not be detected and has not appeared in the Raman spectrum due to its low content. What is worth to be mentioned in the Raman spectrum is the broad peak with maximum at 1500 cm^{-1}. This peak is assigned to C–H chains and is a typical feature of the Raman spectrum of nanocrystalline diamond (Fabisiak et al. 2006). The absence of other amorphous carbon or sp2 graphite peaks that usually appear in the range between 1450 cm^{-1} and 1600 cm^{-1} is an evidence that BDD layers can be considered as phase-pure (high quality) diamond material.

3.2. Veratric acid removal using anodic oxidation with BDD

Effect of supporting electrolyte on COD abatement during electrolysis

The efficiency of electrolysis depends on the generation of oxidants during the process time. Therefore, adding supporting electrolytes along with raw wastewater is a common approach to increase electrolyte conductivity, reduce energy consumption, and organize mass transport by electrical migration [23]. Supporting electrolytes, such as Na_2SO_4, KCl, Na_2NO_3, H_3PO_4 and NaCl are frequently reported in literature [7, 10].

The aim of this part is to study the effect of three supporting electrolytes namely, Na_2SO_4, H_2SO_4, and H_3PO_4 on the rate of COD abatement. **Fig.3** shows that complete COD removal was attained using Na_2SO_4 at an applied current density of 50 mA/cm^2 which is in accordance with the findings of Medeiros de Araújoa et al. [18] and Ealaoud et al. 2011 [20]. Therefore, Na_2SO_4 is a more active supporting electrolyte than either H_2SO_4 or H_3PO_4. H_2SO_4 was less effective than Na_2SO_4 although both can produce peroxodisulfate under anodic oxidation conditions. Serrano et al. [24] reported that peroxodisulfate can be importantly electrogenerated in concentrated H_2SO_4 (>2 M) upon reaction of HSO_4^- with hydroxyl radicals on BDD [23]. The later condition is not valid for the present study since H_2SO_4 was used with a concentration of 50 $mmol/dm^3$. In addition, it can be observed that H_3PO_4 is the weakest among the three tested electrolytes. Cañizares et al. [25] indicated that peroxodiphosphate is optimally electrogenerated at pH in the range of 12-13. Fig. 2

also illustrates that COD decreases exponentially with the electrochemical oxidation time following a pseudo first order model as was also reported by [15].

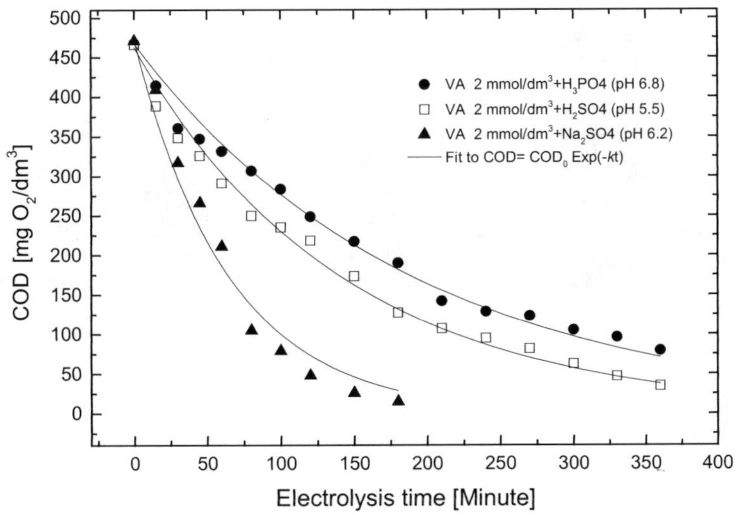

Fig. 3. *Chemical oxygen demand (COD) variation with electrolysis time of three supporting electrolytes (H_2SO_4, H_3PO_4 and Na_2SO_4) for veratric acid (VA) electrolysis on boron doped diamond anode. The electrolyte concentration was 50 mmol/dm^3 and veratric acid concentration was 2 mmol/ dm^3, the applied current density was 50 mA/cm^2.*

The pseudo-first order rate constant (k) values of the electrolysis process of veratric acid with the three different supporting electrolytes are illustrated in Fig.4. The rate constants for Na_2SO_4, H_2SO_4, and H_3PO_4 were 0.016, 0.007, and 0.005 min^{-1} respectively. The correlation coefficient was greater than 0.95 in all cases. The sulfate-containing supporting electrolytes showed higher rate constants comparing to phosphate-containing electrolyte. Consequently, sulfuric acid and sodium sulfate are more efficient than phosphoric acid in stimulating COD removal. The highest k value (0.016 minutes^{-1}) was obtained using Na_2SO_4 with a COD removal of almost 100%. This finding could be attributed to the strong oxidizing power of electrogenerated persulfate which are most probably formed due to the important overpotential for oxygen evolution on BDD anode [26].

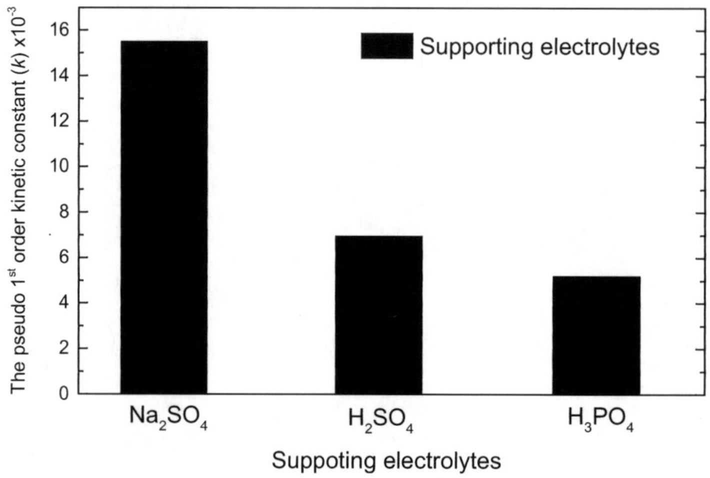

Fig.4. *The pseudo first order rate constants (k) of the 2 mmol/dm^3 veratric acid samples with three supporting electrolytes: Na_2SO_4, H_2SO_4, H_3PO_4.*

Fig. 5. Shows the effect of supporting electrolyte nature on energy consumption during electrolysis time. The changes in specific energy consumption indicated that conducting electrochemical oxidation of VA using Na_2SO_4 as conductive electrolyte has the lowest energy requirements. In other words, the Na_2SO_4 showed the lower energy consumption with the highest % COD removal. Consequently, a reduction in the treatment cost using Na_2SO_4 was expected, as shown in Table 1.

Taking into consideration the electrical energy cost of about US$ 0.15 (U.S. currency) per kWh, the monetary value reported in Table 1 and required to degrade a unit volume (1 dm^3) of effluent.

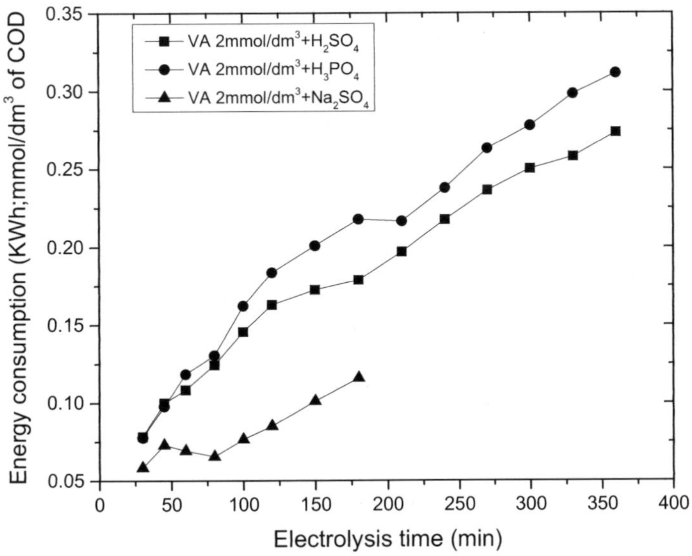

Fig. 5. *The effect of supporting electrolyte nature on specific energy consumption*

Table 1 Cost required for electrochemical treatment of water contaminated with VA (2 mmol/dm^3) using different supporting electrolytes.

Type of electrolyte	Cost (US\$/m^3 treated)
H_2SO_4	16.2
H_3PO_4	16.2
Na_2SO_4	8.1

Effect of veratric acid initial concentration on COD removal during electrolysis on B

As organic aromatic compounds exist in wastewater at different concentrations, a second stage of the experiment was conducted to explore the impact of veratric acid initial concentration on the rate of oxidation on BDD anode. As depicted in Fig. 6. it is obvious that higher veratric acid initial concentration necessitates longer treatment times. However, the pseudo-first order rate constant k, which is determined as previously mentioned by exponential regression and least square method, has almost the same average value (0.016 minutes^{-1}) for the three veratric acid concentrations. Hence, The COD kinetics are independent on VA initial concentration.

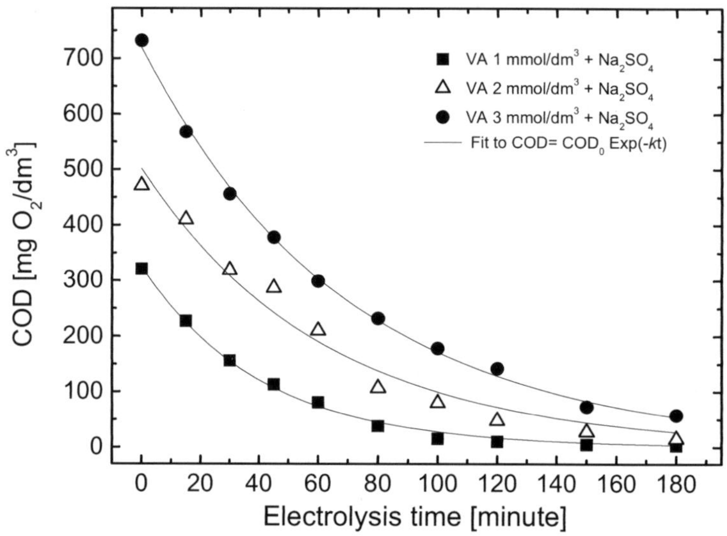

Fig. 6. *COD variation with electrolysis time at three different initial concentrations (1 mmol/dm^3, 2 mmol/dm^3, 3 mmol/dm^3) of VA. Conditions: Na_2SO_4 concentration was 50 mmol/dm^{3}and the applied current density was 50 mA/cm^2. Solid line represents the fitting to the pseudo-first order equation ($COD=COD_o\ e^{-kt}$).*

Effect of the presence of chloride ions on VA oxidation

The impact of chloride ions on COD abatement rate was investigated by adding the NaCl (10 mmol/dm^3) to three solutions containing different VA concentrations (1 mmol/dm^3, 2 mmol/dm^3, 3 mmol/dm^3) with Na_2SO_4 as supporting electrolyte. Fig. 6 (a, b and c) shows the effect of NaCl on COD variation with electrolysis time with the three VA concentrations. By comparing the curves of the COD with and without Cl^-, it is obvious that the chloride ions increased the rate of COD abatement, which means that Cl^- enhanced the kinetics of VA oxidation. The most reasonable explanation is that the Cl^- generates supporting oxidants as hypochlorite ions and enhance the oxidation of organic compounds via electrolysis [6, 15, 26]. Moreover, the allowable concentration of Cl^- (7 mmol/dm^3) set by EPA regulations for treated wastewater is not even exceed in this work (less than 6 mmol/dm^3). Under the current experimental conditions namely acidic medium and low working current density (50 mA/cm^2) , the occurrence of perchlorate can be fully neglected and should not be a concern. It has been reported in the literature [27] that the generation of perchlorates is non promoted at pH less than 10 and current densities as low as 30 mA/cm^2 for 5844 mg/dm^3 NaCl solution. In our case, the NaCl concentration is 10 mmol/dm^3 which is ten times less than the aforementioned concentration and the medium is acidic which does not favorite the perchlorate formation as stated. Finally, the current density used was in the range of 50 mA/cm^2, which is considered as previously mentioned a low density that does not support perchlorate ions production [27].

IOP Conf. Series: Materials Science and Engineering **305** (2018) 012021 doi:10.1088/1757-899X/305/1/012021

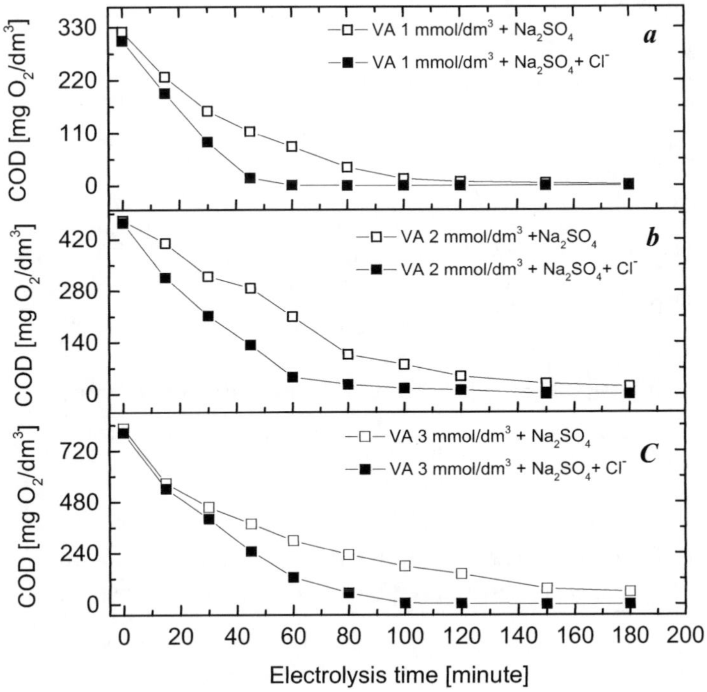

***Fig. 7.** COD variation with increasing electrolysis time of veratric acid solutions in three concentrations (1 mmol/dm^3, 2 mmol/dm^3 and 3 mmol/dm^3) with and without presence of chloride anions (Cl^-). Each solution contains 600 mg/dm^3 of Cl.$^-$*

In order to verify that VA was completely mineralized by electrolysis on BDD with no accumulation of aromatic intermediates in water, the UV spectra time evolution of VA during electrolysis was studied with UV-VIS spectrophotometry. The results are shown in Fig. 8. First, It has been shown that VA has three different λ_{max} when it was analyzed using UV-VIS spectrophotometer which are: 218, 262, 295 nm [28]. In the range of analysis illustrated in Fig. 8, only one peak (295 nm) of VA appears in the figure. This absorption peak disappears completely and the curve is flattened after almost 3 hours of electrolysis. Hence, this is also an additional indicator of the removal of VA from water under the operating conditions mentioned in Fig. 7. In other words, during electrolysis the absorbance peak has decreased with time until it was disappeared completely after 4 hours which suggests that there was no remaining intermediates in water. This observation is consistent with the results reported by Elaoud et al. 2011 [20]

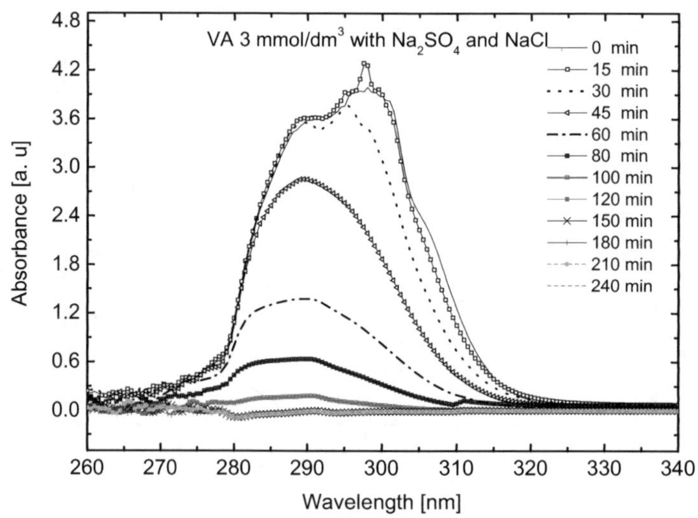

Fig. 8. *Time evolution of the UV-VIS spectrum during electrolysis of 3 mmol/dm^3 of veratric acid at a current density of 50 mA/cm^2 measured with UV-VIS spectrophotometer.*

4. Conclusion

Applications of boron doped diamond (BDD) as efficient anode in electrochemical treatment of organics-contaminated water are studied. It can be reported that electrochemical oxidation using BDD anode is an excellent method for wastewater treatment namely for effluents contaminated with veratric acid. Moreover, the experimental results indicated that the use of supporting electrolytes could considerably enhance the oxidation process and shorten the treatment time needed for complete oxidation of veratric acid, which in turn has direct effect on the total energy consumption. The sulfate-containing supporting electrolytes showed the fast kinetics. The maximum kinetic constant (0.016 minute^{-1}) was observed when Na_2SO_4 was added to water samples. Finally, it was found that further acceleration in the oxidation rate of VA can be achieved by adding chloride to the supporting electrolyte. The addition of chloride had greatly increased the process kinetics as complete oxidation of VA was achieved after about 60 min where 100 min were needed in the absence of chlorine.

References

[1] Panizza M., Bocca C., Cerisola G. 2000 Electrochemical treatment of wastewater containing polyaromatic organic pollutants. wat. res. 9, 2601.

[2] US EPA, Title 40: Protection of Environment: Part 432- Meat and poultry products point source Category, e-CFR Data, 2009.

[3] Bevilaqua J.V., Cammarota M.C., Freire D.M., Santa'Anna G.L. 2002 Phenol removal through combined biological and enzymatic treatments. Braz. J. Chem. Eng. 19, 151.

[4] Siringi D., Home Dr., Chacha J., Mulimi L. 2013 Wastewater Remediation Using Electrochemically Produced Layered Double Hydroxides. International Journal of Engineering and Innovative Technology. 2, 61.
[5] Asselin M., Drogui P., Benmoussa H., Blais JF. 2008 Effectiveness of electro coagulation process in removing organic compounds from slaughterhouse wastewater using monopolar and bipolar electrolytic cells. Chemosphere. 72, 1727.
[6] Matínez-Huitle C. A., Ferro S. 2006 Electrochemical oxidation of organic pollutants for the wastewater treatment: direct and indirect processes. Chem. Soc. Rev. 35, 1324.
[7] Martinez-Huitle C.A., Brillas E. 2009 Decontamination of wastewaters containing synthetic organic dyes by electrochemical methods. A general review. Appl. Catal. B: Environ. 87, 105.
[8] Carlas A., Martinez H., Brillas E. 2008 Electrochemical Alternatives for drinking water Disinfection. Angandeta chemi. 47, 2.
[9] Martínez-Huitle C. A., Rodrigo M.A, Sirés I., Scialdone O. 2015 Single and Coupled Electrochemical Processes and Reactors for the Abatement of Organic Water Pollutants: A Critical Review. Chem. Rev. DOI: 10.1021/acs.chemrev.5b00361.
[10] Brillas E., Martinez-Huitle C. A. 2015 Decontamination of wastewaters containing synthetic organic dyes by electrochemical methods. An updated review. Appl Catal B: Environ. 166-167, 603.
[11] Alfaro M.A., Ferro S., Martínez-Huitle C.A., Vong Y. M. 2006 Boron Doped Diamond Electrode for the Wastewater Treatment. J. Braz. Chem. Soc. 17, 227.
[12] Martinez-Huitle C.A., Brillas E. 2008 Electrochemical Alternatives for Drinking Water Disinfection. Angew. Chem. Int. Ed. 47, 1998.
[13] I. Jum'h , M. AL-Addous , H. AL-Taani , M.S. ABD El-Sadek , N. Ayoub, 2017 Effect of boron concentration on nano-crystalline diamond deposited on niobium substrate. Digest Journal of Nanomaterials and Biostructures Vol. 12, No. 2, 589.
[14] Inshad Jum'h, Arwa Abdelhay, Hussein Al-Taani, Ahmad Telfah, Mohammad Alnaief, Stefan Rosiwal, 2017. Fabrication and application of boron doped diamond BDD electrode in olive mill wastewater treatment in Jordan, Journal of Water Reuse and Desalination, Vol. 7,No. 4, 502.
[15] Souza A., Beraldo R., Ruotolo M., Augusto L. 2013 Phenol Electrooxidation in Different Supporting Electrolytes Using Boron-Doped Diamond Anodes. Int. J. Electrochem. 8, 643.
[16] Carvalho de Almeida C., Fernandes da Costa P. R., Jucilene de Macedo Melo M., Vieira dos Santos E., Martínez-Huitle C. A. 2014 Application of Electrochemical Technology for Water Treatment of Brazilian Industry Effluents. J. Mex. Chem. Soc. 58(3), 276.
[17] Errami M., Salghi R., Zarrouk A., Zougagh M., Zarrok H., Hammouti B., Al-Deyab S. 2013 Electrochemical Treatment of Wastewater Industrial Cartons. Int. J. Electrochem. 8, 12672.
[18] Medeiros de Araújoa D., Sáezb C., Martínez-Huitlea C.A., Ca˜nizaresb P., Rodrigob M.A 2015 Influence of mediated processes on the removal of Rhodamine with conductive-diamond electrochemical oxidation. Appl Catal B: Environ. 166-167, 454.
[19] De Heredia J.B., Torregrosa J., Dominguez J.R., Peres J.A. 2001 Kinetic model for phenolic compound oxidation by Fenton's reagent. Chemosphere. 45, 85.
[20] Elaoud S. C., Panizza M., Cerisola G., and Mhiri T. 2011 Veratric acid treatment by anodic oxidation with BDD anode . J Chem Techno Biotechnol. 87, 381–386.
[21] Dedkov Y. M., Elizarova O. V., Kei'ina S. Y. 2000 Dichromate Method for the Determination of Chemical Oxygen Demand. Journal of Analytical Chemisto. 55, 777.

[22] Mouden O., Errami M., Salghi R., Zarrouk A., Assouag M., Zarrok H., Al-Deyab S., Hammouti B. 2012 Electrochemical degradation of difenoconazole on BDD electrodes. J. chem. pharm. res. 4, 3437.
[23] Ajeel M. A., Aroua M. K., Ashri Wan Daud W.M. 2015 Anodic Degradation of 2-Chlorophenol by Carbon Black Diamond and Activated Carbon Composite Electrodes. Electrochimica Acta. 180, 22–28
[24] Serrano K., Michaud P.A., Comninellis C., Savall A. 2002 Electrochemical preparation of peroxodisulfuric acid using boron doped diamond thin film electrodes. Electrochimica Acta 48, 431.
[25] Ca˜nizares P., Larrondo F., Lobato J., Rodrigo M. A., Sáez C. 2005 Electrochemical Synthesis of Peroxodiphosphate Using Boron-Doped Diamond Anodes J. Electrochem. Soc. 152, 191.
[26] Bouhssine I., Tazi A., Azzi M. 2013 Treatment of olive mill wastewater by electrolysis on boron doped diamond (BDD) electrode. J. Mater. Environ. Sci. 4, 354.
[27] Sánchez-Carretero A., Sáez C., Ca˜nizares P., Rodrigo M.A. 2011 Electrochemical production of perchlorates using conductive diamond electrolysis. Chem. Eng. J. 166, 710.
[28] Robbins R. J. 2003 Phenolic Acids in Foods: An Overview of Analytical Methodology. J. Agric. Food Chem. 51, 2866.

IOP Conf. Series: Materials Science and Engineering **305** (2018) 012022 doi:10.1088/1757-899X/305/1/012022

Field Electron Emission Characteristics of Single-Walled Carbon Nanotube on Tungsten Blunt Tip

Marwan S. Mousa and Samer Daradkeh

Surface Physics and Materials Technology Lab, Department of Physics, Mu'tah University, AL-Karak, Jordan.

marwansmousa@yahoo.com & mmousa@mutah.edu.jo

Abstract: Recent investigations that are presented here illustrate the initial results that were obtained from a modified technique for holding the CNT on a W clean blunt tip. Field Electron Emission (FEE) has been investigated for single walled carbon nanotube (SWCNT) mounted on tungsten tip under ($\sim 10^{-8}$ mbar) vacuum conditions. The measurements recorded presented results showed that the CNT mounted on the W tip could emit electron current of at (0.7 V/μm) and reach up to (24 μA) of emission current at normal emission conditions. Such electron field emission tip was fabricated by electrolytically etching the high purity tungsten wire of (0.1 mm) in diameter in NaOH of (0.1) Molar solution, then mounting the single-walled carbon nanotube on the tip to be nearest to the tin oxide-coated and phosphorus glass anode. Such process was possible to be carried out under the microscope. A field electron microscope with a tip-screen separation at (~10mm) was used to characterize the electron emitter. The system was evacuated to an ultra-high vacuum level obtained after initial backing the system at up to (~180 °C) overnight. The emission characteristic has been investigated employing the I-V characteristics with Fowler-Nordheim plots and recording the emission images

1. Introduction

Carbon nanotube (CNT) have attracted much attention since their first synthesis by arc discharge method by lijima in 1991,[1] and because their own properties like small-radius, high aspect ratio, and remarkably high electrical conductivity, CNTs offer several advantages over other material in electron field emissions like enhanced current stability, lower threshold voltage, and long life time.[2] So there are a lot of potential applications of CNTs including electron field emitters, quantum wires, transistor, molecular filter, and hydrogen storage.[3]

The electron emission from metals can be driven by temperature and applied an electric field. For high electric field and low temperature, the high field narrows the vacuum potential barrier such that electron with energies below Fermi level tunnel through the barrier and predominate the emission current[4] i.e. quantum mechanical tunneling. Which can be described by the Fowler Nordheim theory, and from the theory, the emission current *I* can be expressed in simplified form

$$I = 1.54 \times 10^{-6} \frac{S\beta^2}{\Phi} V^2 \exp(-6.83 \times 10^7 \frac{\Phi^{\frac{3}{2}}}{\beta V})$$

Here, the emission current I and the applied voltage V are expressed in A and V, respectively. The work function Φ is expressed in eV and the emitting area S is expressed in cm2. The field enhancement factor β corresponds to the geometry shape of the emitter, and inversely proportional to the diameter of the emitting tip.[5]
To obtain quantitative information about the microscopic properties of the emitting regime, the total electron emission current I, has been plotted as a function of the applied voltage V between the emitter and the anode. Emission characterization will be presented as $I - V$ plots and as the related FN plots. Through the experiments, electron emission images have been recorded by means of a digital camera, in order to study the spatial distribution and stability of the emission current.[6]

2. Materials and Methods

In our experiments we use SWCNTs are produced by catalytic conversion of high-pressure CO over Fe particle (HiPCO) processed at CNI, Houston, TX, mounted over blunt tungsten tip. This method showed that CNTs/W tip could emit electron at lower applied voltage comparing with individual W tip because the CNTs/W tip has a very high value of enhancement factorβ.
Preparing the emitter has two stages, stage one is preparing the tungsten by electrolytically etching a 0.1 mm diameter wire at the meniscus of a 2 mol/liter NaOH solution .i.e. by the same technique used for tungsten emitters[7, 8]. Stage two is applying a thin layer of epoxy resin, which is necessary for a firm attachment of the nanotube, and then mechanically by approaching the tungsten tip to the CNTs, the CNTs will be stuck on the tungsten tip.
The CNT/W emitter was characterized using the conventional field electron emission apparatus with a tip-screen separation standardized at (10 mm). The input anode is phosphored and acts as the screen of recording emission images using a digital camera. The system can be operated at pressure (10^{-10} mbar) when baked to (~200 °C) for (~12 h), with liquid nitrogen added to the trap, though the operational pressure during the present study was (~10^{-8} mbar). A high tension (EHT) power supply is used to apply a negative voltage to the cathode, and the emission current is measured with a Kiethley 485 auto-ranging Pico ammeter.[9]

3. Results and Discussion

We performed the tests over two samples, (C5\W) and (B5\W) under same conditions. Voltage was applied to the SWCNTs/W tip and I-V characteristics were recorded. For SWCNT\W C5, the electrons emission started at (300 V) with emission current (4 pA) to (1750 V) with emission current (6.6 μA). By increasing the applied voltage slowly caused an increase in emission current, and by further increasing, the emission current suddenly increased to value (I_{SW}=3.10 μA) at applied voltage value (V_{SW}=1100 V), as shown in the figure (1A); figure (1B) show the related Fowler-Nordheim (FN) plot, as well as the field electron microscope image at V_{SW}, V_{SAT}, 1700 V, and 700 V, which is shows a single spot image except at 1700 V.

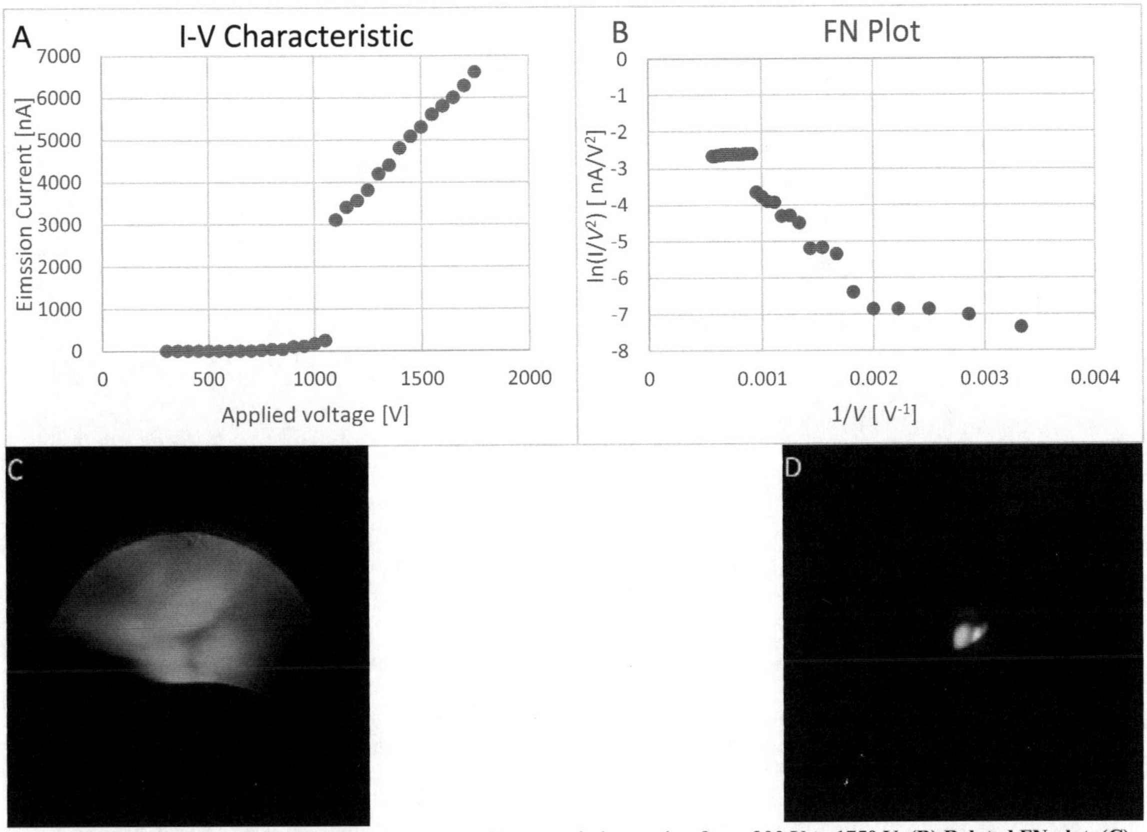

Figure 1.C5\W emitter. Increasing Voltage (A) I-V Characteristic ranging from 300 V to 1750 V. (B) Related FN plot. (C) Field electron image at 1700 V, 6.28μA. (D) field electron image at V_{SW}=1100 V, 3.10 μA.

By decreasing the applied voltage slowly, the saturation region extends down to value V_{SAT}=750 V, with emission current value (1.3 μA), and by further reduction of the applied voltage, the threshold voltage (V_{TH}) has been reached which is the cathode-voltage value necessary to generate the emission current.[10]. Figure (3) shows I-V characteristic, FN plot, and field electron image at V_{SAT}, and at (700 V)

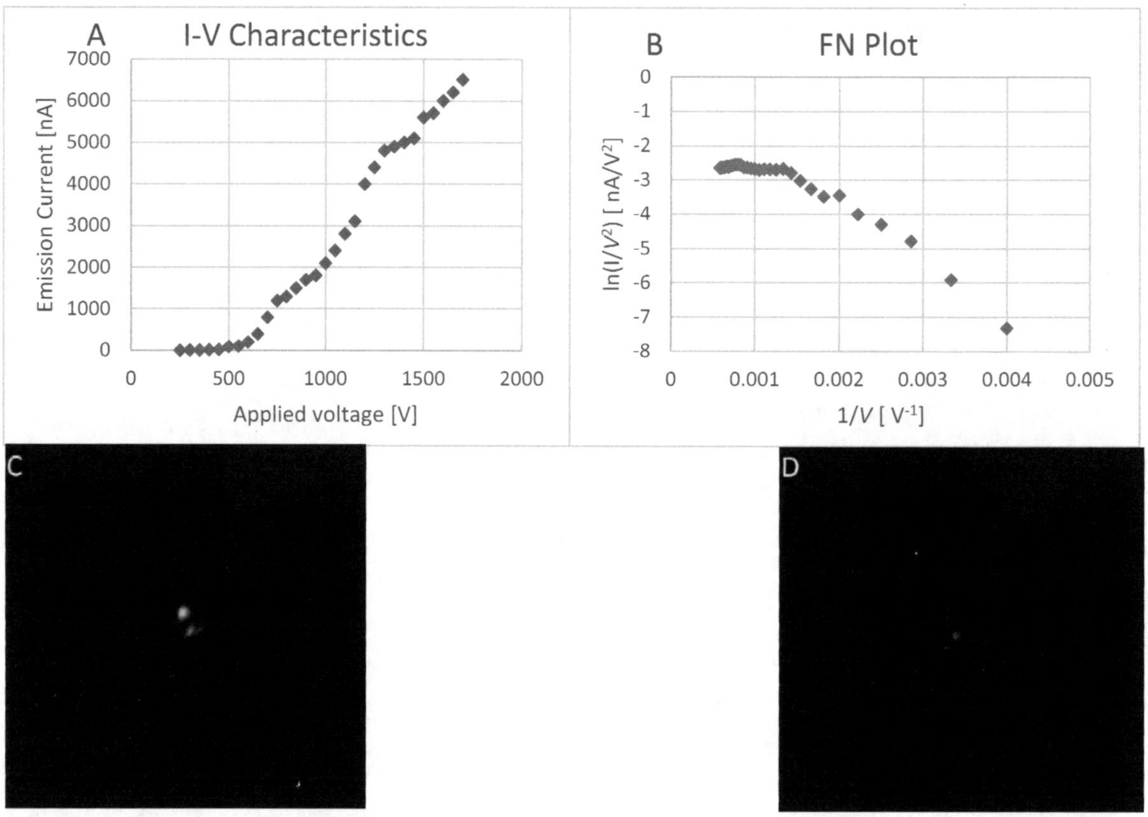

Figure 2. C5\W emitter. Decreasing Voltage (A) I-V Characteristic. (B) Related FN plot with slope -1896.79 decayed V. (C) Field electron image at V_{SAT}= 750 V, 1.2 μA. (D) field electron image at V=700 V, 0.8 μA.

The experiment was repeated second time, under the same condition and the result as following, by increasing the applied voltage the emission current increased as well until V_{SW}=1300 V was reached, the emission current suddenly raised from nA range to μA range specifically I_{SW}=1.6 μA, and by further increase voltage the emission current reach to (5.2 μA) at applied voltage value (1900 V), during this period of increased voltage there are small fluctuations in the emitted current, due to two causes, variation of local field F ($F = \beta V$, where V is the applied voltage and β is the enhancement factor) and variation of work function Φ, for which emission current I is extremely sensitive due to the exponential nature of FN equations, so 1% change in F can readily lead to 15% change in I. Local change of F are likely to occur due to geometric changes of the tip caused by surface migration in the extremely large electric field at the tip and form ion bombardment. On the other hand Φ is sensitive to adsorbates on the surface, these fluctuations can exhibit in electron field emission current in ultra-high vacuum conditions.[11, 12] Figure (3) shows I-V characteristic, its related FN plot, and emission image at V_{SW}. Then we begin slowly decreasing the applied voltage, the saturation region extends down to V_{SAT}=950 V, with emission current I_{SAT}=1.2 μA, and then threshold value reached at V_{TH}=210 V, with emission current V_{TH}=10 pA, as figure (4) showed the I-V behavior at decreasing voltage, with FN plot for this stage.

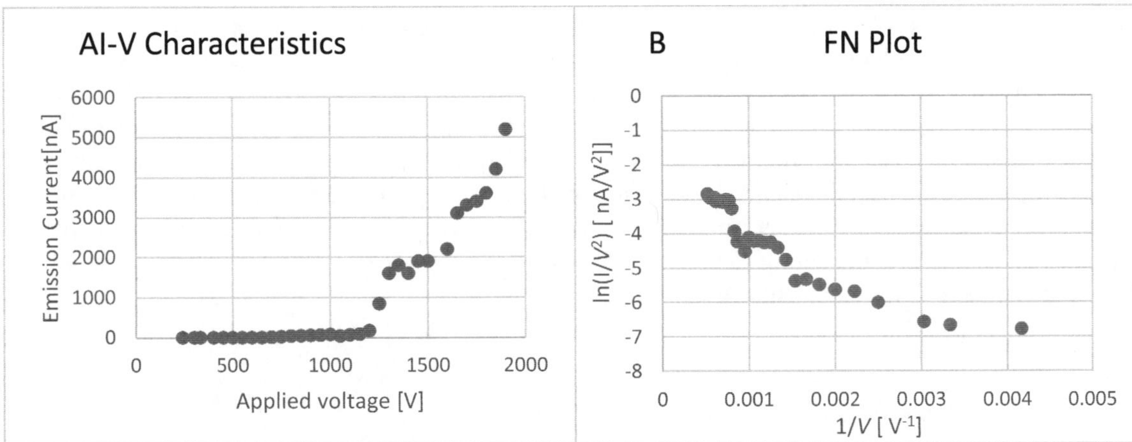

Figure 3. Cyc.2, C5\W emitter. Decreasing voltage. (A) I-V characteristic. (B) FN plot with slope -1766.14 decayed V.

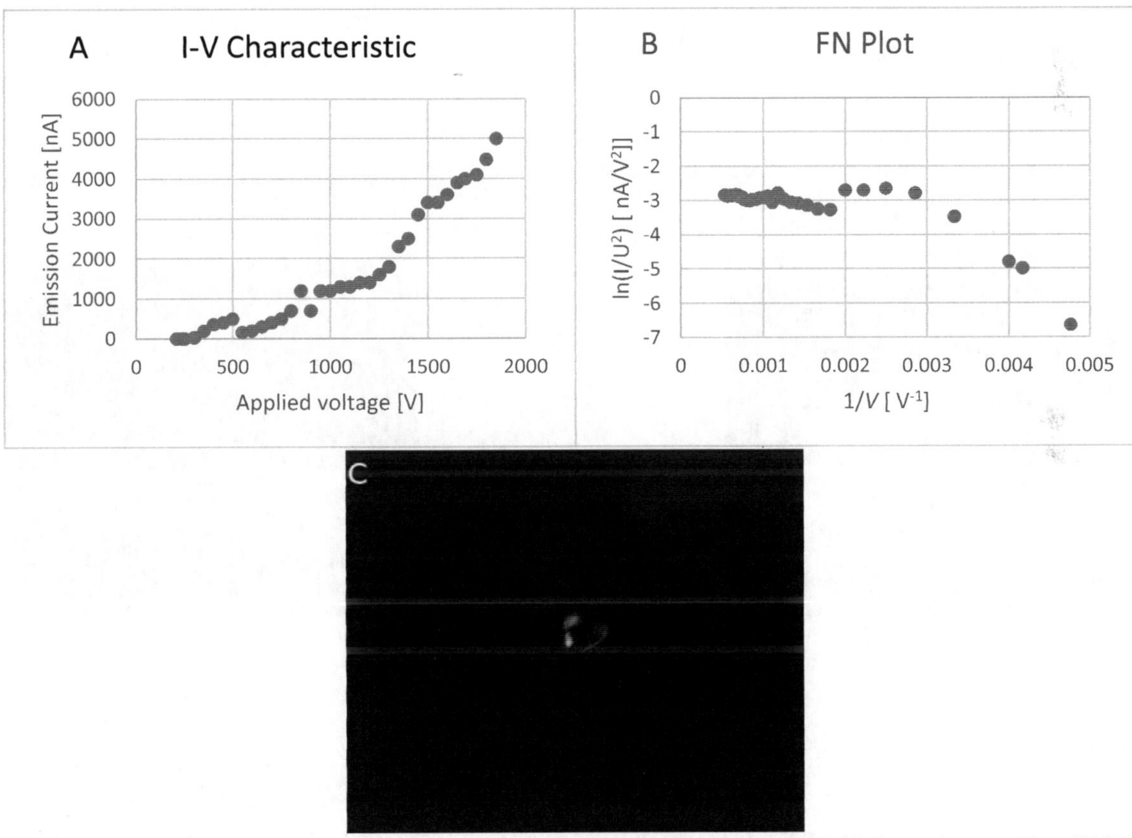

Figure 4. Cyc.2, C5\W emitter. Increasing voltage. (A) I-V characteristic. (B) FN plot. (C) Field electron image at V_{SW}=1300 V, 1.6 μA.

IOP Conf. Series: Materials Science and Engineering **305** (2018) 012022 doi:10.1088/1757-899X/305/1/012022

The second emitter (B5\W) was investigated under same conditions, with same way of cycling the applied voltage up and down two times, and the result as following; in the first cycle, by increasing the applied voltage, the emission current increased until it reaches V_{SW}= 1250 V, where its value has become (2.9 μA), we continue increase the voltage until it reaches V=2850 V, with emission current value (17.9 μA), after that we decrease the applied voltage, while the value of the emission current remained in range of (μA), until the applied voltage reaches V_{SAT}=650 V, where the emission current was (1.06 μA), then it falls to (nA) range. The emission current vanishes at V_{TH}=350 V, I_{TH}=1 pA, the figure (5) shows the behavior of I-V in first cycle, and its related FN plot, and field electron emission at V_{SW}, and V_{SAT}.

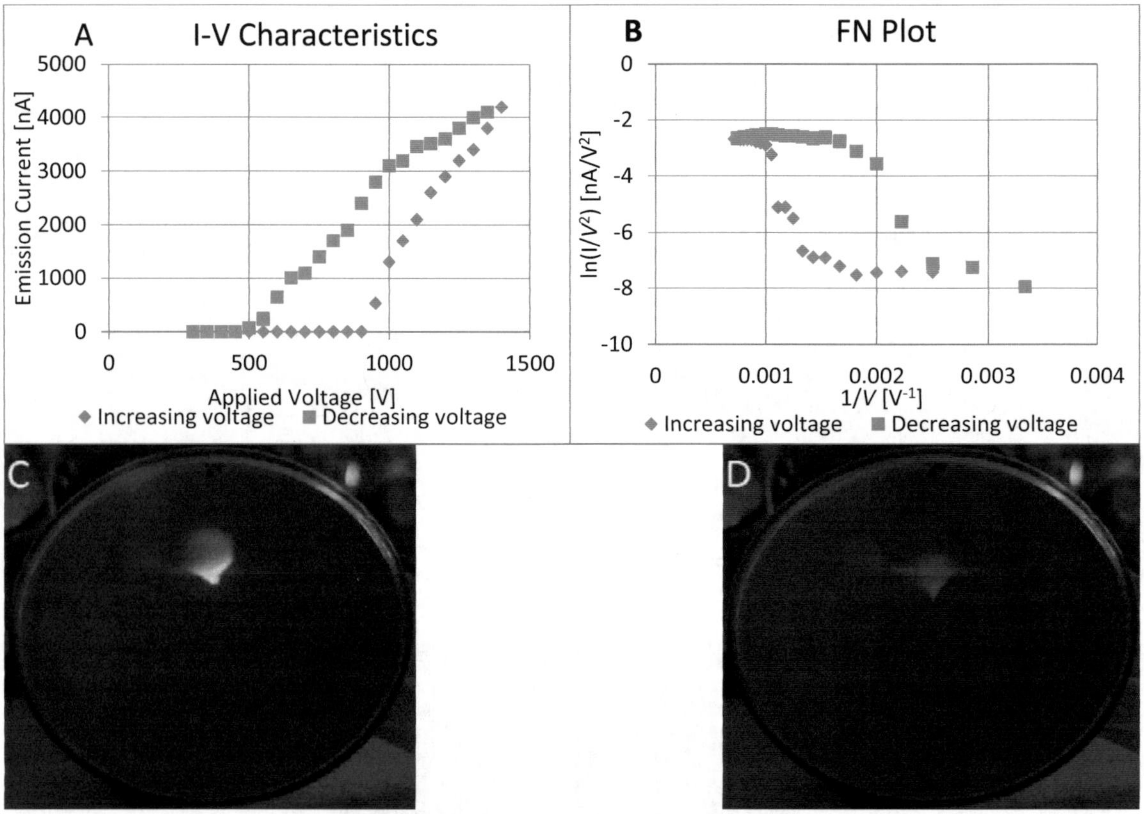

Figure 5.First cycle. B5/W emitter. Increasing and Decreasing voltage (A) I-V characteristics. (B) FN plot with slope -1872.79 decayed V. (C) field electron image at VSW. (D) Field electron image at V_{SAT}.

The same procedure of testing followed with B5/W emitter (second cycle), as we increase the applied voltage, the emission current increase with it from (6 pA) at applied voltage value (400 V) to (4.2 μA) at applied voltage value (1400 V), during this interval of increasing the applied voltage V_{SW} was (1000 V), with emission current (1.3 μA), in decreasing voltage phase, the saturation extends to V_{SAT}=650 V, with I_{SAT}=1.01 μA, and the lowest emission current value was (1 pA) at (300 V). Figure (6) shows the I-V characteristics, FN plot during increasing and decreasing applied voltage, and field electron image at V_{SW}.

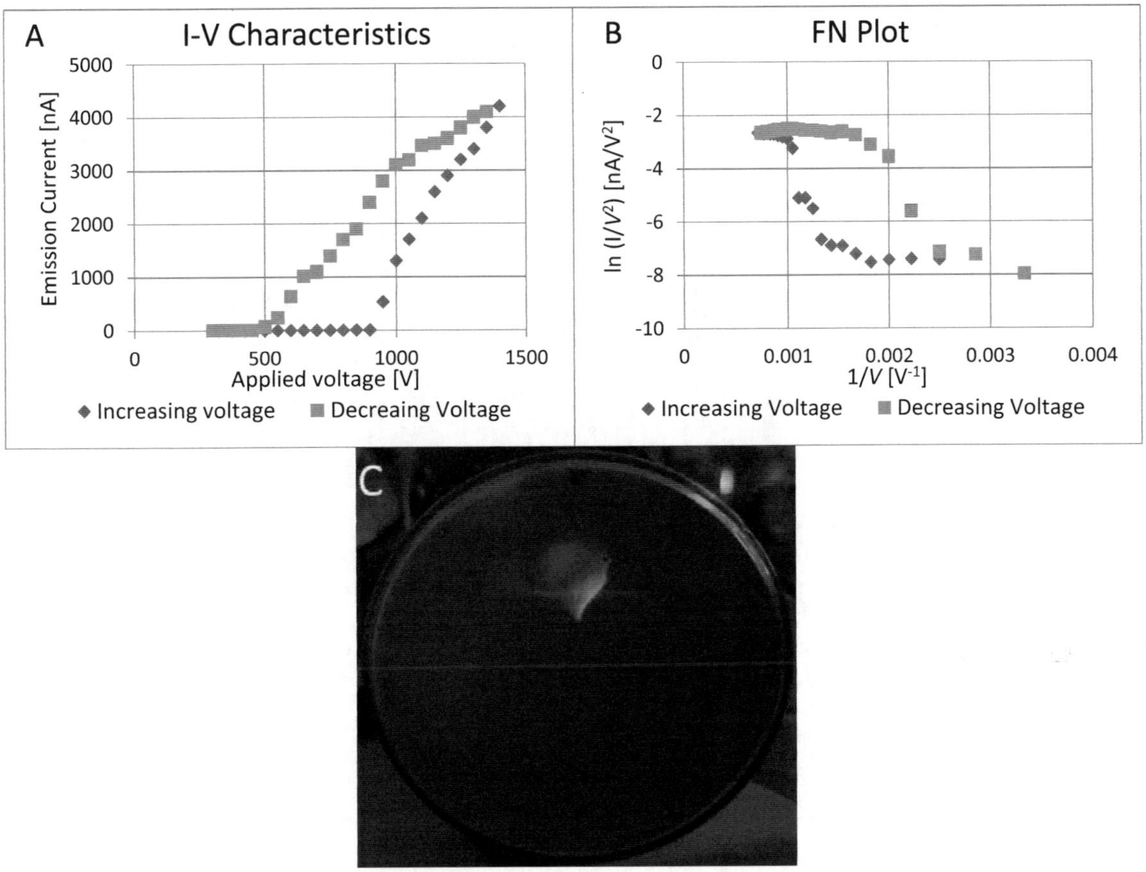

Figure 6. Second cycle. B5/W emitter. Increasing and Decreasing voltage (A) I-V characteristics. (B) FN plot with slope -1766.14 decayed V. (C) field electron image at V_{sw}.

From figure (5C, 5D, and 6C), it shows that the field electron emission image formed in approximately single spot, it seems with higher applied voltages, it will activate a new emission sites on the tip, and that's will make the electron emission image spread on screen.

We see that there are a deviation from FN theory, such deviation are usually attributed to space-charge effect, which induce a diminution of the FN slope at high field,[11] and the electronic properties of the emission behavior may be significantly influenced by the electronic properties of the SWCNT, and to non-metallic density of states at the tip of the tubes.[13]. The emission current fluctuations of conventional nanotube emitter have been found to be caused by head shaking, temperature change, vacuum alteration, or adsorption/desorption of gas molecules, all of these can be divided to two major kinds, the first is a variation in local field F, the second is a variation in local work function Φ, and I is very sensitive to the changes in these two factors[14].

IOP Conf. Series: Materials Science and Engineering **305** (2018) 012022 doi:10.1088/1757-899X/305/1/012022

4. Conclusion

We investigate the electron field emission properties of SWCNTs mounted on blunt tungsten substrate. The Fowler-Nordheim analysis for the emission current-voltage curves provided that we could get single electron field emission bright spot from SWCNTs/W in low applied voltage. In our experiment we use the mentioned tools, and that effect on the quality of the results, like; to perfectly mount the individual SWCNT on the tungsten or other suitable material, using a Nano-manipulator in SEM that could be very precise, and the carbon material is extremely strong and it can only be broken by brute force, and the breaking occurs by joule heating with a current through the tube of more than 20mA, or by apply a mechanical force, and this way would not give you a uniform length to the nanotubes and that will effect on the enhancement factor $\boldsymbol{\beta}$ and in turn effect on the local field and the emission current I.

References

[1] S. Iijima, Helical microtubules of graphitic carbon, Nature (1991) 56-58.

[2] T.T. D. Ferrera, I. Matsuya, G. Zhong, S. Okamoto, and H. Kawarada, Enhancement of field emission characteristics of tungsten emitters by single-walled carbon nanotube modification, Applied Physics letter, vol.88, issue 3 (2005).

[3] C.-s.L.a.S.-H.S.a.G.-T.C.a.Y.-H.L.a.M. Yokohama, the synthesize of carbon nanotube on tungsten wire for field emission cathode, proceedings of asia display, vol.2 (2007).

[4] a.L.C. Shi-Dong Liang, Theories of field and thermionic electron emissions from carbon nanotubes, Journal of Vacuum Science & Technology B, Nanotechnology and Microelectronics: Materials, Processing, Measurement, and Phenomena 28(2) (2010).

[5] R. Gomer, Field Emission and Field Ionization, Harvard University Press, Cambridge 1961.

[6] E.S.B.A.n.M.S. Mousa, Switch-on phenomena and field emission from muti-walled carbon nanotubes embedded in glass, Applied Microscopy 46 (2016) 244-252.

[7] E.W. Müller, Electron microscope observation of field cathodes, Zeitschrift für Physik1937.

[8] M.S.Mousa, Field electron emission studies on zinc oxide coated tungsten microemitters, Surface Science 266(1-3) (1992) 110-120.

[9] K.O.M.M.S.M.a.A. Fisher, M.S. Mousa, Information extraction from FN plots of tungsten microemitters, Ultramicroscopy 132 (2013) 48-53.

[10] N.E. Evgeny Sheshin, Field Emission Electronics, Springer2017.

[11] W.W.D. J. P. Barbour, J. K. Trolan, E. E. Martin, and W. P. Dyke, Space-Charge Effects in Field Emission, physicsal review 92(1) (1953).

[12] a.J.T.L. K.SYeong, Effects of adsorbates on the field emission current from carbon nanotubes, Applied Surface Science 233(1-4) (2004) 20-23.

[13] J.-P.S. Jean-Marc Bonard, Thomas Stöckli, Walt A. de Heer, László Forró, and André Chātelain, Field emission from single-wall carbon nanotube films, Applied Physics Letters 73(7) (1998).

[14] F.A. Lixin Dong, T. Fukuda, Field emission of telescoping multi-walled carbon nanotubes, 4th IEEE Conference on Nanotechnology, IEEE, Munich, Germany, 2004.

Dynamic NMR Study of Model CMP Slurry Containing Silica Particles as Abrasives

In memory of Professor Yuzhuo Li

F Odeh[1]*, A Al-Bawab[1] and Y Li[2]

[1] *Department of Chemistry, The University of Jordan, Amman 11942, JORDAN*
[2] *Department of Chemistry, Clarkson University, Potsdam, NY 13699*

f.odeh@ju.edu.jo

Abstract: Chemical mechanical planarization (CMP) should provide a good surface planarity with minimal surface defectivity. Since CMP slurries are multi-component systems, it is very important to understand the various processes and interactions taking place in such slurries. Several techniques have been employed for such task, however, most of them lack the molecular recognition to investigate molecular interactions without adding probes which in turn increase complexity and might alter the microenvironment of the slurry. Nuclear magnetic resonance (NMR) is a powerful technique that can be employed in such study.

The longitudinal relaxation times (T_1) of the different components of CMP slurries were measured using Spin Echo-NMR (SE-NMR) at a constant temperature. The fact that NMR is non-invasive and gives information on the molecular level gives more advantage to the technique. The model CMP slurry was prepared in D_2O to enable monitoring of T_1 for the various components' protons. SE-NMR provide a very powerful tool to study the various interactions and adsorption processes that take place in a model CMP silica based slurry which contains BTA and/or glycine and/or Cu^{+2} ions. It was found that BTA is very competitive towards complexation with Cu^{+2} ions and BTA-Cu complex adsorbs on silica surface.

1. Introduction:

Silica slurries are of great importance in applications ranging from computer industry to medical fields [1, 2]. However, their use in microelectronic industry is considered one of the most important achievements, specifically in the chemical mechanical planarization (CMP). Several processes take place in CMP including physical and chemical processes [3]. A good understanding of these physicochemical processes associated with CMP (such as surface kinetics, complexation, and corrosion) is critical [4]. This understanding is not also essential for the CMP itself, but is also crucial for post CMP operations such as cleaning and environmental effect of discharged slurries [5]. A wide range of physical and spectroscopic techniques have been applied for the characterization of these slurries such as dynamic [6] and static light scattering [7], scanning tunneling microscope (STM) [8], x-ray photoemission spectroscopy

(XPS) [9], UV- photoemission spectroscopy (UPS) [10], low-energy electron diffraction (LEED) [11], electrochemical impedance spectroscopy [12] and FTIR [13].

Slurries used in CMP are multi-component systems. In such slurries different compounds will be present such as the oxidizing agents and anticorrosion agents. In Cu-CMP silica based slurries in particular, there will be an oxidizing agent (e.g. H_2O_2), complexing agent (e.g. glycine) and anticorrosion agent (e.g. benzotriazole (BTA)) in addition to silica particles and the solvent (water). Due to presence of the oxidizing copper ions (Cu^{+1} and Cu^{+2}) will also be in the slurry. These copper ions will form complexes with the various components. Probable complexes presents are Cu-H_2O, Cu-BTA, and Cu-glycine in addition to copper oxides [4]. The free and complex components will also be available to adsorb on silica surface. This adsorption will modify the silica particles' surface by changing the charge, hydrophobicity and size among other changes. These changes will modify the polishing characteristics of these particles. In addition to changing the surface nature, adsorption might also affect the amount available of these components for action in the CMP process. An understanding of these interactions on the molecular level will provide valuable information about the different components involved in the Cu-CMP process and enables huge advancements in the CMP field.

Nuclear magnetic resonance (NMR) provides a very powerful tool for studying interactions and other dynamic properties on the molecular level. One of the most advantages of NMR is that it is not invasive which is important in keeping number of components in the system low and no disturbances to the present equilibria will happen due to introduction of probe molecules. Also NMR is flexible and can cover a wide spectrum of atoms that can be studied depending on the system under study. For example it can be used to observe ^{1}H, ^{13}C, ^{29}Si and many other different nuclei in the system. Therefore, valuable information about the type and strength of interactions and conformations present in the system will be gained according to the nucleolus observed. Another big advantage for NMR is there are so many different types of experiments that can be performed ranging from simple 1D experiments to 3D ones, and from decoupling experiments to coupling ones, and the coupling between various types of nuclei. Dynamic NMR experiments using Spin Echo (SE-NMR) add to this wide range of different experiments [14]. Since the development of SE-NMR spectroscopy, valuable information were obtained such as solvation effects and solvent-solute interactions [15], internal hydration in proteins [16], probing active sites in enzymes and binding modes [17], interactions in biological systems [18, 19], pharmaceutical and drug release [20] and study of self assembly and organized structures [21, 22].

One of the most important values studied for a system is the relaxation of the observed nucleous in a compound at a certain environment. Relaxation processes are caused by interaction of the magnetic dipole of the nuclear spins with neighboring nuclei [23]. These interactions are stochastic in nature and are modulated by molecular trumbling and translation and dependent on several factors such as intermolecular collisions, Van der Waals binding, and proton exchange with other molecules. there are two pathways for relaxation, longitudinal (spin-lattice relaxation T_1) and transvers (spin-spin relaxation T_2). Longitudinal relaxation time is mainly affected by fast motion at ω_o and $2\omega_o$ frequency orders (ω_o is Larmor frequency of the proton in the NMR instrument), whereas T_2 additionally sensitive to slow motions near zero frequency components resulting on the fast loss of the transverse magnetization without affecting the populations of the spin states. T_1 is the time required for a nucleus to relax from the excited state (after excitation by a pulse of energy) to its ground state. T_1 occurs by getting rid of the excess energy to the lattice (or the environment) through radiationless processes (e.g. heat). The values of T_1 are very

sensitive to any changes of the nucleus environment; these changes could be due to molecular interactions such as adsorption, H-bonding or complexation or to micro-environmental ones such as temperature, pH and ionic strength. It is more convenient to calculate relaxation rate R_1 ($1/T_1$). Relaxation times have been used as an analytical and probing tool in many fields [22, 25-28].
In this paper, a systematic study for the effect of each component in model CMP slurries will be studied using SE-NMR as an attempt to better understand most aspects affecting the CMP process.

2. Experimental:

2.1 General:

Deuterium oxide (CIL D, 99.9%) was used for all experiments and was used without any further treatment. The silica used was Aerosil 200 (Degussa). Silica dispersions were prepared from a stock solution and sonicated for at least 30 minutes. The stock solution was prepared as 10% in D_2O then it was sonicated for at least one hour in a water sonicator to have a homogenous and stable suspension. The Cu^{+2} solution (1000 ppm) was prepared using Copper (II) nitrate,2.5-hydrate (J.T. Baker) in D_2O. 1H-Benzotriazole (ACROS), Glycine (Fisher) were of general laboratory grade and used as purchased and were prepared as stock solutions of 1% and 10% respectively. All solutions prepared were in weight percentages unless it was indicated otherwise.

2.2 NMR measurements:

The relaxation time T_1 was measured using Bruker 400 AVANCE with a BBO probe equipped with z-gradient and the temperature was controlled using a VTU and held constant at 300 °K, the relaxation delay used varied from 10 seconds (in presence of 100 ppm Cu^{+2} ions in the solutions) to 60 seconds depending on the relaxation time of the solution's components. Each sample was prepared and measured at least three times, and the T_1 for each sample was repeated at least three times. The standard deviation was measured for all data presented in the work and was 0.98%.
All samples for the T_1 experiments were prepared by placing the solutions to be studied into NMR (5 mm) tubes. The parameters used to obtain T_1 were the same for all samples. A relaxation delay of 60 seconds was used for all samples that does not contain Cu^{+2} ions with 12 delay times in the t_1-delay list ranging from 0.01 sec to 120 seconds and number of scans ranging from 4 to 16 according to the sample's concentration and receiver gain (rg) values ranging from 64 to 512 also depending on the sample's concentration. The spectral width used was 12 ppm and was not optimized to shorter spectral width to minimize the time for each T_1-experiment.

3. Results and Discussion:

3.1 Effect of D_2O/H_2O changing ratio on T_1 of water:

The relaxation time (T_1) of water is a very useful marker of the environment it is present in. It has been used to study many microenvironments [22, 25, 28], All our measurements were performed in deuterated water to enable the measurements of T_1 for the different components of the model CMP slurry at their respective chemical shift (δ). Hence, it was important to study the relaxation time of water and the effect of increasing percentages of the deuterated form on the relaxation time of these mixtures. A series of samples were prepared with increasing amounts of D_2O and were studied for the relaxation time of the water signal at 4.78 ppm, results are shown in Figure 1. From the data listed below we observe (as expected) that increasing the ratio of D_2O

will increase the relaxation time. This is expected since we are reducing the effective relaxation by adding D atoms instead of protons. Protons relax in a totally different way than D atoms. Also this study is essential to see the effect of the protons added with the various components such as glycine and BTA and not confuse any changes in T_1 due to H-exchange with adsorption or complexation processes. However, T_1 values were smaller than expected for samples with more H_2O ratio. This is caused by the presence of dissolved oxygen in the samples, and since oxygen is paramagnetic it result in a sharp drop of T_1 values [29]. By examining the data in Figure 1 we see that only when H_2O% is more than 20% that T_1 drops to less than 15 seconds and R1 is less than one second. Since the maximum amount of glycine added is 1% and BTA is 0.1%, it is not probable that H-exchange is the cause of any drop in T_1 observed in this study. So Figure 1 can be used effectively to distinguish between T_1 drop due to proton exchange or other processes present.

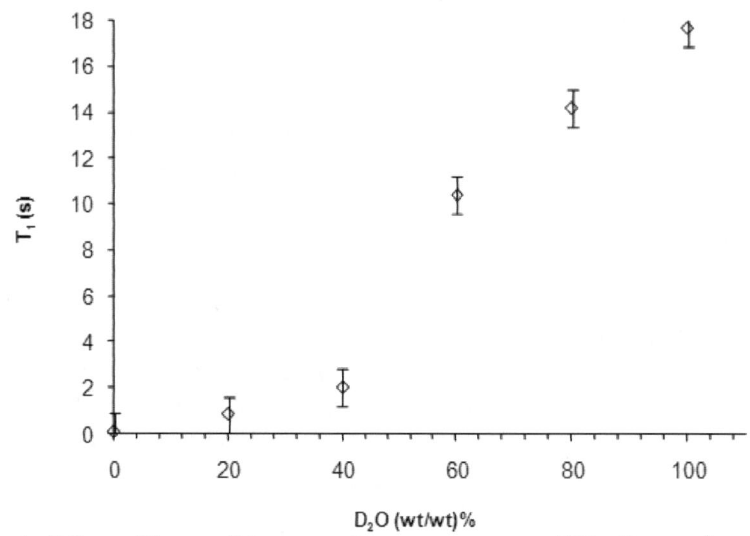

Figure 1: The effect of increasing amounts of D_2O on the relaxation time of water at 300° K.

3.2 Effect of changing silica ratio on T_1 of water:

The relaxation time of water in a series of samples with increasing amounts of silica was measured at a constant temperature of 300 °K (Figure 2). Our analysis of water-silica interaction was based on a two-site exchange model. A fast exchange between surface-bound water and bulk water was assumed since no trace of confined water signal was observed at this temperature. Studies using ^{1}H- and ^{2}H-NMR spin-lattice relaxation studies (T_1) at various magnetic fields and much lower temperatures have been performed [30]. These studies anticipated the presence of a three site exchange between protein and water, bound and free water (bulk), but in general a fast-exchange is seen between bound and free water. Such behavior can also be attributed to the interaction between water and silica particles Increasing silica leads to a decrease in T_1 of water via two pathways of interactions. Firstly, increase of adsorption of water on silica surface due to increase in surface area available for adsorption. The observed T_1 values are the average of T_1 of adsorbed and T_1 of bulk water. Since we are working on a constant magnetic field, the deciding factor affecting T_1 observed is the population of each state. Increasing the surface area available for adsorption by increasing silica% will increase the population of adsorbed water and therefore the observed T_1 will decrease. Secondly, the interaction between water protons and silanol group protons (-Si-OH) on the silica particle's surface [26, 27]. From Figure 1, it can be concluded that adsorption to silica particle's surface and not the exchange between hydroxyl groups of silica and D_2O is the main source for T_1 drop since the amount of protons added due to the addition of silica is less than 10% and 20% proton exchange is needed to drop T_1 of water to less than 15 seconds. These results are in agreement with previous studies of the water structure and dynamics in presence of silica particles conducted using both solution and solid NMR [31] were the contribution of H-bonding contributed less to T_1 decrease than adsorption on silica. These conclusions were supported by a recent NMR study of the orientation of water near silica surfaces [32].

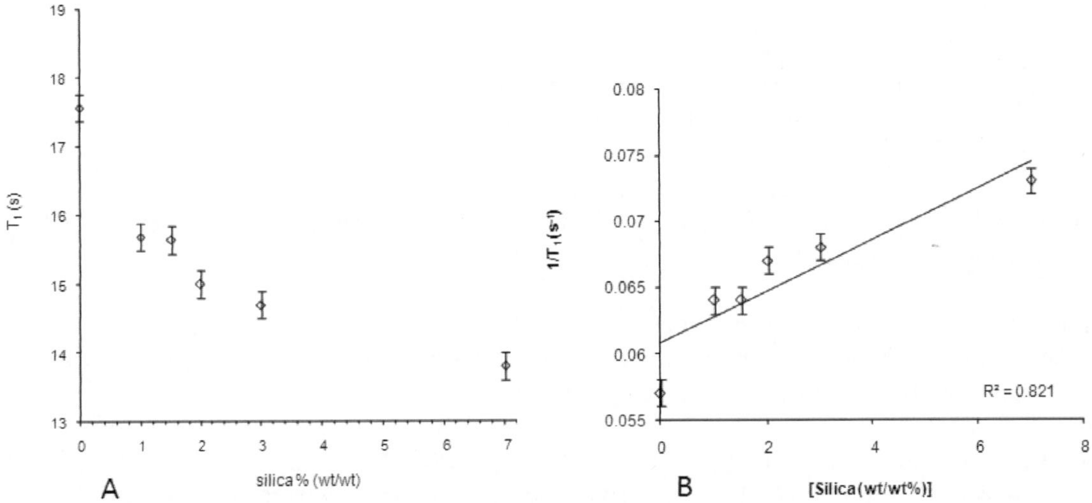

Figure 2: The effect of increasing amounts of silica on (A) the relaxation time and on (B) relaxation rate of water at 300° K.

3.3 Effect of Cu^{+2} on T_1 of water:

The addition of increasing amounts of Cu^{+2} ions to D_2O samples is expected to decrease the values of T_1 of water sharply. This is due to interaction with the paramagnetic ion in two ways. The first way is through short range interactions when water molecules are in the inner-sphere of the complex (very short T_1); the second way is by a long range interaction of water molecules with the metal ion if these molecules are in the outer-sphere of the complex [33]. However, in both cases we have a very fast equilibrium between these sites and the averaged signal is observed by NMR. In this study, constant and increasing amounts of Cu^{+2} were added to D_2O samples and T_1 was measured at a constant temperature of 300 °K. The data are shown in Figure 3, the relaxation time of water dropped sharply from a value of 17.5 s to less than one second when a total final concentration of 100 ppm of Cu^{+2} was reached. This huge effect was anticipated due to the complexation of water molecules with the Cu^{+2} ions, which will eventually decrease T_1.

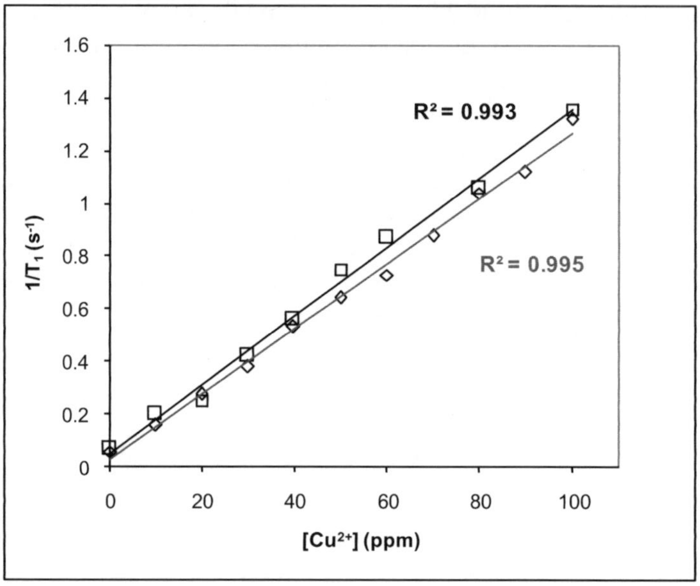

Figure 3: The effect of increasing amounts of Cu^{+2} on the relaxation time of water at 300° K in D_2O (◊) and in 3%Silica (□).

The effect of Cu^{+2} ions was also studied in 3% Silica dispersions, which is the silica ratio that will be used throughout this study in the model silica-based dispersions (Figure 3); indicate that complexation with Cu^{+2} ions cause a large drop in T_1. There was no big difference in presence or absence of silica on T_1 of water in these samples which is a further proof of a complexation mechanism. The addition of transition metal ions to solid surfaces has been the focus of many studies especially in presence of silica particles and many methods were used such as titration methods [33] and ESR [34]. It was found that Cu^{+2} adsorb on oxide surfaces via coordination with the oxygen of the silanol groups but this interaction is weakest if the ligand available was water, while in presence of other ligands chelating to Cu^{+2} ions, the adsorption of these Cu-complexes on the silica surface tend to be stronger. This is very important since the presence of Cu^{+2} ions will participate in occupying some of the adsorption sites on the silica surface but even though, the relaxation time of water is expected to decrease due to the paramagnetic nature of the Cu^{+2} ions. In addition, the fact that relaxation time (T_1) of water is very sensitive to the presence of Cu^{+2} ions is very useful in determining fate of these ions in CMP slurries [5].

3.4 Effect of glycine and benzotriazole on T_1 of water:

It is known that glycine is very soluble in water and can form H-bonding with it, so it is expected to have a decrease in T_1 of both water and glycine which is exactly what we observed (Figure 4A). Addition of Glycine to D_2O causes a moderate decrease in T_1 for both water and glycine indicating the formation of H-bonding between glycine and water as expected. Also the addition of glycine causes changes in the pH of the solutions (more acidic). However, such changes in pH have small effect on T_1 [31]. The addition of glycine to 3% Silica solution was also studied (Figure 4B). By examining the data obtained we see a very small change in T_1 of both water and glycine upon addition of glycine compared to the case in absence of silica particles. This could suggest that adsorption of water and/or glycine to silica particles is more effective than the H-bonding and other interactions between glycine and water.

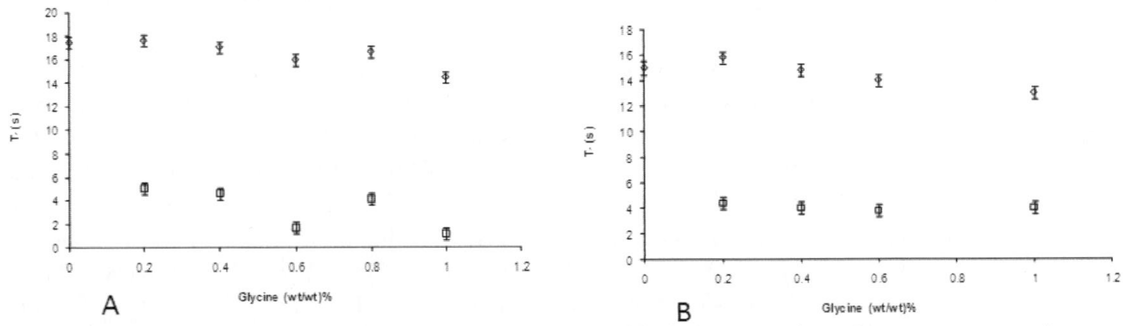

Figure 4: The effect of increasing amounts of glycine on the relaxation time of glycine (□) and water (◊) at 300 °K in (A) D_2O and (B) 3% silica.

The addition of benzotriazole (BTA) to water was not expected to have a large impact on the relaxation time of both. Although BTA has N-H bond and a triazole ring which might interact with water through H-bonding but on the overall BTA is a hydrophobic compound and moderately soluble in water. In Figure 6, the relaxation data for both water and BTA protons (BTA1 at δ 7.9 ppm, BTA2 at δ 7.6 ppm, the third proton in BTA which is NH appeared at δ 16 ppm but was not monitored due to its broad and labile signal) are summarized. When BTA was first added to water in small quantities, no big change was observed. However, when the amount of BTA reached more than 0.2%, T_1 for both BTA and water protons start to decrease suddenly then reached a constant value. Since T_1 reached a steady state even though BTA was still being added, H-bonding was not thought to be the main cause. The cause of this change is most probably due to BTA-BTA interaction and the formation of some sort of aggregation in the BTA system [Figure 5] which was studied in more details by this group [21, 31].

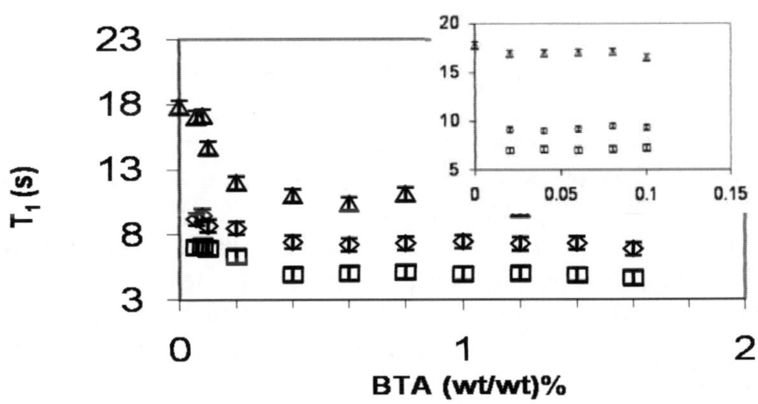

Figure 5: The effect of increasing amounts of BTA in D_2O on the relaxation time of BTA (◊ BTA1, □ BTA2) and water (Δ) protons (A) over a large concentration range, (B) a clearer plot over the small concentration range of BTA at 300 °K.

IOP Conf. Series: Materials Science and Engineering **305** (2018) 012023 doi:10.1088/1757-899X/305/1/012023

The behavior of BTA in 3% silica dispersions was also studied. In presence of silica particles (Figure 6), the relaxation time of both BTA protons started to increase and for water protons, it was an increase upon the first introduction of BTA then it was stable through out the whole additions. This behavior was rather strange since a drop was expected due to adsorption of BTA molecules on silica surface and some sort of interaction (even small) with water. This could be explained by strong adsorption of BTA and forming a film at the same time covering the adsorption sites, hence any further addition of BTA will be marked with increase in T_1 (averaged) but still lower than T_1 for BTA in D_2O solutions. This assumption is strengthened by the way T_1 was changing for BTA1 and BTA2 protons. In D_2O, the BTA protons behaved differently and usually BTA1 (at ~7.9 ppm) has a higher T_1 value than BTA2 (at ~7.6 ppm). However, when BTA was present in silica dispersion, both protons had the same T_1 value indicating very similar effect on both of them which might be an indication of flat adsorption of BTA on silica surface.

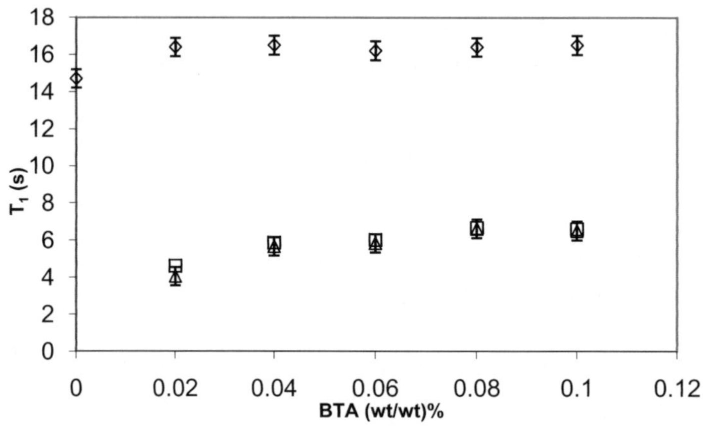

Figure 6: The effect of increasing amounts of BTA on the relaxation time of BTA (□ is BTA1), (Δ is BTA2) and water (◊) at 300 °K in 3% silica.

In presence of Cu^{+2} ions, the relaxation time of glycine, BTA and water dropped significantly (Figure 7) in presence and absence of silica. As was expected, Cu^{+2} ions caused a sharp drop in T_1 values for all components but with different extents. The largest drop was in BTA1 protons (~6 seconds) which is understandable since BTA1 protons are close to the chelating centers in BTA molecule so they will be closer to Cu^{+2} ion. For BTA2, they were a little far away on the edge of BTA molecule. Glycine was also affected by Cu^{+2} ions presence and experienced a drop of ~3.5 seconds in T_1 values. The effect on water was less due to the high competition from BTA and glycine molecules. BTA and glycine are very strong complexing agents compared to water and any drop of T_1 of water is mainly due to its partial participation in the Cu-Gly complex and occupying the axial positions.

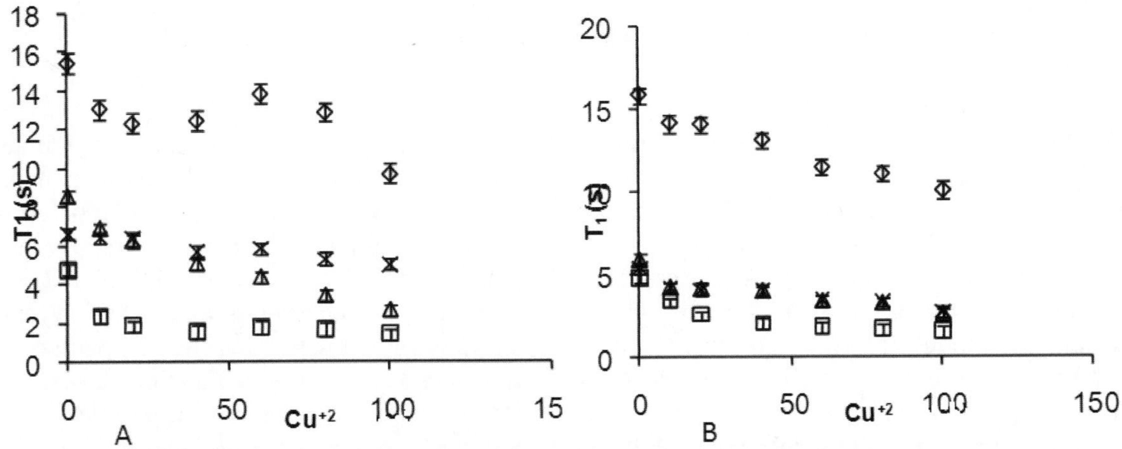

Figure 7: The effect of increasing amounts of Cu^{+2} to D_2O solutions containing BTA (0.1%) and glycine (1.0%) on the relaxation time of BTA (Δ BTA1, x BTA2), glycine (□) and water (◊) in (A) D_2O and (B) 3% silica at 300 °K.

In order to understand the competition between BTA and glycine over complexation with Cu^{+2} ions, BTA was added to solutions containing glycine and Cu^{+2} in presence and absence of silica particles. The results are shown in Figure 8. BTA molecules were able to react with Cu^{+2} ions and remove both glycine and water from their complexes with Cu^{+2} in D_2O at 300 °K (Figure 8A); this is clearly seen from the T_1 data obtained after adding increasing amounts of BTA to prepared solutions containing glycine and Cu^{+2} ions. This addition caused the T_1 for both water and glycine to increase.

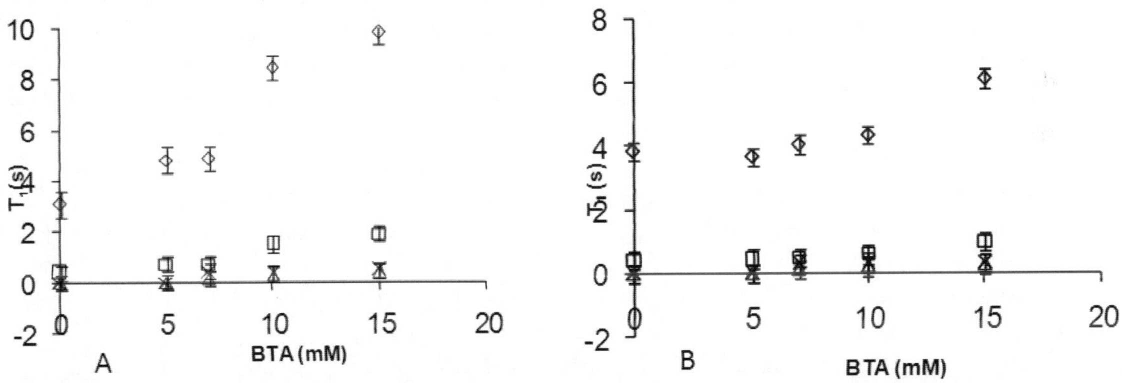

Figure 8: The effect of increasing amounts of BTA to samples containing BTA glycine (20 mM) and Cu^{+2} (1.0 mM) in D_2O on the relaxation time of BTA (Δ BTA1, x BTA2), glycine (□) and water (◊) in (A) D_2O and (B) 3% silica at 300 °K.

The same controlled study was performed in a silica dispersion to see the effect of silica particles on these equilibria. Figure 8B summarizes the data obtained. The same behavior was observed, when BTA was added to a silica dispersion containing 20 mM of glycine and 1 mM of Cu^{+2} ions.

T_1 values for both glycine and water were increased while T_1 for BTA protons were very small and comparable with BTA-Cu complex values obtained earlier.

4. Conclusions:

From the data shown above, we see that BTA behavior has many consequences on the dynamics of these systems in presence and absence of silica particles. The presence of silica particles make BTA's behavior more smooth which is clear from T_1 values of both BTA protons compared to their values in D_2O solutions. This is an indication of both complexation and adsorption when both silica and Cu^{+2} ions are present; several studies mentioned an increased adsorption tendency of Cu^{+2} ions on metal oxide surfaces in presence of ligands other than water. For glycine behavior, although it forms well known and stable complexes with Cu^{+2}, but the competition of BTA towards complexation with Cu^{+2} is very high and since glycine is added to silica based Cu-CMP slurries for chelation with Cu^{+2} ions produced from the oxidation reaction, its role should be studied more. Even though BTA is highly competitive towards complexation but the BTA-Cu complex itself adsorbs strongly on the silica surface and also adsorbs on the Cu surface giving BTA its famous anticorrosion behavior.

References:

[1] Pajonk G.M., (2003), *Colloid Polym. Sci.*, 281, 637-651.
[2] Cousins B.G., Doherty P.J., Williams R.L., Fink J. and Garvey M.J., (2004), *J. Mater. Sci. Mater. Med.*, 15(4), 355-359.
[3] Cooper K., Cooper J., Groschopf J., Flake J., Solomensev Y. and Farkas J., (2002), *Electrochem. Solid State Lett.*, 5(12), G109.
[4] Kirshnan M., Nalaskowski J.WW. and Cook L.M., (2010), *Chem. Rev.*, 110, 178-204.
[5] Speed D., Westerhoff P., Sierra-Alvarez R., Draper R., Pantano P., Aravamudhan S., Chen K.L., Hristovski K., Herckes P., Bi X., Yang Y., Zeng C., Otero-Gonzalez L., Mikoryak C., Wilson B.A., Kosaraju K., Tarannum M., Crawford S., Yi P., Liu X., Babu S.V., Moinpour M., Ranville J., Montano M., Corredor C., Posner J. and Shadman F., ((2015), *Environ. Sci.: Nano*, 2, 227-244.
[6] Gun'ko V.M., Mironyuk I.F., Voronin E.F., Turov V.V., Pakhlov E.M., Goncharuk E.V., Nychiporuk Y.M., Vlasova N.N., Gorbik P.P., Mishchuk O.A., Chuiko A.A., Kulik T.V., Palyanytsya B.B., Pakhovchishin S.V., Skubiszweska-Zieba J., Janusz W., Turov A.V and Leboda R., (2005), *J. Colloid Interface Sci.*, 289, 427-445.
[7] Abarkan I., Doussineau T. and Smaïhi M., (2006), *Polyhedron*, 25(8), 1763-1770.
[8] Bozso F. and Avouris P., (1987), *Phys. Rev. Lett.*, 57, 1185.
[9] Hamers R.J., Avouris P. and Bozso F., (1987), *Phys. Rev. Lett.*, 59, 2071.
[10] Bozso F. and Avouris P., (1988), Phys. Rev. B, 38, 3937.
[11] Yu M.L. and Meyerson B.S., (1984), *J. Vac. Sci. Technol. A*, 2, 446.
[12] Turk M.C., Walters M.J. and Roy D., (2017), *Electrochimica Acta*, 224, 355-368.
[13] Shan J., Wang Y. and Hamer R.J., (1996), *J. Phys. Chem.*, 100, 4961.
[14] Bagno A., Rastrelli F. and Saielli G., (2005), *Prog. Nucl. Magn. Reson. Spec.*, 47(1-2), 41.
[15] Venu K., Denisov V.P. and Halle B., (1997), *J. Am. Chem. Soc.*, 119, 3122.
[16] Fasano M., Baroni S., Aime S., Mattu M. and Ascenzi P., (2003), *J. Inorg. Biochem.*, 95, 64.

[17] Yilmaz A., Tez M., Gorat V., Boylu S., Kaplan A. and Kavak G., (1996), *Phys. Med. Biol.*, 41, 539.
[18] Loth K., Pelupessy P. and Bodenhausen G., (2005), *J. Am. Chem. Soc.*, 127, 6062.
[19] Baumgartner S., Lahajnar G., Sepe A. and Kristl J., (2002), *AAPS Pharm. Sci. Tech.*, 3(4), 1.
[20] Gao Z., Wasylishen R.E. and Kwak J.C.T., (1990), *J. Phys. Chem.*, 94, 773.
[21] Odeh F., Al Bawab A. and Li Y., (2010), *J. Disp. Sci. Tech.*, 31(2), 162-168.
[22] Tu C., Osborne E.A. and Louie A.Y., (2011), Annals Biomed. Eng., 39(4), 1335-1348.
[23] Tadimalla S. and Momot K.I., (2014), *Plos One*, 29, 1-19.
[24] Helpern J.A., Ordidge R.J. and Knight R.A., (1994), *Cerebral Ischemia and Basic Mechanisms*, Editors Hartmann A., Yatsu F. and Kuschinsky W., Springer-Verlag Berlin Heidelberg.
[25] Gavish M. And Vega S., (1993), *Phys. Rev. B*, 48(4), 2191-2199.
[26] Totland C., Steinkopf S., Bolkhus A.M. and Nerdal W., (2011), *Langmuir*, 27, 4690-4699.
[27] Totland C. and Nerdal W., (2016), *J. Phys. Chem. C*, 120, 5052-5058.
[28] Mutisya S.M., Kirch A., de Almeida J.M., Sanchez V.M. and Miranda C.R., (2017), J. *Phys. Chem. C*, 121, 6674-6684.
[29] Anderson W. A. and Arnold J.T., (1956), *Phys. Rev.*, 101(2), 511-512.
[30] Yilmaz A., Tez M., Gorat V., Boylu S., Kaplan A. and Kavak G., (1996), *Phys. Med. Biol.*, 41, 539-549.
[31] Li Y., Gong M., Ramji K. and Li Y., (2009), *J. Phys. Chem. C*, 113, 18003-18013.
[32] Ma Y., Foster A.S. and Nieminen R.M., (2005), *J. Chem. Phys.*, 122, 144709.
[33] Roose P., Craen J.V. and Eisendrath H., (1998), *Colloids and Surfaces A*, 145, 213-227.

IOP Conf. Series: Materials Science and Engineering **305** (2018) 012024 doi:10.1088/1757-899X/305/1/012024

Stability of Benzotriazole Derivatives with Free Cu, Zn, Co and Metal-Containing Enzymes: Binding and Interaction of Methyl-benzotriazoles with Superoxide Dismutase and Vitamin B_{12}

R A Abudalo [1*], M A AbuDalo [2], and M T Hernandez[3]

[1]*Department of Clinical Laboratory Sciences, King Saud bin Abdulaziz University for Health Sciences, P. O. Box. 2477, Al Ahsa 31982, Kingdom of Saudi Arabia*
[2]*Chemistry Department, Jordan Universities of Science and Technology, P.O. Box 3030, Irbid 22110, Jordan*
[3]*Department of Civil and Environmental Engineering, University of Colorado at Boulder, Campus Box 428, Boulder, CO 80309-0428, USA*

abudalo-hullru@ngha.med.sa

Abstract: Benzotriazole derivatives form very strong bonds with transition metals, and are the most widely used type of industrial corrosion inhibitor. Some benzotriazole derivatives have been implicated as hormone regulators which also carry the ability to induce uncoupling responses or otherwise inhibit respiration processes in some microorganisms. However, the mechanisms associated with benzotriazole toxicity and inhibition are unknown. Using Differential Pulse Polarography, the stability constants of commercially significant corrosion inhibitors, 4-and 5-methylbenzotriazole, coordinated with free Cu (II) and Co (III), were determined to be 10^{15} and 10^{8}, respectively. Polarographic analyses were extended to confirm that methylbenzotriazole also binds the copper center(s) in the ubiquitous enzyme superoxide dismutase, and the Corrin site in the coenzyme cobalamin (vitamin B_{12}). These results suggest that the metal-chelating ability of this unique class of compounds may confer inhibition to certain enzyme systems.

1. Introduction

Benzotriazoles (BTs) and its derivatives are widely used in industry, agriculture, and in medicinal chemistry. They have been used as corrosion inhibitors, deicing/anti-icing fluids, radio protectors, and rubber and chemical fiber [1- 6]. They have been also used as UV stabilizers in plastics, and polymer stabilizers in photographic industry [7]. Beside the enormous industrial use, BTs and its derivatives demonstrated many biological activities, such as, antibacterial, antifungal, antineoplastic, anti-inflammatory, analgesic, antimalarial and antiviral activity [8-12].

Apart from industrial and biological importance, acute and chronic toxic responses to BT and its derivatives have been observed in several fish species, zooplankton, and bacteria [1,13-16]. Previous study indicated that BT and methyl-benzotriazoles (MeBT) are toxic to the luminescent bacteria used in the Microtox® assay [14-15]. Acute toxic response to benzotriazole and its commercially significant derivatives have been observed in several fish species, zooplankton, and bacteria at concentration as low as 6 mg/L [1,13,15]. Same response of pure bacterial cultures (Microtox®) to unmethylated benzotriazole and 4- methylbenzotriazole was observed, whereas 5-methylbenzotriazole was more toxic [1]. The mechanism of toxicity however, remains unknown.

Some benzotriazole derivatives have been implicated to inhibit metal-containing oxido/reductase enzymes that are critical to respiration, or these involved with the scavenging of toxic respiration intermediates. O'Brien was the first to isolate a significant benzotriazole toxicity mechanism to pure culture of *Pseudomonas aeruginosa* bacterial cells that was associated with respiratory chain activity: uncoupling oxidative phosphorylation from bacterial respiration [17]. Author suggested that benzotriazole toxicity may be associated with the complexation of membrane bound enzymes involved in the respiratory process. The results were with agreement with De Wever and Verachtert who also suggested that the toxicity of 2-mercaptobenzothiazole (MBT) to bacteria and other organisms is due to the complexation between metal-containing, membrane bound enzymes and benzothiazole [18]. Benzotriazoles are theorized to be successful corrosion inhibitors because of their ability to form stable complexes with different yellow metals, especially copper. Their metal chelating properties may be directed towards metals that serve as critical oxidation/reduction sites in some enzymes. Furthermore, the methyl-benzotriazole binding may significantly change the electron transport abilities such that an enzyme's catalytic properties are adversely affected.

Superoxide dismutase enzyme (SOD) and Vitamin B_{12} are necessary for cells in nearly all living cells. Superoxide Dismutase is found almost in all eukaryotic cells and in many bacteria and forms a crucial part of the cellular antioxidant defense mechanism in many different types of cells. Several research reports suggest that some benzotriazole derivatives form stable, nontoxic, low molecular weight complexes of copper that mimic SOD activity, and have been proposed as substituted for SOD in clinical application. Skorda and coworkers reported that certain copper-methylbenzotriazole complexes can serve as good functional models for the active copper site of Cu-Zn superoxide dismutase [19]. Vitamin B_{12} (Cobalamin) is a coenzyme that is necessary for red blood cells synthesis, maintenance of the nervous system, and growth and development in children. It is unique among all vitamins in that it contains not only a complex organic molecule, but one of the few enzymes to use the essential trace element, cobalt. Vitamin B_{12} as usually isolated is called cynocobalmin because it contains a cyano group attached to the cobalt in the sixth coordination position. A critical coordination position of cobalt in vitamin B_{12} is filled by dimethylbenzimidazole ribonucleotide, bound covalently by its 3'-phosphate group to one of the side chains of the corrin ring through aminoisopropanol.

In this study, the interaction of MeBT with the copper centers of superoxide dismutase enzyme (SOD), and the cobalt center of vitamin B_{12} were characterized. Characterization was done to provide a better understanding for their interactions with methylbenzotriazole (MeBT), and determine if there is potential for the subsequent inhibition of their function. Determination of stability constants of MeBT interacted with free copper and cobalt ions was essential for this characterization. Differential Pulse Polarography (DPP) were used for the determination of consecutive and overall stability constants of metal complexes that can be reversibly reduced. Such an approach was applied to estimate the strength of metal-benzotriazole interactions, as well as for interactions with metal incorporated in the active sites of common (co)enzymes.

2. Differential Pulse Polarography and Stability Theory

The current that flows during a single life time of a dropping mercury electrode (DME) is held at a given potential in a polarography and is described by the Ilkovic Equation:

$$I_d = 708nD_o^{1/2}C_0^*m^{2/3}t^{1/6} \tag{1}$$

In which,

I_d= diffusion current (A)
n = number of electrons transferred/molecule,
m = mercury flow rate (mg/s).
t = Sampling interval(s).
D_o= diffusion coefficient (cm2/sec)
C_o*= Concentration of oxidized species in bulk solution (mol/cm^3)

The overall electrode reaction can be generally represented by:

IOP Conf. Series: Materials Science and Engineering **305** (2018) 012024 doi:10.1088/1757-899X/305/1/012024

$$M^{x+} + nL^{y-} \leftrightarrow ML_n^{(x+y)}$$
$$M^{x+} + xe^- \leftrightarrow M \qquad (2)$$

Where, M^{x+} represent some metal ion, L^{-y} a ligand, and $ML^{(x+y)}$ a complex formed.
In their classic report, Deford and Hume (1951) defined a series of functions, where the change in reduction potential of a target ion, measured with a polarograph, can be directly related to its complexation stability by the progressive addition of a known ligand [20].

$$F_0(x) = \sum \beta_n C_L^n = anti\log\left[0.434\frac{nF}{RT}\Delta E_P\right] + Log\frac{(I_p)_s}{(I_p)_s} \qquad (3)$$

In which,

β_n =overall stability constant of the nth complex,
C_L = ligand concentration assuming $C_L >> C_M$,
$(I_p)_s$ = diffusion current constant of free metals.
$(I_p)_c$ = diffusion current constant of complexed metal.

The shift in DPP potential on the addition of complexing ligand.

$$\Delta E_P = (E_p)_s - (E_p)_C \qquad (4)$$

Related to Heath and Hefter and under most of the common testing conditions, $(I_p)_s$ for most inorganic species is equal to $(I_p)_c$ [21]. For this reason, equation (3) condenses to the following equation:

$$F_0(x) = \sum \beta_n C_L^n = anti\log\left[0.434\frac{nF}{RT}\Delta E_P\right] \qquad (5)$$

Stability constants were obtained as follows

$$F_0(x) = \sum \beta_n C_L^n = \beta_0 + \beta_1 C_L + \beta_2 C_L^2 + \dots\dots\dots + \beta_n C_L^n \qquad (6)$$

Where β_0 has a value of unity. The function $F_n(X)$ is introduced as follow

$$F_n(X) = \frac{F_{n-1}(X) - \beta_{n-1}}{C_L} \qquad (7)$$

In order to obtain β1,…., βn, a widely-accepted graphical extrapolation method derived by Laden (1941) was applied in this research [22]. Sequential plots of (F_n (X) vs. C_L) yielded a family of curves, through which coordination chemistry between metals and benzotriazoles was determined

An independent voltammetric approach was used to verify the maximum coordination number, and magnitude of stability constants determined by DeFord-Hume regression. The maximum coordination number was determined using the Lingane method reported by Crow (1969) by plotting of $\Delta E_{1/2}$ Vs. Log C_x relationship, using the Lingane equation [23]:

$$(E_p)_S = (E_p)_C - \frac{0.0591}{n} Log\beta_{mxj} - j\frac{0.0591}{n} LogC_x \qquad (8)$$

β_{mxj} (the overall stability constant), and j (maximum number of ligands coordinated with a metal through a reversible series of complexing interactions), were determined for Cu (II), Co (III) and Zn (II) interactions with benzotriazoles.

This study reported the development and use of polarographic methods for quantitation of benzotriazole complexes to study their behavior in coordinating free copper, cobalt, and zinc ions and metal-containing enzymes, where their ions are integral to the active site specifically superoxide dismutase and cobalamin.

3. Material and Methods

3.1 Polarography

Polarographic and voltammetric measurements were performed with a CV-50W Voltametric analyzer fitted with a controlled growth mercury electrode. The multimode electrode was used in static mercury drop electrode (SMDE) mode. A three-electrode system was multiplexed with a platinum auxiliary electrode and an Ag/AgCl (3.0 M KCl) reference electrode. All potentials were relative to the Ag/AgCl electrode. A system scan rate of 5mV s^{-1} was used with a drop time of 1.0 sec. For DPP, the pulse amplitude was 100 mV with a sample width of 17 msec.

3.2 Reagents

Britton-Robinson buffer was the supporting electrolyte to determine the stability of methylbenzotriazoles interactions with Cu, Co, and Zn. Phosphate buffered saline (pH 7) was used as supporting electrolyte to observe methylbenzotriazole complexation interactions with superoxide dismutase and vitamin B_{12}.

All chemicals used for polarographic analysis were of analytical grade and in high purity (>97%). Copper, and zinc sulfate, and hexamine cobalt chloride (for DPP analysis) were obtained from Fisher Scientific (Pittsburgh, PA, USA). When added to the solution used in this study, these salts liberated Cu (II), Zn (II), and Co (III) as the predominant forms of these metals. Superoxide dismutase and Vitamin B_{12} were obtained from ICN Biomedicals, Inc. (Aurora, OH, USA). Britton-Robinson buffers were prepared in the polarographic vessel, by mixing 0.04 M phosphoric acid, 0.04 M acetic acid and 0.04 M Boric acid to give a pH 2.5. MeBT standards were obtained from PMC Specialty Group Inc. (Cincinnati, OH, USA) or bought from Sigma-Aldrich Corporation (St Louis, MO).

3.3 Procedures

3.3.1 Differential Pulse Polarography (DPP).

A 15 ml supporting electrolyte solution was pipetted into the voltammetric cell, and a metal ion (Co, Zn, or Cu) was added to give a concentration between 10^{-4}-10^{-5}M. The ligand (methybenzotriazole) was then added in increasing concentrations to a maximum of 3.0 x10^{-3} M. The pH and the temperature were maintained at pH=2.5; room temperature respectively. The shift in the half wave potential was recorded after each addition and polarograms were compiled. All DPP measurements were made in triplicate, and each complexation experiment was independently replicated at least three times. Since commercially significant application of corrosion inhibitors often include a mix of 4- and 5- methybenzotriazoles, a consecutive set of complexation experiments were extended with each isomers alone, as well as with the mix of isomers (ratio of 4-MeBT to 5-MeBT was 45:55) that are commercially available in the North American market.

4. Results and Discussion

4.1 Reversibility

DeFord-Hume analysis is based on an assumption that the complexation observed on the electrode surfaces is reversible, which is the case for many benzotriazole-metal interactions. According to a widely accepted metric developed by Parry and Osteryoung (1965), the reversibility of a polarographic wave is most accurately assessed by its half peak width [24]. These researchers have derived a general expression for the half peak width of DPP polarogram, which predicts a value of 62mV for a fully reversible two electron reduction at an applied potential pulse of 50 mV. In the present work, DPP polarograms in Britton-Robenson buffer had half peak widths of 62 ± 2mV for both free copper, cobalt, and complexed copper and cobalt, indicating a high degree of reversibility. To confirm the reversible reduction, reversibility was further tested using Tomes criterion [25]. According to this criterion, $E_{3/4}$-$E_{1/4}$ for a two electron reduction process resulted in a 29 mV response with applied pulse at 50 mV. $E_{3/4}$-$E_{1/4}$ represents the potential at ¾ and ¼ of the diffusion current respectively. In this present DPP work, the $E_{3/4}$-$E_{1/4}$ for Cu (II) and Co (II) was found to be 30 ± 1 and 29 ± 1 mV respectively when applied pulse at 50 mV.

4.2 Cu (II): 4, and 5-Methylbenzotriazole Systems

IOP Conf. Series: Materials Science and Engineering **305** (2018) 012024 doi:10.1088/1757-899X/305/1/012024

A relatively large potential shift was observed when 4-MeBT and 5-MeBT were added separately to solution containing Cu (II); this shift pattern validated the use of DPP for formation constant determination. Atypical DPP of those associated with a copper-4-Methylbenzotriazole and 5-Methylbenzotriazole systems is shown in Figure 1. The conditional stepwise stability constants (i.e. non-saturated copper complexes) as well as the overall (conditional) stability constant β_2 for Cu (II) were determined.

Polarographic results obtained for Cu (II): 4 and 5-methylbenzotrizole systems are presented in Table 1 and in Figure 1. Plotting $F_o(X)$ vs. [MeBT] using the data in Table 1 yielded a rising curves with an intercept of β_o, which was equal to one for both systems. A first derivative of the previous plot ($F_1(X)$ vs. [MeBT]) yielded a straight line indicating that copper ions are coordinated with benzotriazole ligands at a maximum stoichiometric ratio of 2 for both systems as well. Regression analysis for Cu (II): 4-Methylbenzotriazole and 5-Methylbenzotriazole systems showed that the intercept of the first derivative plot, $F_1(X)$ corresponded to a value of β_1 (first complex stability) that approximately equal to 3.0 $x10^{10}$ and 6.0 $x10^{12}$ respectively. The slope corresponded to a value of β_2 (overall stability constants) approximately equal to 5.0 $x10^{13}$ and 7.0 $x10^{15}$ respectively. A confirmation for the value for β_2 was provided by the intercept of the second derivative plot, F_2 (X) vs. [MeBT], which yielded a straight line parallel to the [MeBT] axis, indicating that the highest ligand: metal ratio of Cu-4 or 5-MeBT interactions was 2. The intercept was found to equal 4.6 $x10^{13}$ and 7.0 $x10^{15}$ respectively.

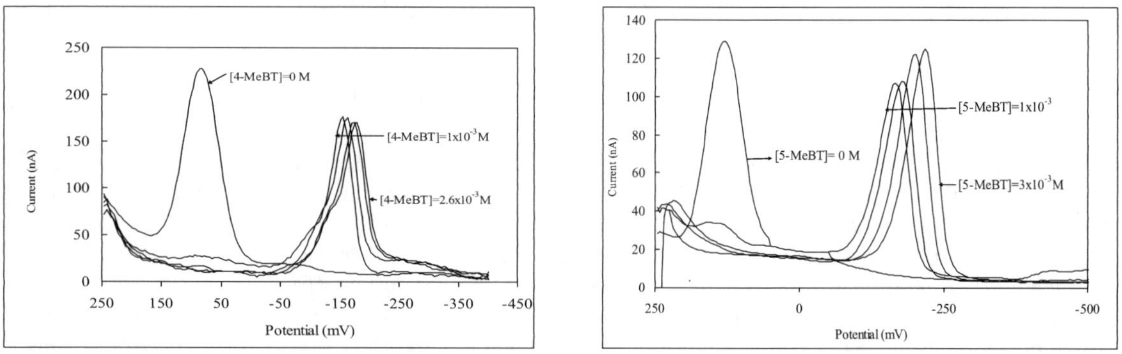

Figure 1: Differential pulse polarogram of Cu (II): methylbenzotriazole systems. A significant shift in the potential occurred in both systems (4-methylbenzotriazole (left), 5- methylbenzotriazole (Right)) systems as a result of an increase in the benzotriazole. [Cu^{+2}] is equal to $5.3x10^{-5}$M.

Table 1: Polarographic data for a Cu (II):4 and 5-methylbenzotriazole systems

Added Ligand Conc. (M)	ΔE_p (mV)	I_s/I_c	$F_0(CL)$	$F_1(CL)$	$F_2(CL)$
Cu(II):4-methylbenzotriazole system					
0.00E+00	0	-	1.00E+00	-	
1.12E-03	235	1.03	8.97E+07	8.01E+10	4.47E+13
1.30E-03	240	1.02	1.24E+08	9.56E+10	5.06E+13
1.47E-03	241	1.03	1.42E+08	9.63E+10	4.51E+13
1.64E-03	244	1.05	1.73E+08	1.06E+11	4.6E+13
1.98E-03	248	1.03	2.30E+08	1.17E+11	4.39E+13
2.61E-03	255	1.08	3.94E+08	1.51E+11	4.62E+13
2.77E-03	257	1.02	4.30E+08	1.55E+11	4.53E+13
Cu (II):5-methylbenzotriazole system					
0.00E+00	0	-	1.00E+00	-	-
1.09E-03	292	2.05	1.47E+10	1.34E+13	6.77E+15

1.44E-03	298	1.78	2.30E+10	1.60E+13	6.99E+15
1.77E-03	304	1.69	3.25E+10	1.84E+13	7.02E+15
1.93E-03	306	1.63	3.69E+10	1.91E+13	6.80E+15
2.09E-03	308	1.60	4.21E+10	2.01E+13	6.76E+15
2.40E-03	312	1.54	5.47E+10	2.28E+13	6.98E+15
2.55E-03	314	1.50	6.22E+10	2.43E+13	7.18E+15

The maximum coordination number was confirmed for both systems using Lingane method [23]. Half wave potential shifts in response to Log [4 and 5-MeBT] shown in Figure 2 confirmed the previous

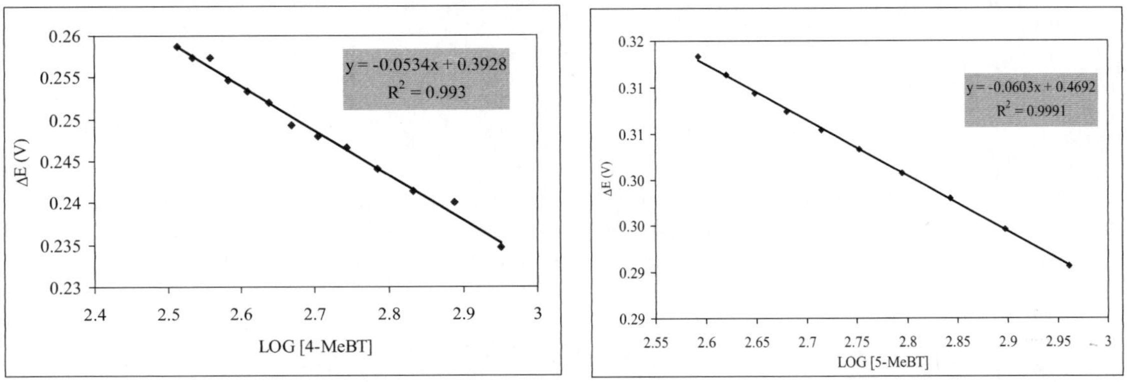

Figure 2: Half wave potential shift for the reduction of Cu (II) in response to increasing concentrations of 4-methybenzotriazole (left) and 5-methybenzotriazole (right) systems. pH=2.5; T=25°C. Symbols represent the averages of triplicate observations; error bars represent standard deviation.

results with the slope approaching the maximum ligand number of 2.

4.3 Zinc (II)-Methylbenzotriazole System

No significant potential shifts were observed when increasing 4-or 5-methybenzotriazole masses were added to zinc-containing solutions. Those results indicating that either weak, no complexes formed, or DPP could not be used in this application. Since DPP was unsuitable for formation constant determination of Zn(II): MeBT complexes, further electrochemical characterization was not required for zinc-MeBT interactions.

4.4 Co (III): 4 and 5-Methylbenzotriazole Systems

A potential shift was observed when 4 and 5-MeBT were added to solution containing Co (III); this shift again validated the use of DPP for formation constant determination. Figure 3 shows the typical polarograms associated with cobalt: 4 and 5-methylbenzotriazole systems respectively after the incremental additions of MeBT. As before, conditional stepwise stability constants as well as the overall stability constant β_2 for Co (III) were determined. Polarographic results obtained for Co (III): 4 and 5-methylbenzotrizole systems are presented in Table 2 and Figure 3. Plotting $F_o(X)$ vs. [MeBT] yielded a rising curve with an intercept of β_o, which was equal to one. Also, first derivative of the previous plot, $F_1(X)$ vs. [MeBT], yielded a straight line indicating that copper ions are coordinated with benzotriazole ligands at a maximum stoichiometric ratio of 2.

Regression analysis for both 4 and 5-methylbenzotriazole systems indicated that the intercept of the first derivative plot, $F_1(X)$ corresponded to a value of β_1 approximately equal to 1.0 $x10^6$ and 9.0 $x10^5$ respectively while the slope of this line corresponded to a value of β_2 approximately equal to 2.0 $x10^8$ and $5.0x10^7$ respectively as well. A confirmation for the value for β_2 was provided by the intercept of the second derivative plot, $F_2(X)$ vs. [MeBT], which yielded a straight line parallel to the [MeBT] axis,

IOP Conf. Series: Materials Science and Engineering **305** (2018) 012024 doi:10.1088/1757-899X/305/1/012024

indicating that the highest ligand: metal ratio of Co-MeBT interactions was 2 for both systems. The intercept was found to equal 2.4 $x10^8$ and 5.0 $x10^7$ respectively.

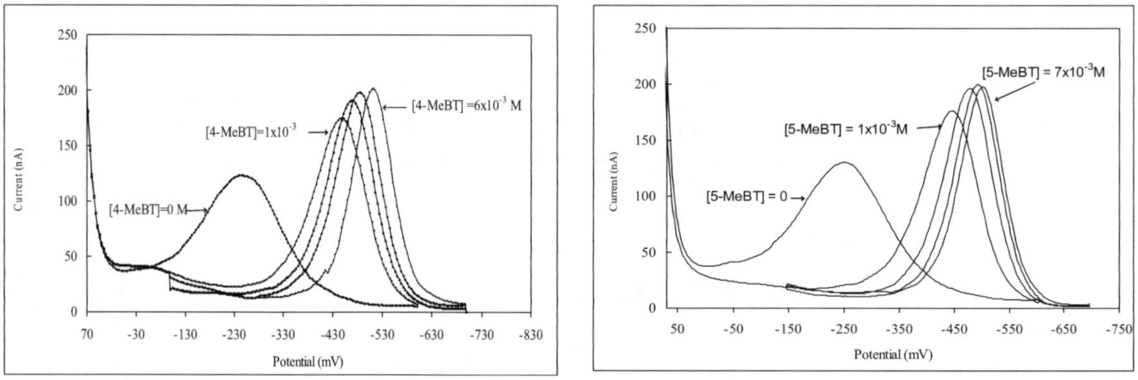

Figure 3: Differential pulse polarogram of Co(III): methylbenzotriazole systems. A significant shift in the potential occurred in both systems (4-methylbenzotriazole (left), 5- methylbenzotriazole (Right)) systems as a result of an increase in the benzotriazole. [Co III] is equal to $1.5x10^{-5}$M.

Table 2: Polarographic data for a Co(III): 4 and 5-methylbenzotriazole systems

Added Ligand Conc. (M)	ΔE_p (mV)	I_s/I_c	F_0(CL)	F_1(CL)	F_2(CL)
Co(III):4-methylbenzotriazole system					
0	0	-	1.00E+00		
1.095E-03	200	0.59	1.42E+03	1.30E+06	2.71E+08
2.094E-03	224	0.51	3.16E+03	1.51E+06	2.42E+08
3.008E-03	240	0.47	5.44E+03	1.81E+06	2.69E+08
4.621E-03	256	0.42	9.12E+03	1.97E+06	2.11E+08
5.337E-03	264	0.41	1.21E+04	2.26E+06	2.36E+08
6.618E-03	272	0.40	1.59E+04	2.39E+06	2.11E+08
Co(III):5-methylbenzotriazole system					
0.00E+00	0		1.00E+00		
1.07E-03	188	0.63	9.61E+02	8.97E+05	-5.57E+06
2.94E-03	228	0.51	3.17E+03	1.08E+06	5.96E+07
3.76E-03	236	0.49	4.12E+03	1.10E+06	5.13E+07
5.22E-03	248	0.45	6.09E+03	1.17E+06	5.08E+07
5.87E-03	252	0.44	6.92E+03	1.18E+06	4.72E+07
7.03E-03	260	0.42	9.14E+03	1.30E+06	5.64E+07

The maximum coordination number was confirmed for both systems using Lingane method [23]. Half wave potential shifts in response to Log [4 and 5-MeBT] shown in Figure 4 confirmed the previous results with the slope approaching the maximum ligand number of 2 for both systems.

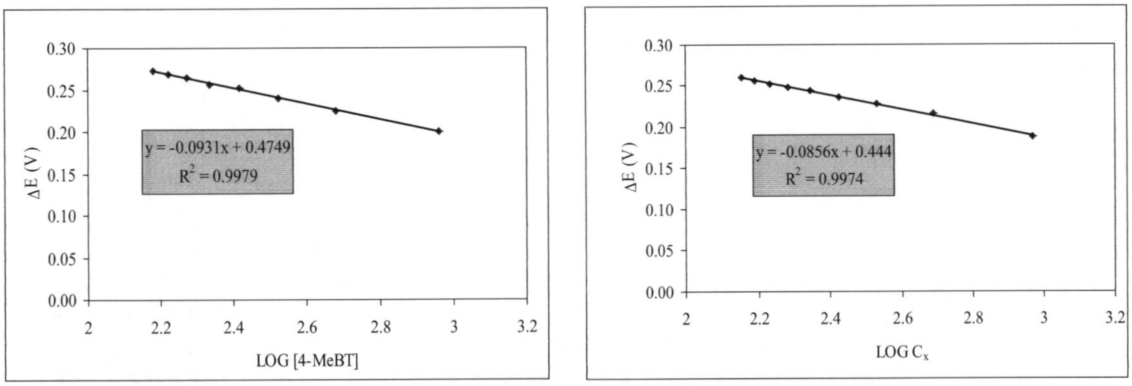

Figure 4: Half wave potential shift for the reduction of Co (III) in response to increasing concentrations of 4-methybenzotriazole (left) and 5-methybenzotriazole (right) systems. pH=2.5; T=25°C. Symbols represent the averages of triplicate observations; error bars represent standard deviation.

Summary of stability constants for methyl benzotriazole isomers complexes with Cu(II), Co(III) and Zn(II) derived from DeFord-Hume and Lingane polarographic analysis is shown in Table 3.

Table 3: Stability constants for 4-methylbenzotriazole and 5-methylbenzotriazole complexes with Cu(II) and Co(III) derived from DeFord-Hume and Lingane polarographic analysis.

	Concentration (M)		**Method**			
System	Metal	MeBT	Lingane	DeFord-Hume		**Max. coordination number**
			Log β_n	**Log β_1**	**Log β_2**	
Cu(II):5-MeBT	5.2 x10^{-5}	0 – 3.0 x10^{-3}	15.9	12.8	15.8	2
Cu(II):4-MeBT	5.3 x10^{-5}	0 – 3.0 x10^{-3}	13.7	10.5	13.7	2
Co(III):5-MeBT	1.5 x10^{-5}	0 – 7.0 x10^{-3}	7.8	6.0	7.7	2
Co(III):4-MeBT	1.5 x10^{-5}	0 – 7.0 x10^{-3}	8.0	6.0	8.3	2
Zn(II):5-MeBT	5.5 x10^{-5}	0 – 2.0 x10^{-3}	-	-	-	-
Zn(II):4-MeBT	5.5 x10^{-5}	0 – 2.0 x10^{-3}	-	-	-	-

4.5 Interactions of Superoxide Dismutase with MeBT

An Addition of benzotriazole isomers to solutions containing active superoxide dismutase (SOD) causes a potential shift (Figure 5). This shift is related to the formation of a complex between the copper metal in the SOD enzyme, and methylbenzotriazole isomers. The behavior was the same as that between free copper metal and MeBT, meaning that the complex formed between SOD and MeBT is strong nearly as strong as that between MeBT and free Cu and as judged by -200 mV potential shifts observed upon MeBT addition to solution containing active SOD. However, unlike MeBT interactions with free copper, the reduction was found to be irreversible once the MeBT was bound to SOD. Such that DPP could not be used to calculate the conditional stability constant.

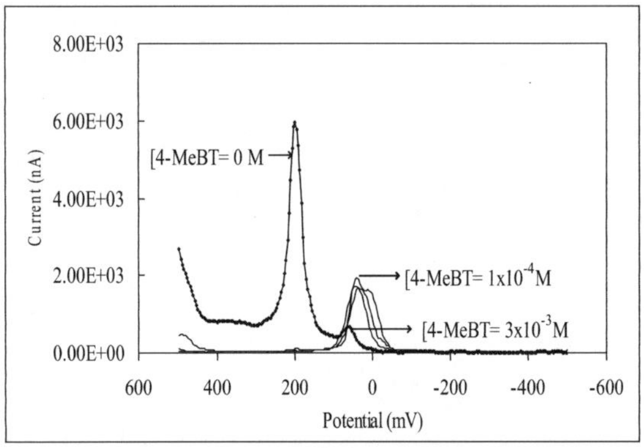

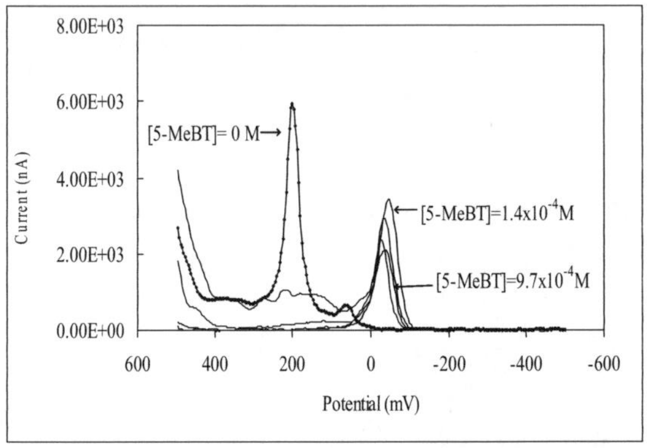

Figure 5: Differential pulse polarogram of SOD: 4-methylbenzotriazole (Left), 5- methylbenzotriazole (Right) systems, showing the potential shift and streaming current reduction resulting from the addition of increasing amount of benzotriazole. [Cu(II)] in SOD= 6.0 x10^{-7}M.

4.6 Interaction of Vitamin B_{12} with MeBT

The interaction between methybenzotriazole and free cobalt ions was observed, however, no significant shift in potential occurred with the increase of the MeBT concentration in a solution containing cobalamin (Figure 6). Decreasing in the current were observed, and the decrease were proportional to incremental masses of MeBT added (Figures 7). The results suggest that a MeBT-cobalmin complex indeed formed, and the amount formed was proportional to the amount of ligand added; however, the inability to measure a potential shift indicated that DPP techniques are unsuitable for formation (stability) constant determination between MeBT and cobalamin.

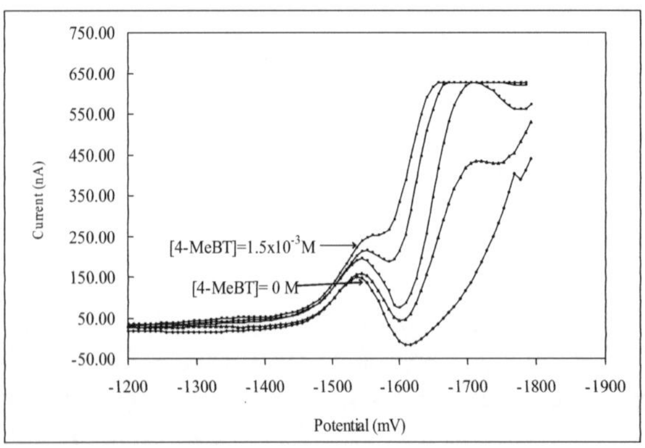

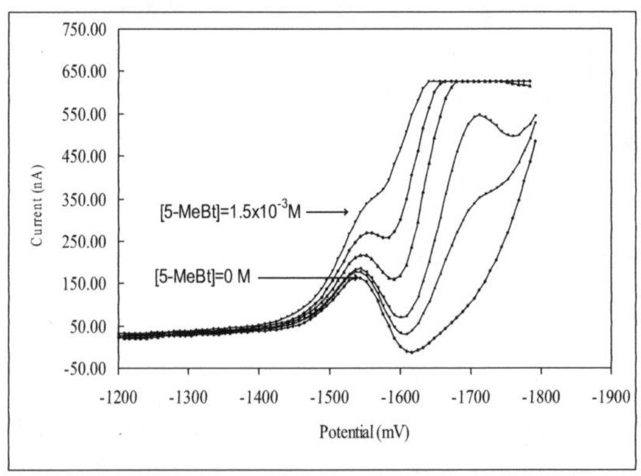

Figure 6: Differential pulse polarography of Vitamin B_{12}: 4-methylbenzotriazole (left) and 5-methylbenzotriazole (right) systems showing streaming current reduction resulting from the addition of increasing amounts of benzotriazole. [Co(III)] in Vitamin B_{12}=5.7x10^{-7}M.

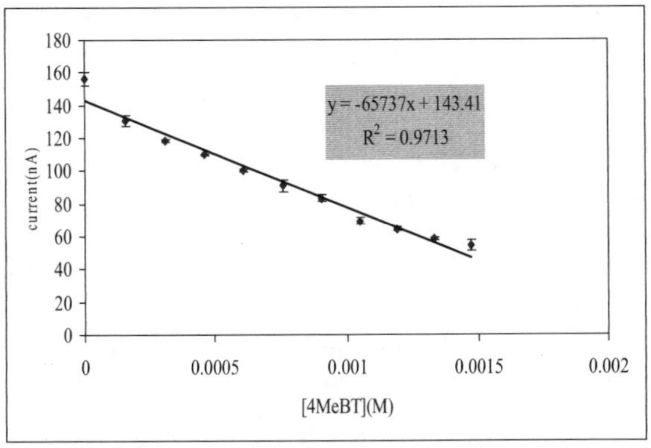

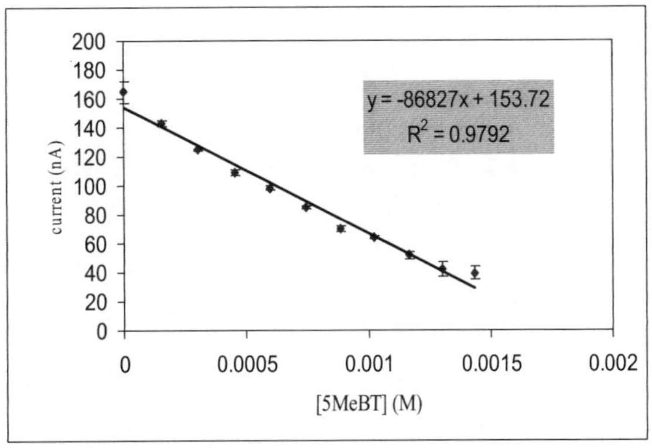

Figure 7: Streaming reduction current response of vitamin B_{12} (Co(III)) in response to the addition 4-MeBT (left) and 5-MeBT (right) at pH 7 (phosphate buffer). Results are averages of 3 independent observations. Error bars represent 1 standard deviation.

5. Summary and Conclusion

Results from this study suggested that commercially significant methylbenzotriazole isomers are able to form a very strong complexes with free copper and cobalt for both benzotriazole isomers tested. The conditional stability constants were approximately 10^{15} and 10^{8} respectively for copper and cobalt. Complexation of MeBT with Zn (II) could not be detected by differential pulse polargraphic techniques (DPP). It was found that Lingane and DeFord-Hume analysis were in a very good agreement under all conditions tested. The differences did not exceed $10^{0.3}$ difference.

Methylated benzotriazoles coordinated and bound copper in the active site of a superoxide dismutase, as well as with the redox-active cobalt conjugated in a corrin group in the ubiquitous co-enzyme cobalamin (vitamin B_{12}). The strength of binding (i.e. stability constant) could not be directly determined by polarography; however, the reduction spectra suggest that like free copper and cobalt, the stability constants and associated binding energy is quite large. This type of strong selected coordination may represent a novel mode of toxicity associated with corrosion inhibitors: changing the redox potential of a transition metal responsible for electron transfer in the active site of essential oxido-reductase enzymes.

While concurrent assays of benzotriazole-enzyme coordination and enzyme activity were not experimentally feasible during these investigations, the large negative potential shift observed when benzotriazoles were added to active enzyme-containing solutions suggested that once it bound, some catalytic inhibition would result.

References

[1] Pillardb DA, Cornell JS, et al 2001. "Toxicity of Benzotriazole and Benzotriazole Derivatives to Three Aquatic Species." *Water Research* **35**(2): 557-560.

[2] Cancilla DA, Baird JC, Geis SW, & Corsi SR 2003. Studies of the environmental fate and effect of aircraft deicing fluids: detection of 5-methyl-1H-benzotriazole in the fathead minnow (Pimephales promelas). E*nviron. toxic. and chem.*, **22**(1), 134.

[3] Deshpande S, Kuiry SC, Klimov M, Obeng Yand Seal S 2004. *J. Electrochem. Soc.* **151** G788.

[4] Antonijevic MM and Petrovic MB 2008. Copper Corrosion Inhibitors. A Review. *Intern. J. of Electrochem. Science*, **3**, 1-28.

[5] Maisuradze M, Ugulava G, Bolkvadze N, Phalavandishvili G, Gaxokidze N, and Matnadze M 2013. Some New Derivatives of 3H-Benzo (b) Furo Benzotriazole. *Journal of Chem. Chem. Eng.*, **7**: 621-625.

[6] Singh VK et al 2017. Benzotriazole: a heterocyclic molecule with diversified pharmacological activities. *Intern. J. of Pharm. Sci. and Res.,* Vol. **8**(2): 446-456.

[7] USEPA 1977. Investigation of Selected Potential Environmental Contaminants: Benzotriazoles. Washington D.C., *Environmental Protection Agency*: 191.

[8] Al-Soud YA, Al-Masoudi NA, Ferwanah AS, 2003. Synthesis and properties of new substituted 1,2,4-triazoles: potential antitumor agents. *Bioorg. Med. Chem.* **11**, 1701–1708.

[9] Katarzyna KZ, Andzelika N, Justyna Z, Lidia C, Janusz P, Przemysaw M, Maria B 2004. Synthesis and activity of 1H-benzimidazole and 1H-benzotriazole derivatives as inhibitors of acanthamoeba castellanii. *Bioorg. Med. Chem.* **12**, 2617–2624

[10] Wu CY, King KY et al 2006. Stable benzotriazole esters as mechanism-based inactivators of the severe acute respiratory syndrome 3CL protease. *Chem. Biol.* **13**, 261–268.

[11] Suma BV, Natesh NN, Venkataramana CHS, Jays J, Madhavan V (2012) Synthesis and antibacterial of some new 1, 2, 3 benzotriazoles derivatives containing pyrazolidinedione moieties. *Int J Pharm Pharm Sci* **4**:16.

[12] Mishra SS, Sharma CS, Singh HP et al 2015. A Review on Biological Activities of Benzotriazole Derivatives. *Intern. J. of pharm. Tech. and biotech.*; **2**(2):10-20

[13] Hartwell SI 1995. "Toxicity of Aircraft Deicer and Anti-icer Solutions to Aquatic Organisms." *Environ. Toxic.and Chem.* **14**(8): 1375-1386.

[14] Cancilla DA, Holtkamp A, et al 1997. "Isolation and Characterization of Microtox(R)-Active Components from Aircraft De-Icing/Anti-Icing Fluids." *Environ. Toxic. and Chem* **16**(3): 430-434.

[15] Cornell J, Pillard DA, et al 2000. "Comparative Measures of the Toxicity of Component Chemicals in Aircraft Deicing Fluid." *Environ. Toxic. and Chem.y* **19**(6): 1465-1472.

[16] Seeland A, Oetken M, Kiss A, Fries E, Oehlmann J 2012. Acute and chronic toxicity of benzotriazoles to aquatic organisms. *Environ. Science and Poll. Res.*, **19** (5), pp 1781–1790

[17] O'Brien, I 2002. The Environmental Chemistry of Aircraft Deicing Fluid (ADF) Component Chemicals. Civil, Environmental and Architectural Engineering. Boulder, University of Colorado.

[18] De WeverH and Verachtert H 1997. "Biodegradation and Toxicity of Benzothiazoles." *Water Research* **31**(11): 2673-2684.

[19] Skorda K, Per leapes SP, et al 1999. "A structural model for the copper (II) site of Cu-Zn superoxide dismutase: preparation, crystal structure and properties of Cu (Meta)4(H2O) (CLO4)2 center dot 0.4 EtOH(Mebta= 1-methylbenzotriazole)." *Trans. Metal Chem.* **24**(5): 541-545.

[20] Deford D D and Hume DN 1951. "The Determination of Concecutive Formation Constants of Complex Ions from Polarographic Data." *J. AM. Chem. Soc.* **73**: 5321-5322.

[21] Heath GA and Hefter G 1977. "The Use of Differential Pulse Polarography for the Determination of Stability Constants." J. Electroanal, Chem. **84**: 295-302.

[22] Leden J. 1941. *Z. phys. chem.* A, **188**, 160

[23] Crow DR 1969. Polarographic of Metal Complexes. *London and New York*, Academic Press.

[24] Parry EP and Osteryoung RA 1965. "Evaluation of analytical pulse polarography." *anal. Chem.* **37**: 1634.

Comparison between Single-Walled CNT, Multi-Walled CNT, and Carbon Nanotube-Fiber Pyrograf III

Marwan S. Mousa

Surface Physics and Materials Technology Lab, Department of Physics, Mu'tah University, AL-Karak, Jordan

marwansmousa@yahoo.com & mmousa@mutah.edu.jo

Abstract: Single-Walled CNT (SWCNTs), Multi-walled Carbon Nanotubes (MWCNTs), and Carbon Nanotube-Fibers Pyrograf III PR-1 (CNTFs) were deposited by chemical vapor deposition under vacuum pressure value of (10^{-7}mbar). Their structures were investigated by field emission microscopy. Carbon Nano-Fibers Pyrograf III PR-1 showed an average fiber diameter within the range of 100-200 nm and a length of (30-100) μm. Single-walled Carbon Nanotubes were produced by high-pressure Carbon Monoxide process with an average diameter ranging between (1-4) nm and a length of (1-3) μm. Thin Multiwall Carbon Nanotube of carbon purity (90%) showed an average diameter tube (9.5 nm) with a high-aspect-ratio (>150). The research work reported here includes the field electron emission current-voltage (I-V) characteristics and presented as Fowler-Nordheim (FN) plots and the spatial emission current distributions (electron emission images) obtained and analyzed in terms of electron source features. For the three types of emitters, a single spot pattern for the electron spatial; distributions were observed, with emission current fluctuations in some voltage region.

1. Introduction

Multi-Walled Carbon Nanotubes (MWCNTs) and Single Wall Carbon Nanotubes (SWCNTs) [1-5] have a growing attention due to their perfect properties that make them the best candidate to be the next generation of field emitters such as, low turn-on fields, high emission current, stable emission current, long lifetime, low energy spreading, and high brightness [5,6]. In addition, Stacked-Cup Carbon Nanotubes (SCCNTs) or Carbon Nanotube-Fibers (CNTFs) have unique structures with highly graphitic as shown in figure 1. The graphene planes in the structures of (SCCNTs) or (CNTFs) are canted from the fiber axis, resulting in exposed edge plane on the interior and exterior surfaces of the fiber; CNTs, on the other hand, typically resemble an assembly of concentric cylinders of graphene [7-9].

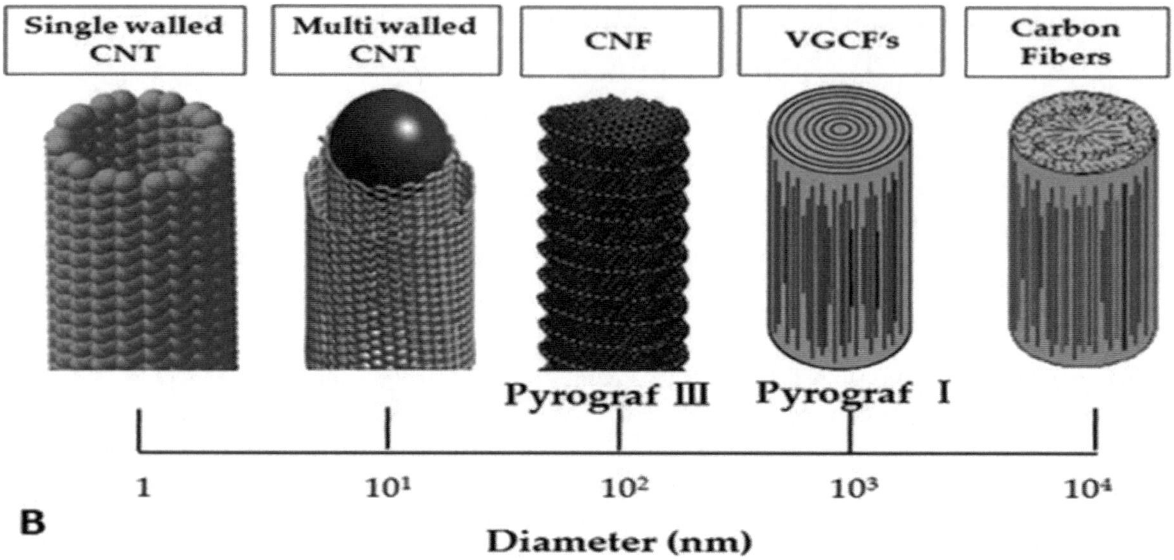

Figure 1 Nanotube-Fibers sizes from Single Wall Carbon Nanotubes to conventional Carbon Fibers.

Fowler-Nordheim (FN) first derived a semiclassical theory of FE based on physical assumptions and mathematical approaches; the theory has been modified many times over years to describe Cold Field Emission (CFE) from bulk metal where it is known as Fowler-Nordheim (FN) equation which is given by [10,11]:

$$i = A_n\{ a\phi^{-1}(\beta_V V)^2\}\exp(-\frac{b\phi^{\frac{3}{2}}}{\beta_V V})$$

Where i is the current emission, where it is exponentially dependent on the barrier width and, when graphically expressed on $\ln(I/V^2)$ vs V^{-1} scales, will exhibit a linear relationship, ϕ is the local work function, V is the applied voltage, a and b are the first and second FN constants, β_V is the local voltage-to-surface field conversion factor at some specific reference point on the emitter surface, and A_n is the national emission area defined by considering the local emission current density at the reference point. This equation uses the macroscopic field as an independent variable have often been used to describe CFE from Carbon Nanotubes (CNTs) including closed single-walled (CNTs) of small apex radius, particularly in technological context [12]. The Nanotube tip does not necessarily behave like a metal; electrons are not completely free and the effect of the tips atomic structure can be dominant. Moreover, an SWCNT is a three-dimensional structure (3D) and not an infinitely wide surface; thus, a more detailed treatment than the FN model is required. In this work there are three type of emitters: Single-Walled CNT, Multi-walled CNT, and Carbon Nanotube-Fiber Pyrograf III will be investigated and tested using a technique of the glass puller.

2. Experimental Details

Figures 2 and 3 showed the produced sample and the glass puller, respectively. Three types of CNTs were used: (a) Single-Walled Carbon Nanotubes (SWCNTs) were produced by high-pressure CO over Fe particles (HiPCO: High-Pressure Carbon Monoxide process), having an average diameter ranging between (1-4) nm with length (1-3) μm. (b) Multi-Walled Carbon Nanotubes (MWCNTs) Nanocyl™ NC7000 with 9.5nm diameter and high aspect ratio (>150), fabricated by chemical vapor deposition. (c) Carbon Nanotube-Fibers pyrograf III PR-1 (CNTFs), having average fiber diameter that is ranging between (100-200) nm, with the length of (30-100) μm.

Both tips prepared by employing a drawing technique using a glass puller apparatus (see figure 3), in this technique, two bearings are located accurately on plates supported by three stainless steel rods fixed rigidly to the frame of the control, this frame is strong enough to serve as a stable base for the instrument.

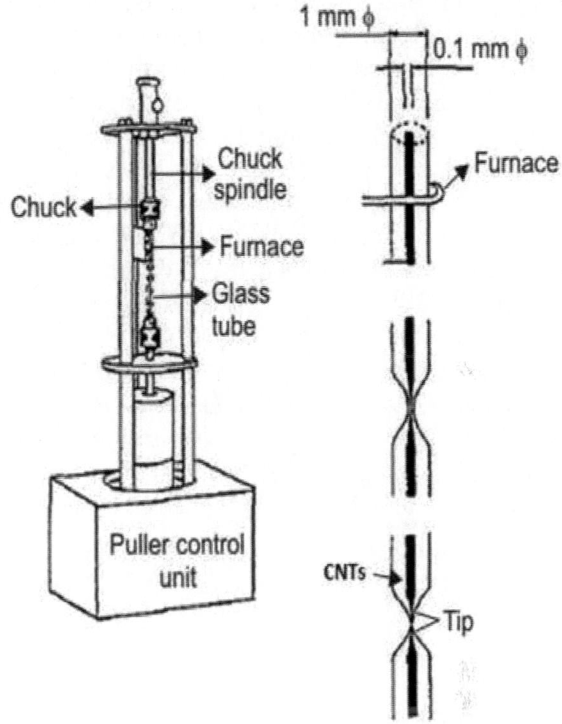

Figure 3 Schematic diagram of the glass puller.

A glass tube (outside diameter=1 mm, internal diameter=0.1 mm) fits inside these bearings between the upper and lower chunks, with a furnace loop located around it. The CNTs were entered into the opposite end of each glass tube so they would protrude at the tip with wire plunger [13] the lower chunk spindle can slide vertically in order to pull down the glass tube under gravity, after preparing the emitters, they are mounted in a standard Field Emission Microscopy (FEM) with an emitter-screen distance of ~10 mm [14], and current limiting resistor of (100 MΩ) is used. The system was evacuated to pressure ~10^{-8} mbar after baking at ~170°C overnight. The FEM applied voltage V is increased slowly until a "switch-on voltage" V_{SW} is reached, at which point the emission current suddenly "switch-on", at V_{SW} the current increase rapidly from about a Nano-

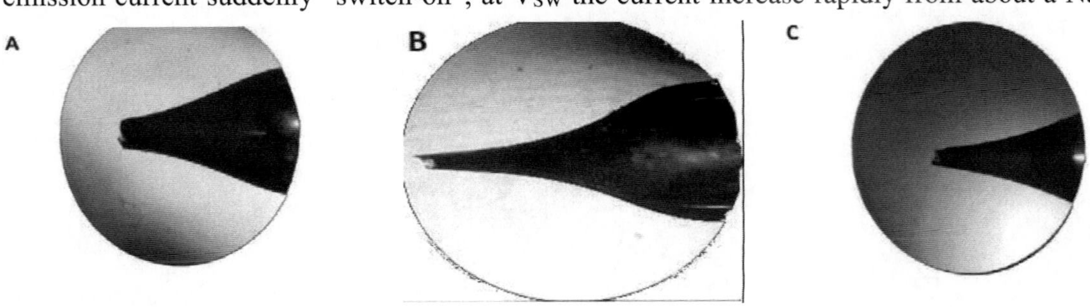

Figure 2 Optical micrograph at magnification 50X times for: (A) SWCNT-15. (B) CNTF-25. (C)MWCNT-A4.

IOP Conf. Series: Materials Science and Engineering **305** (2018) 012025 doi:10.1088/1757-899X/305/1/012025

ampere to a much greater saturated value I_{SW}, by slowly decreasing the applied voltage to a lower value the emission current starts to be decreasing until limit reached, beyond which the emission current falls smoothly to zero as the applied voltage is decreased to the threshold value V_{TH}.

3. Results and Discussion

This section describes the results of experiments on the emission characteristics of the Carbon Nano-Fibers and Single-Walled Carbon Nanotubes. These results include the I-V characteristics, Fowler-Nordheim (FN) plot, and emission images. Emission current measurements are made using a pico-ammeter connected between the cathode and earth. As the applied electric field is slowly applied, the emission is only observed after a threshold field is reached, and emission suddenly jumps to a constant saturated value I_{SAT}, further increasing the applied electric field the V_{MAX} will be reached, after that the tip explode, as the voltage slowly decreasing, the emission current fall to nearly zero, reaching the threshold value V_{TH}.

In the first set of specimens, we perform two sweeps of increasing voltage then decreasing voltage.Figure (4) showed the I-V characteristics for the three specimens, its related Fowler-Nordheim (FN) plot, and emission images (Fig. 4 (C), (D), and (E)) at V_{SW}. From the figure, it is shown that the V_{SW} for SWCNTs, CNTFs, and MWCNTs are 800 V, 1100 V, 4300 V with emission current 4 μA, 1.2 μA, and 25 μA, respectively, the slope for SWCNTs, MWCNTs, and CNTFs is (-3294.54), (-6484.71), and (-1727.94) decayed V, respectively, and we notice that the MWCNTs gave higher emission current the SWCNTs and CNTFs, but the turn-on field emission at SWCNTs and CNTFs is lower than MWCNTs. The experiment is repeated for another sweep of increasing and decreasing the applied voltage, and the result shown in the figure (5).

IOP Conf. Series: Materials Science and Engineering **305** (2018) 012025 doi:10.1088/1757-899X/305/1/012025

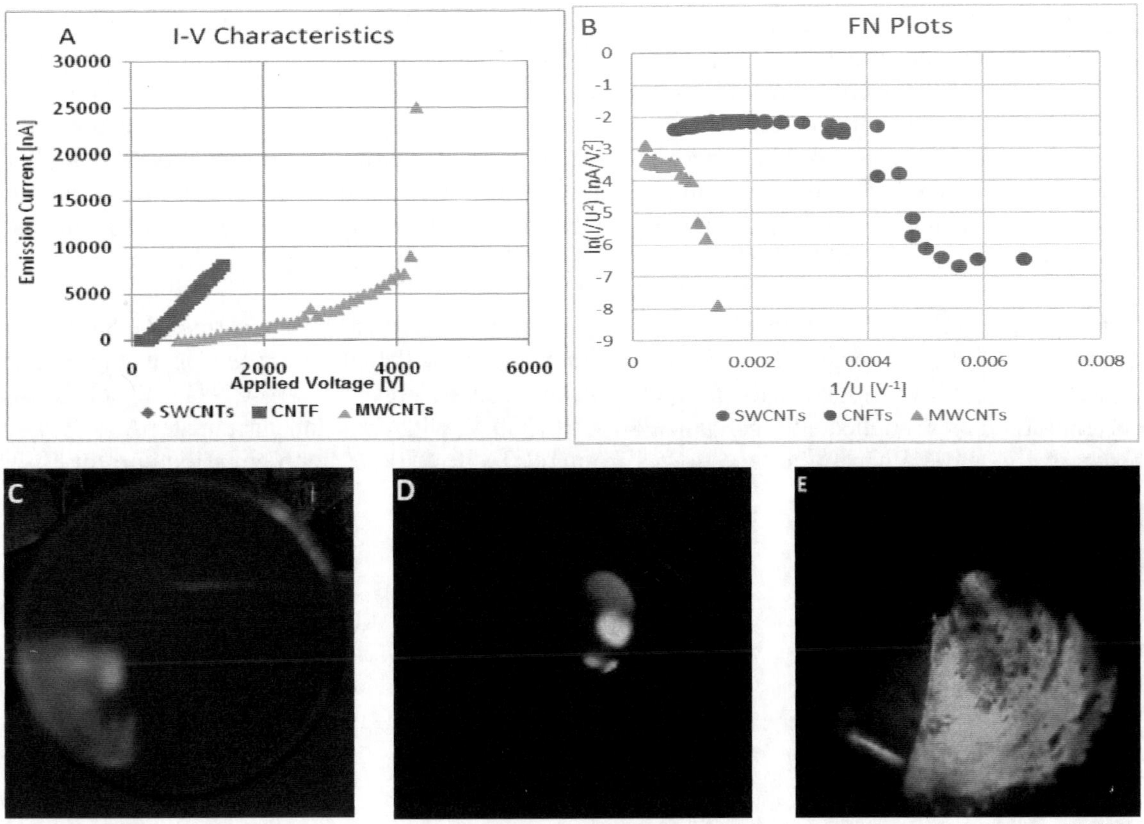

Figure 4. I-V characteristics for the three specimens. (B) Related FN Plot. (C) Emission Image at V_{SW} = 800 V, 4 μA for SWCNTs emitter. (D) Emission Image at V_{SW} = 1100 V, I_{SW} = 1.2 μA for CNTFs emitter. (E) Emission Image at V_{SW} = 4300 V, I_{SW}=25 μA for MWCNTs emitter.

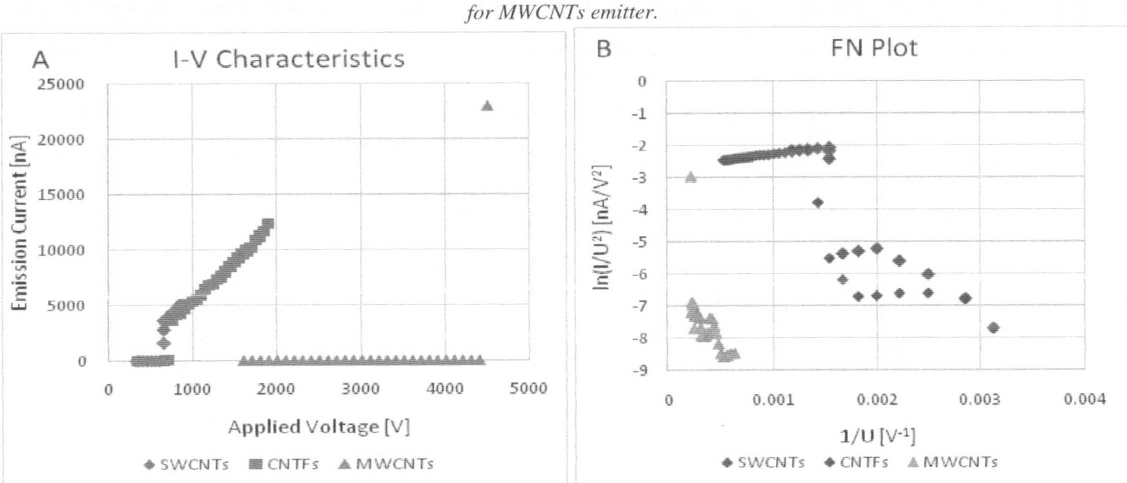

Figure 5. 2nd cycle, increasing voltage. (A) I-V Characteristics. (B) FN plot.

From figure 5, it shows the same result as in figure 4, it seems from the figure and the experimental observation that the value of the V_{SW} of the SWCNTs, MWCNTs, and CNTFs are 650 V, 1100 V, and 4500 V, respectively with emission current value 1.6 μA, 3.6 μA, and 23 μA, and there is

IOP Conf. Series: Materials Science and Engineering **305** (2018) 012025 doi:10.1088/1757-899X/305/1/012025

some emission current fluctuation in the low range in (MWCNTs-A4) emitter, due to their sensitivity to gaseous molecules [15] .As we decrease the applied voltage in the three specimens the saturated region extends down to V_{SAT} value, which is for SWCNTs, CNTFs, and MWCNTs are 380 V, 750 V, and 1300 V with emission current 1.06 μA, 1.2 μA, and 1 μA. From the previous result, the emission current has been reduced by comparing the emission current between the first and second sweep of voltage, that would be referred to the existence adsorbate, and the variation of local electric field (F) [16]. The experiment has been performed on another set of specimens. Figure 6 shows the SWCNTs, CNTFs-26, and MWCNTs-A5 emitters.

The voltage and emission current range in the three emitters as following; in case of the SWCNTs the turn-on applied voltage value is V=750 V and the last applied voltage value in increasing voltage is V=2850 V, with emission current 2 pA, and 18.9 μA, respectively, CNTFs, the emission current turned on at applied voltage value 340 V to 1350 V, with emission current 30 pA, 8.49 μA, respectively, and MWCNTs voltage ranges from 1600 V to 2700 V, with emission current 8 pA

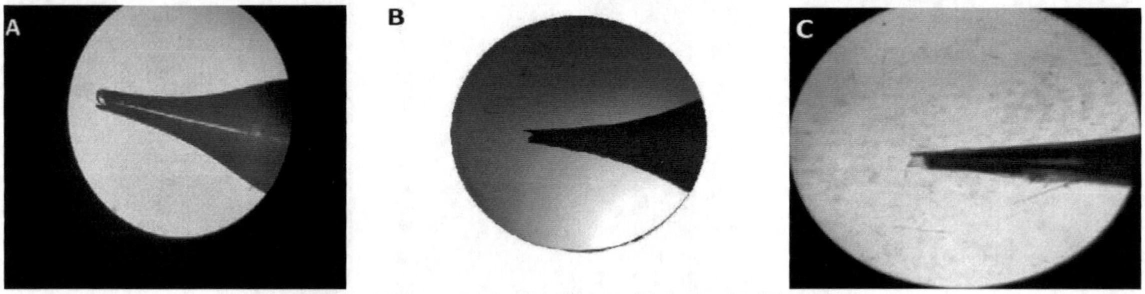

Figure 6 Optical Micrograph for (A) SWCNTs (B) CNTFs-26. (C) MWCNTs-A5.

to 6 nA. Figure 7 shows I-V characteristics and its relative FN plot.
The SWCNTs and CNTFs emitters' shows switch-on phenomenon at V_{SW} value = 1900 V, 1000 V, with emission current value 1.2 μA, 4.8 μA, respectively. As we continue increasing the applied voltage on SWCNTs specimen after switch-on phenomena appeared at V_{SW}=1900 V, I_{SW}=1.2 μA, the emission current suddenly raised from few μA range up to ~20 μA, that behavior appeared because the contribution of the adsorbate in the electron emission [17], as we decreased the applied voltage the saturation region extends down to V_{SAT} value for SWCNTs, CNTFs, and MWCNTs are 1050 V, 1000 V, and 1750 V, with emission current value 3.99 μA, 1.11 μA, and 1 μA, respectively.

IOP Conf. Series: Materials Science and Engineering **305** (2018) 012025 doi:10.1088/1757-899X/305/1/012025

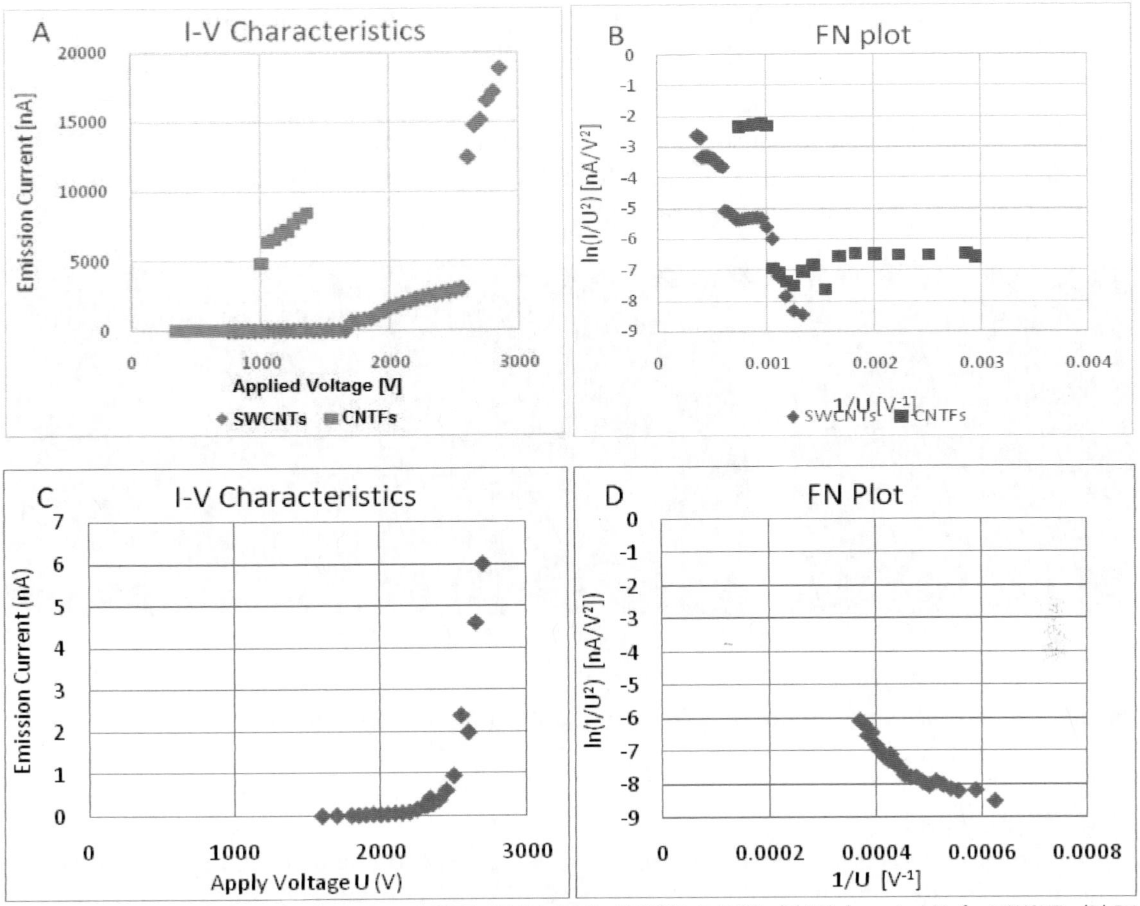

Figure 7(A) I-V Characteristics for SWCNTs and CNTFs (B) FN plot for SWCNTs & CNTFs (C) I-V characteristic for MWCNTs. (D) FN Plot for MWCNTs.

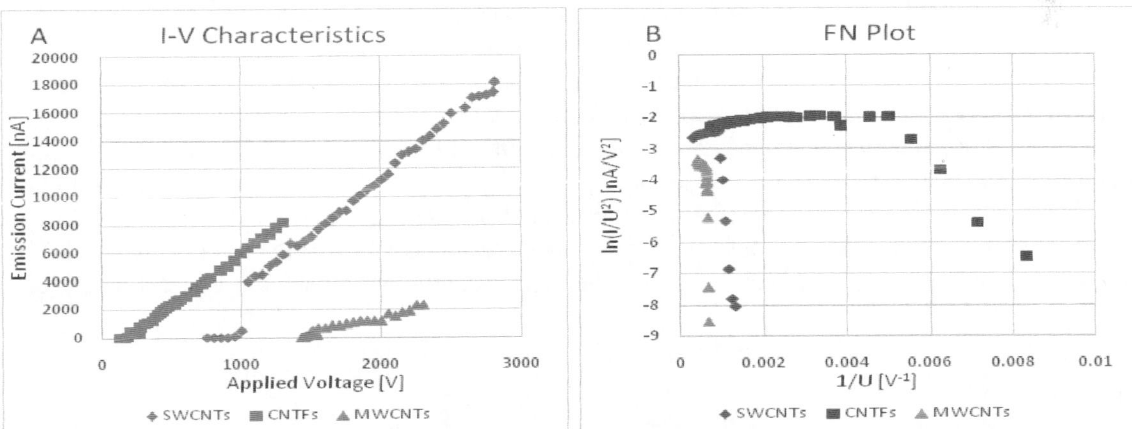

Figure 8. (A) I-V Characteristics for three type for specimen (B) Relate FN plot, where the slope value for SWCNTs, CNTFs, and MWCNTs are -16028.2, -1394.01, and -87187.2 respectively.

figure 8 shows I-V characteristics and FN value during the decreased V. As we see from the figure 8, that the V_{TH} for the CNTFs is less than SWCNTs, and MWCNTs and SWCNTs emitter gave emission current higher than CNTFs, and MWCNTs. According to this equation:

$$\gamma_{phe} = -6.44 \times 10^{9} \phi^{\frac{3}{2}}/K$$

where γ_{phe} is the phenomenological field enhancement factor, and K is the slope of FN plot, the value of γ_{phe} for CNTFs will be higher than SWCNTs and MWCNTs, so that will explain the value of V_{TH} for the three type of the emitter[18, 19].

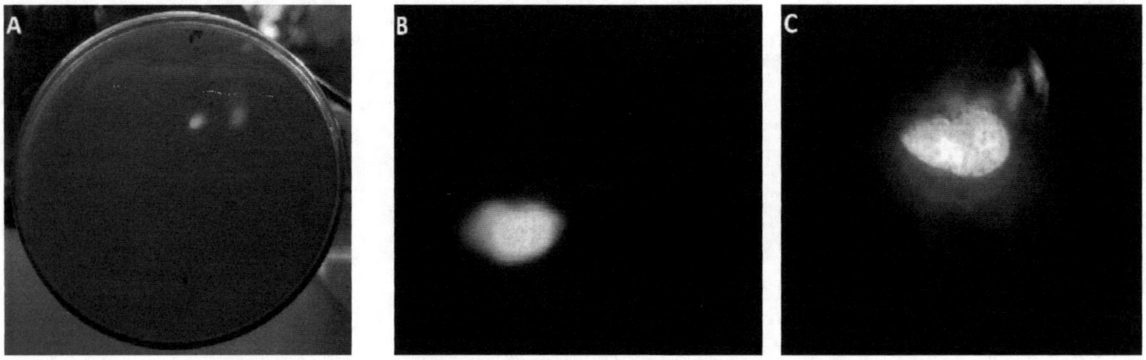

Figure 9 Emission Image for (A) SWCNTs at VSW. (B) CNTFs at VSW. (C) MWCNTs.

Figure 9 shows the emission image at V_{SW} for the second set of the emitter.

4. Conclusion

Three type of emitters have been tested, and all prepared with the same technique, by employing a drawing technique using a glass puller. Theoretically, the field emission from carbon nanotubes is a very challenging topic in many respects. Fundamentally, the field emission phenomena are highly non-equilibrium quantum mechanical processes that require the precise description of electronic structures of the nanotubes, and the equation of FN still in development to fully describe the electron emission from single-walled carbon nanotubes. From the result we can say that the SWCNTs and CNTFs have approximately similar behavior, while SWCNTs emission current turned on earlier, and that referring to the smaller radius of the nanotubes, but there are other influences on this matter, like the effect of the enhancement factor and field penetration, which the electric field can penetrate to CNTs to suppress the charge potential near the tip of tube and induce a few extra electrons [13], and the effective work function where its affected by the penetration of the electric field [18]. The cleanness the Carbon Nanotubes (CNT) surface effects on the current fluctuation, which is can be achieved by wet chemical cleaning and thermal post-treatment [19].

References

[1] Chambers, T. V. H. a. C. R. (1889). USA Patent No. U.S. Patent.

[2] De Jong, K. P., & Geus, J. W. (2000). Carbon Nanofibers: Catalytic Synthesis and Applications. Catalysis Reviews, 42(4), 481-510. doi: 10.1081/CR-100101954

[3] de Jonge, N., Allioux, M., Oostveen, J. T., Teo, K. B. K., & Milne, W. I. (2005). Low noise and stable emission from carbon nanotube electron sources. Applied Physics Letters, 87(13), 133118. doi: 10.1063/1.2058225

[4] Forbes, R. G. (2010). Thin-slab model for field electron emission. Journal of Vacuum Science & Technology B, Nanotechnology and Microelectronics: Materials, Processing, Measurement, and Phenomena, 28(2), C2A43-C42A49. doi: 10.1116/1.3322736

[5] Forbes, R. G., & Deane, J. H. B. (2007). Reformulation of the standard theory of Fowler–Nordheim tunnelling and cold field electron emission. Proceedings of the Royal Society A: Mathematical, Physical and Engineering Science, 463(2087), 2907-2927. doi: 10.1098/rspa.2007.0030

[6] Fowler, R. H., Nordheim, L. W. (1982). Electron emission in intense electric fields. Proceedings of the Royal Society of London, series A: Mathematical, Physical and Engineering science, 119 (781), 137-181.

[7] Lijima, S. (1991). Helical microtubules of graphitic carbon. Nature, 354(6348), 56-58.

[8] Lijima, S., & Ichihashi, T. (1993). Single-shell carbon nanotubes of 1-nm diameter. Nature, 363(6430), 603-605.

[9] Lassagne, B., Raquet, B., Broto, J. M., Cleuziou, J. P., Th, O., Monthioux, M., & Magrez, A. (2006). Electronic fluctuations in multi-walled carbon nanotubes. New Journal of Physics, 8(3), 31.

[10] More, M. A., Kashid, R. V., Patil, S. S., Shinde, D. R., & Joag, D. S. (2012, 9-13 July 2012). Field emission current noise analysis of carbon based materials. Paper presented at the 25th International Vacuum Nanoelectronics Conference.

[11] Mousa, M. S. (1988). Effect of lacomit films on cold-cathode hot-electron emission. Journal de physique, 49(C6), C6-237 - C236-242.

[12] Mousa, M. S., & Hibbert, D. B. (1993). Analysis of some properties of metal-glass microemitters subjected to strong electric fields. Applied Surface Science, 67(1), 59-65. doi: http://dx.doi.org/10.1016/0169-4332(93)90295-M

[13] Peng, J., Li, Z., He, C., Chen, G., Wang, W., Deng, S., . . . Forbes, R. (2008). The roles of apex dipoles and field penetration in the physics of charged, field emitting, single-walled carbon nanotubes (Vol. 104).

[14] S. Alnawasreh, M. S. M., and A. N. Al-Rabadi. (2015). Investigating the Effects of Sample Conditioning on Nano-Apex Carbon Fiber Tips for Efficient Field Electron Emission. Jordan Journal of Physics, 8, 95-101

[15] Peng, J., Li, Z., He, C., Deng, S., Xu, N., Zheng, X., & Chen, G. (2005). Quantum mechanical understanding of field dependence of the apex barrier of a single-wall carbon nanotube. Physical Review B, 72(23), 235106.

[16] De Jong, K. P., & Geus, J. W. (2000). Carbon Nanofibers: Catalytic Synthesis and Applications. Catalysis Reviews, 42(4), 481-510. doi: 10.1081/CR-100101954.

[17] More, M. A., Kashid, R. V., Patil, S. S., Shinde, D. R., & Joag, D. S. (2012, 9-13 July 2012). Field emission current noise analysis of carbon based materials. Paper presented at the 25th International Vacuum Nanoelectronics Conference.

[18] Qiu, H., Joshi, R. P., Neuber, A., & Dickens, J. (2015). A model study of the role of workfunction variations in cold field emission from microstructures with inclusion of field enhancements. Semiconductor Science and Technology, 30(10), 105038.

[19] Tittmann-Otto, J., Hermann, S., Kalbacova, J., Hartmann, M., Toader, M., Rodriguez, R. D., . . . Gessner, T. (2016). Effect of cleaning procedures on the electrical properties of carbon nanotube transistors—A statistical study. Journal of Applied Physics, 119(12), 124509. doi: 10.1063/1.4944835

.

Use of tapered Pyrex capillary tubes to increase the mechanical stability of multiwall carbon nanotubes field emitters

M.S. Mousa[1], E.S. Bani Ali[1], M.J. Hagmann[2]

[1]*Department of Physics, Mu'tah University, Al-Karak 61710, Jordan.*
[1]*Surface Physics and Materials Technology Lab, Physics Dept., Mu'tah University, AL-Karak, Jordan*
[2]*University of Utah, Dept. Electrical and Computer Engineering, Salt Lake City, Utah, USA, 84112*

marwansmousa@yahoo.com & mmousa@mutah.edu.jo

Abstract: In this study, Nanocyl^{TM} NC 7000 Thin Multiwall Carbon Nanotubes (MWCNTs) were used with a high aspect ratio (>150) made by the process of catalytic chemical vapor deposition (CCVD). The field emitter tips were prepared by inserting these MWCT into fine glass capillary tubes that were pulled at high temperatures and then cut. Measurements were carried out under ultra-high vacuum (UHV) conditions with a base pressure of 10^{-9} mbar. The data show the effects of initial conditioning of MWCNT and hysteresis. Compression of the MWCNT by the capillary tubes appears to provide adequate mechanical support without requiring the use of a low-melting point electrically-conductive binder as has been used previously. Emission currents in excess of 1 μA were obtained so this technique shows promise as a reliable, stable, powerful electron source.

Keywords: Cold Field Emission, Carbon Nanotubes, Field electron microscope, Ultra High Vacuum, Fowler-Nordheim plot, Glass Microemitters.

1. Introduction

Efficient electron sources are the fundamental components of miniature vacuum tube sources for high frequency applications [1]. Many analytical instruments are used for elemental and mineralogical analyses [2]. Vacuum microelectronic devices that are radiation-insensitive and high-temperature tolerant for extreme environments electronic applications. Each application requires an electron source with specifically designed beam forming optics and emission densities that span a range of tens to hundreds of amperes per sq. cm. State-of-the art thermionic cathodes are ill-suited for miniature instruments because of their bulkiness, high temperature operation, and high power consumption. Conversely, state-of-the-art cold cathodes [3], based on atomically sharp micromachined tips [4, 5] are highly susceptible to poisoning when operated in non-UHV (10^{-8} to 10^{-9} mbar) environments.

A new class of field emitters was identified when carbon nanotubes (CNTs) were discovered [6]. Carbon Nanotube (CNT) based field emitters have been reported to exhibit high emission current with long-term stability, resulting from their high aspect ratio with small tip radius of curvature and high chemical inertness [7–9]. Thus a large amount of research is being carried out to utilize their excellent properties in practical applications, such as field emission displays (FED) [10, 11] and x-ray sources [12,13]. Another fascinating application of CNT emitters is a field emission backlight unit (FE-BLU) for liquid crystal displays (LCD) [14]. Although various types of lamps are being used as backlights for LCD, a cold cathode fluorescent lamp (CCFL) is by far the most widely used backlight for commercial LCDTVs and monitors [15].

Previous studies of field electron emission from and through glasses have had three aims. These include defining the mechanism for high-voltage breakdown [16, 17], creating stable bright electron sources, and studies of field emission from glass and metal-insulator composites [18]. Previous studies of field electron emission from glass with a conductive coating applied to the opposite side have been reported [19, 20]. In field electron emission electrons are emitted from the surface of a cathode under an intense applied electric field (> 3×109 V/m) [3]. The electrical current depends strongly on the work function of the emitter. In this paper, a new type of emitter is introduced that is based on Multiwall Carbon Nanotubes (MWCNTs) and fabrication techniques are presented with field emission data.

Previously glass and Pyrex capillary tubes have been used by our group to increase the mechanical stability MWCNTs as field emitters [20]. Then, the capillaary tubes were filled with low melting point electrically-conductive materials such as indium, tin, and gallium, and heated the tubes to a temperature just below the melting point so that the tube softened and then necked at several points along its length. After each tube was slowly cooled to room temperature it was cut at these constrictions to form several field emitters, each having an apex at the point where the tube was cut.

2. Experimental

We used apparatus similar to what was used before, as shown in Fig. 1, to prepare agglomerates of MWCNT in tapered PyrexTM tubes having an inside diameter of 0.1 mm and outside diameter of 1.0 mm to provide physical support for the MWCNT [20]. A Nichrome wire was used to heat a single straight tube as shown in Fig. 1.b, to taper the tube to much smaller inner diameters at several locations as shown in Fig. 1.c. Then the tube was cut to form several tapered sections that were filled with MWCNT to form the field emitters. Nanocyl™ NC 7000 Thin Multiwall Carbon Nanotubes (Nancyl S.A., Sambreville, Belgium) were used, which have a typical diameter of 9.5 nm and aspect ratio greater than 150. After each tube was filled with the MWCNT a tungsten wire having a diameter of 0.1 mm was inserted into the end which was not tapered to make a permanent electrical contact with the MWCNT.

Prepared emitters were investigated in a locally built point to-plane field electron microscope (FEM), with an axial tip to-screen distance of 10 mm [21]. The FEM was evacuated to ultra-high vacuum conditions by the combination of a rotary pump (which produces a backing pressure of about 10^{-3} mbar) and an oil diffusion pump system, connected to a liquid nitrogen (LN2) trap. Without baking, the FEM reaches a base-pressure of about 10^{-7} mbar. The pressure reaches about 10^{-9} mbar after baking the system for 12 hours at 180°C, and adding liquid nitrogen to the trap [22]. In operation, the FEM applied voltage V is increased slowly until a "switch-on voltage" V_{SW} is reached, at which point the emission current suddenly "switches on". At V_{SW} the current increases rapidly from about a nano Ampere to a much greater saturated value I_{SAT}, similar to the phenomena that has been previously reported [23-24]. The current-voltage (I-V) data yields a FN plot that has linear segments.

IOP Conf. Series: Materials Science and Engineering **305** (2018) 012026 doi:10.1088/1757-899X/305/1/012026

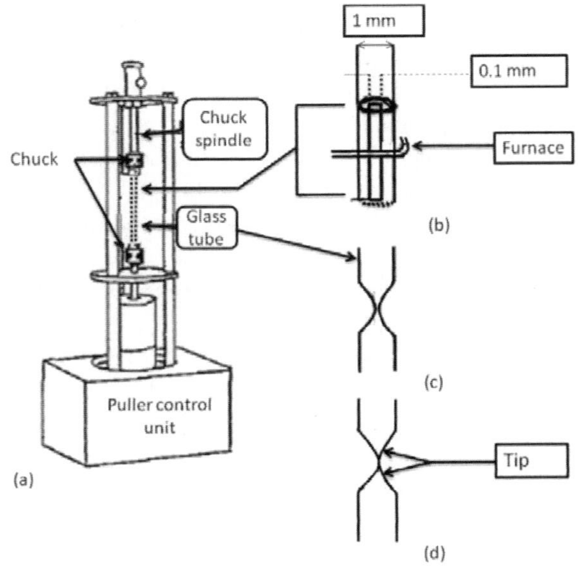

FIG. 1. (a) Apparatus used to prepare the Pyrex tubes that were used to hold the MWCNT; (b) Nichrome wire heater used to heat the tube to taper it to a smaller diameter; (c) Pyrex tube tapered at several locations where it is cut to form the field emitter tips.

Figure 2 is an SEM image of one of these field emitters in which the tapered end has an opening with a diameter of approximately 184 nm. An agglomerate of MWCNT is held securely by pressure at the opening without requiring the use of epoxy, pitch, or other agents for attachment. An EDX (energy dispersive X-ray spectroscopy) analyzer was used to determine the composition of the emitter surface which was primarily carbon, oxygen and sodium.

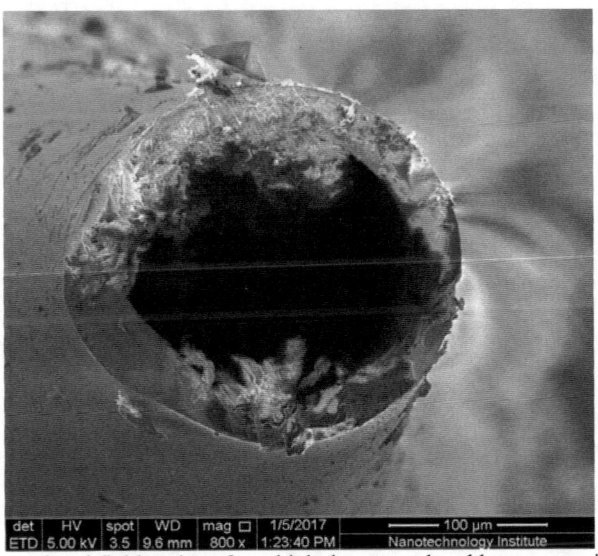

FIG. 2. SEM image of the completed field emitter for which the tapered end has an opening diameter of 184 nm.

3. Results and Discussion

The field emitter shown in Fig. 2 was placed in a vacuum chamber in which a phosphor screen at a distance of 10 mm was used as the anode. The chamber was evacuated to a base pressure of 10^{-9} mbar. Current vs. voltage measurements were manually recorded using a Keithley 485 Picoammeter and a high-voltage power supply with a digital voltage readout. The potential was increased in steps of 100 V. Figure 3 is a Fowler-Nordheim (F-N) plot for these measurements made with the field emitter that is shown in Fig. 2. The plots in the figure represent the I-V measurements which were repeated as the voltage was increased or decreased in steps.

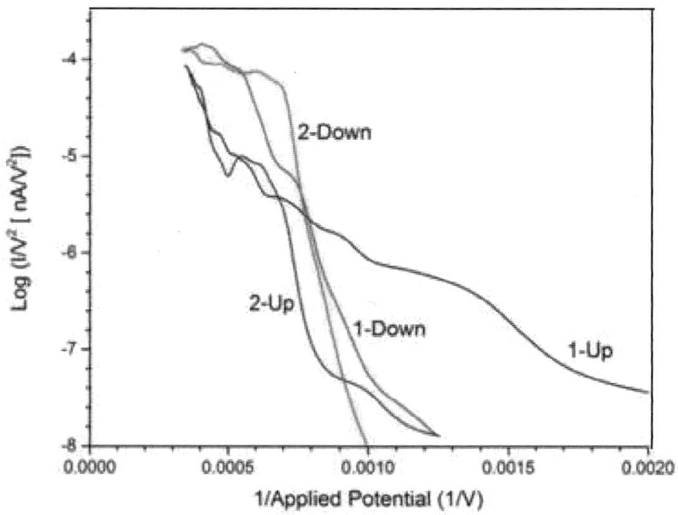

FIG. 3. Fowler-Nordheim plot of current vs. voltage measurements made with the field emitter shown in Fig. 2. The data are plotted for the first and second time that the voltage is increased and decreased.

Figure 3 shows the effects of conditioning of the emitter as well as the hysteresis. Note that as the voltage is increased (1/V decreased) for the first time (1-Up), the initial value of the current, normalized by dividing by the square of the voltage (I/V^2), is greater than its value at same point during later cycles. This effect is referred to as the "initial conditioning" of the MWCNT. Also, as the voltage is lowered in the reverse cycle of the hysteresis loop (1-down), the normalized current is greater than in the former part of the hysteresis loop. Subsequent loops are similar to 2-up and 2-down in the second loop.

In conclusion, using compression by a Pyrex capillary tube to mechanically stabilize the MWCNT made it possible to avoid adding pitch, Epoxy, or other materials that would increase the electrical resistance and thermal resistance required to lower the maximum current, and may also contaminate the emitter. We have already obtained currents in excess of 1 μA (Fig. 3) without requiring such binders.

4. Conclusion

In summary, using compression by a Pyrex capillary tube to mechanically stabilize the MWCNT has made it possible to avoid adding pitch, epoxy resin, or other materials. Such materials might increase the electrical resistance (and hence lower the maximum current), or increase thermal resistance (and hence potentially make the emitter less stable), or contaminate the emitter. Currents in excess of 1 μA (Fig. 3) without requiring such binders, were obtained.

Aknowledgments

We would like to acknowledge the support of the Deanship of the Academic Scientific Research at Mu'tah University for supporting this work through the research project # 120\14\121. Emad S. Bani Ali would like to express gratitude to the Scientific Research Support Fund (SRSF) in the Hashemite Kingdom of Jordan for support extended to him.

References

[1] Siegel, P.H., Fung, A., Manohara, H.M., Xu, J. and Chang, B. (2001). Proceedings of the 12th International Symposium on Space Terahertz Technology, (Jet Propulsion Laboratory, San Diego, CA) 94-103.
[2] Vaniman, D., Bish, D., Blake, D.F., Elliot, S.T., Sarrazin, P., Collins, S.A. and Chipera, S. (1998). J. Geophys. Res. **103** (E13), 31,477–31,489.
[3] Gomer, R. (1993). Field Emission and Field Ionization, (AVS Classics Series, American Institute of Physics, New York).
[4] Spindt, C.A., Brodie, I., Humphrey, L. and Westerberg, E.R. (1976). J. Appl. Phys. **47** (12), 5248–5263.
[5] Trujillo, T. and Hung, C.E. (1993). Maejo Int. J. Sci. Technol. B **11** (2), 454–458.
[6] Ijima, S. (1991). Nature **354**, 56–58.
[7] Seelaboyina, R., Boddepalli, S., Noh, K., Jeon, M. and Choi, W. (2008). Nanotechnology **19**, 065605.
[8] Jung, Y.J., Swastik, K., Sanikat, T., Caterina, S., Gunaranjan, V., Xuesong, L., Zhaoling, Y., Fung, S.O., Aditya, A., Robert, V., Seamus, C., Omkaram, N. and Pulickel, M.A. (2006). Nano. Lett., **6**, 413-418.
[9] Chen, K.C., Chen, C.F., Chiang, J.S., Hwang, C.L., Chang, J. Chang, Y.Y., and Lee, CC. (2006). Thin Solid Films **498,** 198-201.
[10] Choi, W.B., Chung, D.S.,Kang, J.H., Kim, H.Y., Jin, W., Han, I.T., Lee, Y.H., Park, G.S. and Kim, J.M. (1999). Appl. Phys. Lett. **75,** 3129.
[11] Choi, Y.C., Jeong, K.S., Han, I.T., Kim, H.J., Jin, Y.W., Kim, Kim, J.M., Lee, B.G., Park, J.H. and Choe, D.H. (2006). Appl. Phys. Lett. **88**, 263504.
[12] Zhang, J., Yang, G., Cheng, Y., Gao, B., Qiu, Q., Lee, Y.Z., Lu, J.P. and Zhou, O. (2005). Appl. Phys. Lett. **86,** 184104.
[13] Liu, Z., Yang, G., Lee, Y. Z., Bordelon, D. , Lu, J.P., and Zhou, O. (2006). Appl. Phys. Lett. **89**, 101111.
[14] Kim, Y.C. and Yoo, E.H. (2005). Japan. J. Appl. Phys. **44**, L454-456.
[15] Jung, Y.J., Son, G.H., Park, J.H., Kim, Y.W., Berdinsky, A. S., Yoo, J.B., and Park, C.Y. (2015). Diamond Relat. Mater. **14**, 2109-2122.
[16] Murray, J. J. (1960). Glass cathodes in vacuum insulated high voltage systems. Lawrence Radiation Laboratory Report.
[17] Latham, R. V. (1981). High Voltage Vacuum Insulation: The Physical Basis (Academic Press, New York).
[18] Bayliss, K.H. and Latham, R.V. (1986). Proc. R. Soc. London: A, Mathematical and Physical Sciences, **403**, 285-311.
[19] Mousa, M.S. and Hibbert, D.B. (1993). Appl. Surf. Sci. **67**, 59-65.
[20] Mousa M.S. and Hibbert D.B. (1993). Phys. D. Appl. Phys. **26**, 697-703.

[21] AL-Qudah, A. Mousa, M.S. and Fischer, A. (2015). Mater. Sci. Eng. **92**, 012021.

[22] Madanat, M.A., Mousa, M.S. Al-Rabadi, A.N. & Fisher, A. (2015). Jordan. J. Phys. **8**, 79-85.

[23] Latham, R.V. and Mousa, M.S. (1986). J. Phys. D: Appl. Phys., **19**, 699-713.

[24] Bani Ali, E.S. and Mousa, M. S. (2016). Applied Microscopy, **46**, 4, 244-252.

Institute of Physics
Dirac House, Temple Back
Bristol BS1 6BE UK

ISSN: 1757-8981
ISBN 978-1-5108-5951-7

9 781510 859517

CURRAN ASSOCIATES INC.
proceedings.com

8th International Particle Accelerator Conference (IPAC'17)

Journal of Physics: Conference Series Volume 874

Copenhagen, Denmark
14 – 19 May 2017

IOP
Institute of Physics